THE ECONOMIC AND HERITAGE PLANTS OF KASHMIR HIMALAYA

NIPA GENX ELECTRONIC RESOURCES & SOLUTIONS P. LTD.
New Delhi-110 034

About the Author

Dr. Nazir Ahmad Zeerak (M Sc., M Phil., Ph.D. Botany) working as Professor cum- Chief Scientist, Division of Genetics & Plant Breeding, Sher-e-Kashmir University of Agricultural Sciences and Technology of Kashmir, India. He specializes in the field of genetic improvement of crop plants through induced mutations and polyploidy. His major research areas also include Cytogenetics and Plant Biodiversity. Professor Zeerak has a teaching and research experience of more than 30 years and has about 150 research papers to his credit which stand published in national and international journals. He has been the Principal Investigator/Nodal Officer/Co-PI of several research projects sponsored by Horticulture Board of India and Indian Council of Agricultural Research. He is also actively involved as a member of many national and international scientific associations including the Indian Botanical Society and European Conservation Biology. During the last several years, he has been involved in the documentation, conservation and sustainable utilization of Plant Genetic Resources for Food and Agriculture in the Indian Himalayan Regions.

THE ECONOMIC AND HERITAGE PLANTS OF KASHMIR HIMALAYA

Dr. Nazir Ahmad Zeerak
Division of Genetics & Plant Breeding
Sher-e Kashmir University of Agricultural Sciences & Technology of Kashmir
Shalimar, Jammu and Kashmir-190 025, India

NIPA GENX ELECTRONIC RESOURCES & SOLUTIONS P. LTD.
New Delhi-110 034

NIPA GENX ELECTRONIC
RESOURCES & SOLUTIONS P. LTD.
101,103, Vikas Surya Plaza, CU Block
L.S.C.Market, Pitam Pura, New Delhi-110 034
Ph : +91 11 27341616, 27341717, 27341718
E-mail:newindiapublishingagency@gmail.com
www: www.nipabooks.com

For customer assistance, please contact
Phone: + 91-11-27 34 17 17 Fax: + 91-11- 27 34 16 16
E-Mail: feedbacks@nipabooks.com

ISBN: 978-93-90591-64-0

Composed and Designed by NIPA.

In memoriam

Prof. Reayat Khan

&

Prof. P. Kachroo

Learned teachers and humanists

Acknowledgements

My excuse for compiling this book is my deep association with nature and my culture. Infact my mother has been the source of my knowledge about the names and ethnobotanical uses of local plants.

The author acknowledges with gratitude the cooperation and help rendered by farmers and agriculturists in giving access to their farms and orchards and helping in identifying the local economic and heritage plants and providing valuable information about them. The fruitful discussion with some Agricultural scientists, Horticulture experts and local *Hakims / Amchis* also helped in recording the traditional uses of plants and the cultivation status of the crops and their varieties.

I owe a deep sense of gratitude to my family who encouraged and inspired me to compile and enlist the economic and heritage plants of my homeland.

The working facilities provided by Sher-e-Kashmir University of Agricultural Sciences and Technology, Kashmir (SKUAST-K) are gratefully acknowledged.

N.A. Zeerak

Preface

Biological diversity is essential to life on earth. It contributes to the regulation of the climate and functioning of the biosphere. It is also the common heritage of mankind and, since the dawn of human civilization, it has been providing human beings with their means of subsistence, securing their food supplies and health through many traditional medicines and remedies. However the recent human population growth at global level followed by the resulting agents of destruction like pollution, habitat destruction, over-harvesting, invasive species, unplanned industrial activities, social and economic factors have destroyed the habitats of millions of plant and animal species causing extinction of several species and sharp decline in the population sizes of others. As a result, our planet is facing loss of living resources which will have a clear and major impact on the development and welfare of mankind.

Having realized the importance of biodiversity for human survival, great concern is now being felt globally over the loss of biodiversity resources and accordingly an increased recognition has been imparted to the value of biodiversity and its essential role in promoting sustainable development and poverty alleviation. In this regard, International agreements like Convention on Biological Diversity stand passed by United Nation's Food & Agriculture Organization (during 1992) and ratified by member nations including India. The Convention focussed on the importance of biodiversity, reflected the world wide concern for preventing unfair exploitation of biological resources and reaffirmed the sovereign rights of member nations over their biological resources. It also recommended to the member nations to protect their traditional knowledge and customary practices related to the use of biological resources. Accordingly, various measures and policies have been adopted by many countries of the world, including India, to document, evaluate and conserve their bio-resources.

India has been blessed by nature with one of the richest floras on earth, and is the centre of origin of a number of economic plants. Indian Himalayan regions, situated in its north-west and north-east of the country, are noticeably included among the global hotspots of biodiversity. Within the North Western part of Indian Himalayas the state of Jammu and Kashmir, alone occupies about 67% of the geographical area, spread on about 2,22000 Km^2 (including the area under illegal occupation of Pakistan and China) comprising mostly of Hilly terrain formed by a series of mountain ranges like the Shivaliks, the Pir Panchjal, the Great Himalayas, the Karakoram and the Lesser Middle mountain ranges. Though being hilly and a mostly mountainous state, with varied altitudinal gradients and a variety of macro and micro- climatic regimes, a wide range of economically important plants thrive on its soils, in the cultivated and wild forms. Additionally since centuries the native farmer's selections among the useful indigenous plants for biomass, from the natural stands, and the selective introductions from the Mediterranean, the Europe and other adjacent temperate areas, in the past, has also resulted in accumulation and diversification of enormous plant genetic variability. As a result, it is the leading biomass producing region of India and is known for the export of its fruit and nuts, vegetables and their seed, medicinal and aromatic herbs, honey, mushrooms and morels, condiments like saffron and black zeera, fur, silk and woollen garments, woollen and silken carpets, wood carving and wood crafts, paper machie and wicker works, sports goods, joinery items, etc. The vast plant genetic resources of the region, existing in its different gene pools, thus hold vast potential for current and future users for sustained development and well-being of its inhabitants. However, these precious resources are now increasingly being threatened mainly because of human population pressures, climatic changes, urbanization, change in land use pattern, deforestation and unplanned tourism. Use of plant resources for rapid commercial gains, over-harvesting, mono-cultures, rapid replacement of adapted and indigenous cultivars by modern high yielding varieties and the adoption of bio-hazardous production technologies have also resulted in loss of many native plant species and the land races of all major crops, further shrinking the native plant genetic diversity. The losses would prove catastrophic to the ecology of the region, food security and livelihood resources of its farming communities if urgent steps are not taken for safeguarding these biological resources.

Kashmir and Ladakh regions of Jammu & Kashmir state, in particular, are also rich in ethnic folklore, culture and heritage. They have their own floristic composition, geographical peculiarities and social and cultural traditions which are markedly different from any other part of India. The people have their own traditional health care systems, food habits and beverages. The man-plant relationship has been shaped to a great extent by their traditions, culture and history. As a matter of fact, the natural geographic barriers of the valley and Ladakh have indirectly assured such a unique position of isolation to this region from other cultures that its ethnobotany and local traditions have remained largely undisturbed. However, over the past several decades, due the influence of scientific technologies and the use of some national and international languages, as mediums of education, many trusted age old ethnobotanical practices and plant nomenclatures have been abandoned which have narrowed down the vocabularies of regional languages, thus initiating a clear cultural erosion.

Keeping in view the increasing threat to plant biodiversity and culture of the regions, it is, therefore, imperative that a scientific estimation and documentation of their existing plant resources are immediately and urgently taken up. This will help in evaluating the diversity status of economic plants, documentation of their local use, the related indigenous technical knowledge and the preparation of a comprehensive document on the economic cum heritage plants of the region. The activity will also indirectly lead in protecting Plant Genetic Resources for Food and Agriculture and guaranteeing the rights of the farming communities of the regions over them. In this regard, an attempt has been made to enlist and put to record the existing plant diversity, especially in respect to local heritage plants and plants of agri- horticultural importance growing across North-Western Himalayas falling under Ladakh and Kashmir regions.

The work, presented in the form of a book, is partly the result of surveys carried out over the past few years. However, keeping in view the State's vastness and difficult terrain the total picture of the diversity for food and other economic plants, in the areas surveyed, needs much more dedication and time, nevertheless it may be considered a beginning. The information generated on the useful plant species and their uses may, therefore, contain deficiencies or omissions which shall be taken care of in future publications.

In this edition of the book about 1300 native, introduced and locally adapted plant species of economic importance and about 500 varieties of various crops have been listed and briefly described. Indigenous technical knowledge and ethnobotanical information, associated with the available diversity together with phyto-geographical information has also been given. The listed plant species and germplasm lines of all economic plants, and varieties of field and horticulture crops have been classified on the basis of their present status of occurrence or cultivation. Accordingly, the field crop varieties have been referred to like- **Widely Cultivated (WC), Restrictedly Cultivated (RC) and Not Cultivated (NC). Similarly the horticultural crop varieties have been assigned the status like Under Commercial Cultivation (UC), Not Under Commercial Cultivation (NC), Naturalized after having been introduced (NI)** while as the economic plant species found only in the **wild state** are mentioned as **Wild (W)**. The plant species are arranged in alphabetical order of their scientific names, followed by their **synonyms (syn.)**, respective plant families, then by their **English (Eng.), Hindi, Kashmiri(Kash.)** and **Ladakhi** names and lastly by their status of **occurrence /cultivation**.

It is hoped that the book would be very useful to the students and research workers especially those interested in economic botany, agri-horticultural biodiversity, ethnobotany as well as culture of the region. The plant diversity documented shall also prove helpful to conservationists and planners for devising future strategies in the conservation and sustainable utilization of Plant Genetic Resources for Food and Agriculture; and also in guaranteeing livelihood resources and cultural identities of the inhabitants of the regions.

The author appeals to all the conscious citizens, students and plant lovers of this Himalayan region, to rise to the occasion and promote public awareness about the economic and heritage plants and their importance towards development and our cultural identity; and to be a part of collective efforts for conservation of this heritage so that our plant genetic resources as well as our cultural and social identities are preserved for posterity.

April, 2021 **N.A. Zeerak**

Contents

List of Figures

A

Abelia grandifolia Villarreal (Caprifoliaceae): Eng.- Glossy Abelia; Kash.– Abeel [NI]. A vigorous, arching, medium sized semi-evergreen shrub commonly grown in gardens for hedging. Leaves ovate, short petioled, entire, dark glossy-green. Flowers born in axillary cymes and terminal panicles, slightly fragrant, produced over a long period from June to October; sepals 2-5; corolla funnel form, white tinged pale- pink.

Abelia triflora R. Br. ex Wall. (Caprifoliaceae): Eng.- Himalayan Abelia [W]. An erect ornamental shrub, 1-3 m tall, growing in sub-alpine woods at 1600 – 2400 m altitudes. Leaves lance shaped, 4-5 cm long, entire or coarsely toothed, sparingly hairy. Flowers borne in 3 – flowered clusters, from the axils of leafy bracts. Flowers tubular – funnel shaped, tube reddish with five spreading white petals (Fig. A 1).

Abelmoschus esculentus (L.) Moench (syn. Hibiscus esculentus L.) (Malvaceae): Eng.- Ladies' Fingers /Oakra; Hindi – Bhindi [UC]. An erect stout herb cultivated, as a *kharif* crop, for its tender fruits. Leaves palmately lobed. Flowers white. Fruit a capsule. Commonly cultivated varieties include *Pusa Sawni* and *Pusa Makhmali.*

Abies pindrow (Royle. ex D Don.) Royle. (syn. Abies pindrow Brandis) (Pinaceae): Eng.- Himalayan Silver Fir; Hindi- Paludar/Badar; Kash.- Yaere/Budul [W]. A tall evergreen coniferous tree of slender, conical habit, forming an important forest element from 1800 to 3200 m altitudes. Leaves borne on the upper side of the branchlets, erect, loosely overlapping, shining dark green above with two greyish-white, stomatic bands beneath. Flowers monoecious, male cones sessile usually clustered; female cones solitary or in distant pairs, erect, situated a little below the tips of the shoots; mature cones cylindrical, dark purple when ripe. Wood is used in construction and furniture works. In earlier times the wood was also used for making water troughs and filling for boats.

Abies spectabilis (D.Don) Mirbel. (syn. Abies webbiana (Wall. ex D Don.) Lindl.) (Pinaceae)**: Eng.- East Himalayan Silver Fir; Kash.– Rayei Budul [W].** A magnificent large forest tree of slender pyramidal habit with dark grey, deeply longitudinally grooved bark; branches pendulous; branchlets stout, stiff, spreading horizontally, young shoots usually clothed with short brown hairs. Leaves shining dark green above and gleaming silvery white beneath; cones cylindrical, erect, violet purple when young. Wood is used in construction and furniture works.

Abutilon pictum (Gillies ex Hook) Walp. (Malvaceae): Eng.- Red- veined Indian Mallow/Chinese Bell Flower; Hindi – Kanghi [NI]. An erect, evergreen half-hardy ornamental shrub, with ovate to rounded 3 to 5 lobed, pale to mid-green leaves, margins serrate to crenate-dentate, glabrous, and occasionally mottled creamy white. Flowers campanulate, deep orange-red, with blood red veins, open in succession between March to October.

Abutilon theophrasti Medik. (syn. Abutilon avicennae Gaertn.) (Malvaceae): Eng.- Button weed; Kash.- Yechker / Tchaungan [W]. A tall bushy annual growing on sub-alpine slopes and as a weed in croplands and gardens. Leaves heart shaped, velvety. Flowers yellow; fruits button shaped. Branches used for making rugs and baskets. Fruits and seeds edible.

Acantholimon lycopodioides (Girard.) Boiss. (Plumbaginaceae): Ladakhi- Burch [W]. A medium sized prickly and densely tufted shrub growing as a woody component in Ladakh. Branches stout, densely leafy. Flowers pale pink, in dense clusters. Twigs used for roofing of houses and as fuel (Fig. A 2).

Acer caesium Wallich ex Brandis (Sapindaceae): Eng. - Indian Mapple; Kash. – Kanar /Kanzul [W]. A fairly large deciduous tree growing as isolated tress in coniferous forests at 2400- 3000 m. Leaves large with a heart shaped base and with 5 unequal broad long-pointed coarsely saw toothed lobes, usually pale and glaucous beneath, red when young. Flowers greenish-yellow, appearing with the young leaves, in branched flat-topped clusters. Wings of fruit divergent or sometimes overlapping, nuts dark-brown, with a hump-like swelling. Grown in gardens as an ornamental. In earlier times its wood was used in carving and for making bowls, plates and ladles.

Acer negundo L. (Sapindaceae): Eng.- Box Elder [NI]. A fast growing, bushy-headed deciduous ornamental tree of medium to large size. The young shoots are bright green and the leaves are pinnate; leaflets, ovate or oblong-lanceolate, coarsely serrate, bright green above, paler beneath, turning yellow in autumn. Flowers appearing before leaves, yellowish-green. The staminate flowers are borne in pendulous corymbs while the pistillate in pendulous racemes, petals wanting. Fruit glabrous, wings curved inward, acute angled, yellow-white in long racemes.

Acer palmatum Thunb. (Sapindaceae): Eng. - Japanese Mapple [NI]. A shrub or round-headed deciduous tree with glabrous branchlets, petioles and peduncles. Leaves deeply lobed, doubly serrate or incised, glabrous, fresh green underside, lighter carmine- red at fall. Flowers small purple, in glabrous erect corymbs, fruits with wings spread at obtuse angles. Two sub species are found growing in open grasslands and occasionally in gardens as shady ornamentals viz., *atropurpureum* (leaves medium sized, deeply 5 lobed, nearly black- red and remaining so until fall) and *disecetum rubrifolium* (growth dense, branches often twisted; leaves are divided to the base into 5-7 pinnatifid lobes, brownish red, later green; fruit wings spread at obtuse angles).

Acer platanoides L. (syn. Acer platanoides cv. schwedleri K.Koch.) (Sapindaceae): Eng.- Norway Mapple [NI]. A large ornamental tree with round spreading head. Leaves 5-lobed, cordate at base, remotely sharp dentate, lobes acute, bright red when young changing to dark green, light green and lustrous beneath. Flowers yellowish green in erect glabrous corymbs. Fruits glabrous, pendulous with horizontally spreading wings.

Acer sterculiaceum Wallich. (syn. Acer sterculiaceum thomsonii (Miq.) Murray (Sapindaceae): Eng.- Himalayan Mapple; Kash.- Kanar [W]. A deciduous tree, up to 20 m tall, with spreading habit; very large palmately lobed leaves which are green beneath; young shoots are covered with brownish hairs, often gives good autumn colour. Flowers in clusters appearing before the leaves. Fruits large winged, usually rusty-haired and borne on leafless side branches in drooping clusters; nuts pale brown. Grows in northern Indian forests at altitudes of 1800 – 3000 m. Wood used for making agricultural implements.

Achillea filipendulina Lam. (Asteraceae): Eng.- Fernleaf Yarrow/Nose Bleed [NI]. A herbaceous, perennial border ornamental species with mid- green leaves and compact, wide clusters of lemon yellow flowers. *Coronation Gold* is the common variety under cultivation in gardens. Carries flat heads of deep yellow flowers from June to August.

Achillea millefolium L. (syn. Achillea lanulosa Nutt.) (Asteraceae): Eng.- Millfoil / Common Yarrow; Hindi- Gandana,Biranje-sif; Kash.- Berguir /Momudna/ Pahel Gausse; Ladakhi- Saijun [W]. A sparsely hairy erect perennial herb, occurring in West Himalaya in wastelands and borders of cultivated fields. Leaves with 2 -3 pinnatisect angular lobes. Heads corymbose, flowers white. Leaves and flower heads are used as tonic, diaphoretic, for relieving tooth ache and for colds. In Ladakh the flowering tops are taken in the form of decoction to cure liver troubles (Fig. A 3).

Achillea tomentosa L. (Asteraceae): Eng.- Woolly Yarrow [NI]. Herbaceous perennial suitably planted for ground cover, for a rock garden or a border. It has long narrow, downy, grey- green leaves and bright yellow densely packed flower heads.

Achyranthes bidentata Blume. (Amaranthaceae): Eng.- Ox Knee; Hindi- Puthkunda [W]. Perennial herb, 70- 120 cm tall, among scrubs in ravines and rock crevices of sub-alpine areas. Stems green, or tinged purple, angulate, branches opposite. Flowers greenish yellow. Seeds used for abdominal pains.

Acomastylis elata var. elata (syn. Geum elatum Wall. ex G. Don.) (Rosaceae): Eng.- Herb Bennet; Kash.- Gogal mool [W]. A stout clump forming perennial inhabiting alpine meadows at 3500 – 5400 m altitudes. Leaves irregularly toothed; radical leaves interrupted pinnate. Inflorescence terminal, cymose, 2-6 flowered; flowers yellow. Root extract used for dysentery and leucoderma.

Aconitum chasmanthum Stapf. ex Holmes (syn. Aconitum dissectum Watt.)(Ranunculaceae): Hindi- Balnag [W]. An alpine perennial tuberous herb endemic to Kashmir Himalaya; leaves palmate-partite, racemes dense, bluish green.Tuberous roots are the source of drug aconite considered suitable substitute for the drug from *Aconitum napellus* (Fig. A 4).

Aconitum dissectum D Don. (syn. Aconitum napellus var. rigidum Hook. f. & Thoms., Aconitum ferox Wall ex Ser.) (Ranunculaceae): Eng.- Monk's Hood / Wolf's Bane; Hindi- Bish [W]. Hardy herbaceous perennial growing wild in alpine/sub-alpine forest slopes and also cultivated in gardens. Leaves dark-green, deeply divided to segments. Racemes of deep blue coloured flowers with hoods. Flower from July to August. Roots used as antipyretic, diaphoretic and diuretic.

Aconitum heterophylloides (Burhl) Stapf. (syn. Aconitum deinorrhizum Stapf.) (Ranunculaceae): Hindi- Mohra Bish / Safed Bikh [W]. A perennial herb of alpine ranges of Baderwah. Root and leaves used in rheumatism, rheumatic fever and headache. Plant has been used for poisoning arrows for killing wild life.

Aconitum heterophyllum Wall ex Royle (syn. Aconitum atees Royle.) (Ranunculaceae): Eng.- Indian Atis; Hindi- Tis; Kash.- Hongi safed

Patres; Ladakhi- Ponkar / Buana [W]. A tall perennial tuberous herb found on open sub-alpine and alpine zone of N-W Himalayas at 2000 to 3500 m altitudes. Flowers blue, helmet shaped. Roots/ tubers used as astringent, stomachic, digestive tonic and for bleeding piles. Also useful in diarrhoea, cough and vomiting of children and for deworming. Tubers were being used for washing hair.

Aconitum napellus L. (Ranunculaceae): Eng.- Wolf's Bane; Kash.- Mohand Gooje; Ladakhi- Aitvish [W]. A herbaceous perennial, up to 1 m tall, growing in alpine forest slopes. Leaves and stem hairless. Leaves round palmately divided into 5-7 deeply lobed segments. Flowers dark purple, helmet shaped. Used for body aches, coughs and sore throats.

Aconitum rotundifolium Kar. & Kir (Ranunculaceae): Eng.- Roundleaf Monk's Hood; Ladakhi – Charkhan [W]. A biennial herb of Himalayan forest slopes, with paired tubers. Stem erect, with rosette of leaves at the base. Leaves circular or ovate, incised-toothed to deeply cut. Flowers bluish, bilabiate at apex. Herb used for relieving headaches.

Aconitum violaceum Jacq. ex Stapf. (Ranunculaceae): Eng.- Atis /Voilet Monk'sHood; Kash.- Laderwal Posh / Mohand /Mori; Ladakhi- Nyaman / Jimba [W]. Robust erect perennial tuberous herb growing on open alpine slopes of Himalayas at 3600 – 4800 m. Flowers deep blue, shaded with white. Tubers are eaten by hill men as a pleasant tonic, especially after a prolonged illness. Also used medicinally as astringent tonic and in diarrhoea and cough. Dried seeds and roots, mixed with honey and animal fat, are used as tonic during convulsions.

Acorus calamus L. (Acoraceae): Eng.-Sweet Flag /Calamus; Hindi- Safed bach; Kash.- Wai [W]. A perennial aromatic herb growing in north-western Himalayas above 1800 m altitude, chiefly in marshy and moist situations. Leaves long, lanceolate with parallel venation. Aromatic rhizomes are used medicinally as carminative, stimulant to central nervous system and as tonic. Leaves and rhizomes also have usefulness for flavouring drinks and food stuffs and for preparing insecticides. The plant has social significance as its rhizomes used to be placed on a specially prepared rice bowl and be sighted first by all family members (as an auspicious object) in the mornings, especially on New Year Days (Fig. A 5).

Actaea spicata var. acuminate (Wall ex Royle) Hara (Ranunculaceae): Eng.- Himalayan Baneberry; Kash.- Girbare /Mamera [W]. Erect perennial herb, in forests and scrubs, usually in moist shady places above 2400 m. Leaves toothed, bipinnate compound with lance shaped leaflets. Flowers white,in clusters; fruit a berry. The dried roots used as emetopurgative. Powder of the plant used as insecticide. Root/fruit extract used for high fever and skin eruptions (Fig. A 6).

Actinidia chinensis Planch. (Actinidiaceae): Eng. - Chinese Gooseberry [W]. A vigorous twining climber of ravines with densely reddish hairy branchlets. Leaves long petioled, orbicular or oval, cordate at base, usually rounded or emerginate at apex, dark-green above, whitish stellate pubescence beneath. Flowers creamy white, turning to buff-yellow, fragrant produced in axillary clusters. Fruit edible, green turning brown on maturity, resembling a large Goose berry with a similar flavour.

Adiantum capillus –veneris L. (Pteridaceae): Eng.- Maidenhair fern / Venus hair fern; Hindi- Hansraj; Kash.– Gewthir [W]. Perennial fern growing on moist mud walls, and damp shaded places. Fronds arising in clusters, bi-pinnate, light green with purple black wiry stalks. Grown and better suited for cool green house and for growing indoor potted plants. Used as demulcent, expectorant and diuretic.

Adiantum venustum D. Don. (Pteridaceae): Eng.- Himalayan Maiden-hair fern; Hindi- Hansraj; Kash.- Gewthir/ Wushan [W]. Perennial rhizomatous herb inhabiting rock crevices, forest slope of N-W Himalayas from Kashmir to Sikkim at 1800 to 2500m altitudes. Fronds green, triangular with numerous fan shaped segments. Leaves have expectorant, diuretic, purgative and deobstructant properties. Commonly used in diseases of chest as demulcent and expectorant in chronic catarrh. Also used in ritual bath of ladies after child birth.

Adonis aestivalis L. (Ranunculaceae): Eng.-Pheasants' Eye; Kash.-Beauket [W] Annual herb, in cultivated fields. Flowers saffron coloured. Leaves finely divided. Plant used as cardio tonic and diuretic while flowers are used as laxative. Leaf extract is used for boils.

Adonis chrysocyathus Hook. f. & Thoms. (Ranunculaceae): Eng.- Yellow Himalayan Oxeye Daisy; Ladakhi- Tzemetz [W]. Perennial herb, with a long vertical scaly rootstock, in alpine vegetation above 3000 m. Flowers bright yellow. Leaves leathery, radical compound, segments linear. Powdered flowers used against hyperacidity (Fig. A 7).

Aegilops tauschii Cosson. (syn. Triticum aegilops P. Beauv. ex Roem & Schult.; Triticum tauschii (Coss.) Schmalh.) (Poaceae): Eng.-Tausch's Goatgrass [W]. Annual herb growing in grasslands and in cultivated fields. The grass species is a valued fodder plant.

Aschynomene indica L. (Syn. Aeschynomene cashemeriana Cambess. (Fabaceae): Eng.- Indian Joint Vetch; Kash.-Gaente Moughte [W].

Annul herb, grwoing along borders of rice fields. Leaves compound with 15-30 pairs of small leaflets, rounded at tip. Pale yellow flowers are borne on floral racemens. Pods linear jointed. Seed Kidney shaped and brown. Seeds are often found as adult rant in milled rice.

Aesculus hippocastanum L. (Sapindaceae): Eng.- Horse chestnut; Kash.- Hauna doon [W/UC]. A deciduous tree growing as a woody element of sub-alpine forests. Also planted on roadsides and other exposed areas as an avenue tree and for its wood. Leaves long stalked, palmate. Flowers white (red streaked) in narrow- pyramidal clusters. Fruit ovoid, dark brown, shining. Fruit used as a cataplasm for mammary and other cancers. Bark used for healing ulcers. Seed tincture applied against painful piles.

Aesculus indica (Wall. ex Cambess.) Hook. (Sapindaceae): Eng.- Himalayan Horse chestnut; Hindi- Pangla; Kash.- Hauna–doon [W/ UC]. A magnificent spreading, rounded deciduous tree (planted and naturalized) of N W Himalaya. Leaves compound, 7-palmate, composed of obovate to lance shaped leaflets, bronze when young, becoming later dark glossy green, turning yellow in autumn. Bark peeling in long strips. Fruit is a large capsule with smooth leathery walls. *Sydney Pearce* is the common Horse chestnut variety grown in gardens around Srinagar as an avenue tree. It is a free flowering form of upright habit, with dark olive green leaves. Flowers in large panicles, petals white marked yellow and prettily suffused pink. Leaves and twigs used as fodder. Seed oil is used for rheumatic pain, frost bites. Bark/nut infusions used for treating ulcers. Seed tincture is applied to painful piles and as an anti-carcinogen. Wood used as fire wood. In earlier times wood was also used in making water troughs and filling of boats (Fig. A 8).

Agapanthus campanulatus Leight. (Amaryllidaceae):Eng.- African Lily/ Bell Agapanthus [NI]. A hardy perennial growing in campact clumps with fleshy roots. Leaves mid- green and strap shaped. Crowded umbels of blue flowers are borne in late summer. *Iris* is the common variety under cultivation in gardens.

Agaricus bisporous (J.E. Lange) Imbach (Agaricaceae): Eng.- Common mushroom; Kash.- Hedoure [W/UC]. A saprophytic edible fungi usually growing in fields and grassy areas, usually after rains. Pelius pale grey-brown, flattened convex, button shaped; hymenium free, stipe with a ring. Also grown commercially.

Agave americana L. (Asparagaceae): Eng.- Century Plant; Hindi- Cantala [NI] A mono-carpic succulent, producing basal rosettes of spreading, lance shaped, spine-tipped and spiny margined, grey-green leaves.

Inflorescence 5-8 cm high, with 25-30 branches in loose panicles or racemes; flowers greenish with yellow anthers. Planted for hedging in gardens. *Agave americana* var. *marginata* Trel. is having dark green leaves with yellow margins, twisted.

Ageratum houstonianum Mill. (syn. Ageratum maxicanum Sims.) (Asteraceae): Eng.- Floss Flower /Blue weed [NI]. An annual, bushy dwarf ornamental planted in gardens/parks. Leaves hairy and dark green, heart shaped. Blooms from early summer till October. Flowers bright blue in colour in truses. *Blue capis* is the commonly cultivated cultivar.

Agrimonia eupatoria L. (Rosaceae): Eng.- Common Agrimony [W]. A perennial herb, up to 60 cm tall growing in pastures, land clearings, road sides and forests. Stem unbrached with a long flowering spike bearing yellow flowers. Leaves bi-pinnate with dentate margins, hairy all over, more towards midrib. Flower tops used to treat gastrointestinal disturbances.

Agropyron striatum Nees ex Steud. (Poaceae) [W]. The plant grows across grasslands of North-Western India including Kashmir Himalaya. Spikes slender; lemma usually has oppressed hairs on whole surface. A valuable fodder plant.

Agrostis gigantea Roth. (Poaceae): Eng.- Black Bent; Ladakhi- Howakzet [W] A perennial grass, 30- 90 cm tall; stem unbrached, glabrous and terete in cross section. The nodes of culms are red and glabrous.The plant generally inhabits worn-out soils in agricultural fields and is a favoured fodder plant for milking cows.

Agrostis munroana Aitch. & Hemsl. (Poaceae): Eng.- Creeping Bent Grass; Kash. -Kardkani [W]. Annual tufted herb, up to 70 cm tall. The plant, forming a part of alpine pastures of N-W Himalayas, is acceptable to sheep, goat and cattle. Panicles open.

Ailanthus altissima (Mill) Swingle (syn. Ailanthus glandulosa Desf.) (Simaroubaceae): Eng.- Ailanto; Kash.- Bouhdeh [W/UC]. A large spreading deciduous tree with distinct ash like, large, oblong elliptic, pinnate leaves, 40-60 cm long, leaflets to 10-15 pairs which appear reddish green and later turn mid-green, bluish beneath, stalked usually truncate at base, acute or acuminate. Flowers insignificant, greenish in 10-20 cm long panicles; staminate flowers ill-scented. Bark longitudinally striped, smooth with characteristic white grooves; samaras 3-4 cm long, numerous, twisted, light brown. Planted for wood as well as an avenue tree in gardens and public places. Wood used for making fruit boxes and as fuel.

Ajania fruticulosa (Ledeb.) Poljakov. (syn.Tanacetum fruticulosa Ledeb.) (Asteraceae): Eng.- Shrubby Ajania; Ladakhi- Yakzae [W]. Twiggy hoary perennial sub-shrub on dry alpine slopes. Leaves tri-partite. Flower heads yellow, mostly stalked, congested at the top of shoots. In Ladakh the dried plants are used for thatching roofs of houses.

Ajuga integrifolia Buch.- Ham (syn. Ajuga bracteosa var. alba (Gurke) Engl.) (Lamiaceae): Eng- Bugle; Hindi- Bungle; Kash.- Jauni-Audam [W]. A perennial herb growing from Kashmir to Nepal up to 2100 m altitudes in W Himalayan woodlands on rocky slopes. Leaves toothed and reddish-brown beneath. Flowers pale lilac. Whole plant and aromatic leaves used as stimulant and diuretic. Plant yields a bitter astringent given in the treatment of fevers (Fig. A 9).

Alangium chinense (Lour.) Harms (syn. Marlea begonifolia Roxb.) (Cornaceae) [W] Shrubs or small tree inhabiting exposed places or forest margins below 2400 m altitude. Leaf blade ovate or orbicular to cordate, petioles reddish. Drupes ovoid, fleshy, dark purple. In folk medicine the plant is used for snake bites, rheumatism and wounds.

Albizia julibrissin var. rosea (Carriere) Mouillefert (Fabaceae): Eng.- Pink Silk Tree;Kash.- Panjabe Kikar [NI]. An introduced and naturalized, small, spreading, graceful deciduous avenue tree with angular branches when mature; fern-like, light to mid-green leaves, 30-45 cm long, pinnate, with sickle-shaped, oblong, leaflets. Flowers bright pink, on stalked heads at the ends of the higher branches.

Alchemilla alpina L. (Rosaceae): Eng.- Alpine Lady's Mantle [W].A perennial alpine/ sub-alpine herb with woody rhizome, growing under moist situations in meadow lands. Leaves orbicular-cordate, 7-9 lobed; flowers yellow. Weak stems are silkily hairy and grow from a basal rosette. Plant used to treat acute intestinal troubles and irregular mensuration.

Alisma plantago-aquatica L. (Alismataceae): Eng.- Water Plantain [W]. Perennial herb, usually in shallow water ponds, swampy rice fields. Flowers white. Rhizomes and leaves edible.

Allardia tomentosa Decne. (syn. Waldheimia tomentosa (Decne.) Regel) (Asteraceae):Eng.- White leaf Ground Daisy; Ladakhi- Solo-karpo [W]. A loosely clustered perennial herb inhabiting alpine grasslands at 3500- 5000 m altitudes. Flower heads solitary with many pinkish-white linear ray florets. Leaves cut to segments, covered with dense soft white-woolly hairs. Extract of the achenes of the plant in raw form used in acidity. The crushed fresh leaves are applied as a poultice in arthritis.

Allium ampeloprasum *L.* **(Amaryllidaceae): Eng.- Wild leek; Hindi- Jangli Lahsun; Kash.- Wan Rohan [W]**. A wild bulbous herb of W. Himalayas from Kashmir to Kumoan, occurring as a specific weed in saffron fields of Kashmir. Substitute to garlic. Leaves used as minor vegetable and for seasoning dishes. Bulbs used to relieve rheumatism, control skin diseases, kill germs and worms, biliousness and bleeding piles.

Allium-carolinianum DC. (syn. Allium thomsonii Baker.) (Amaryllidaceae): Eng.- Wild onion: Ladakhi- Arum / Skiche [W]. A perennial bulbous herb growing on alpine grasslands and slopes at 2500- 3500m altitudes, near moist situations. Bulbs cylindrical, outer coats coriaceous.Tepals rose or pink. Crushed bulbs are used as poultice on swollen joints to relieve pain. Mature bulbs are fried and used as vegetable in winters. In Ladakh it is fed to ladies after delivery (Fig. A 10).

Allium caesiodes Wendelbo (syn. Allium kachrooi G. Singh) (Amaryllidaceae): [W]. Perennial bulbous herb, growing in sub-alpine moist slopes at 1700-2000m.Used as a vegetable by villagers.

Allium cepa L. (Amaryllidaceae): Eng.- Onion; Hindi – Piyaz; Kash. - Koushre Ganda [UC]. A bulbous biennial herb cultivated for its bulbs. Scape 60-75 cm in height, ventricose; leaves sub- distichous, fistular, acute; flowers greenish white, stellate, in large globular umbles, without bulbils, enclosed by a thin membranous spathe; fruit a capsule. Bulb single roundish.The local cultivar under cultivation has yellowish – white outer scales and dull white flesh, strongly flavoured. Foilage as well as mature bulbs are consumed as vegetable. Plant susceptible to downy mildew and bulbs have low shelf life.

Allium cepa L. var. viviparum (Metz.) Alef. (Amaryllidaceae): Eng- Top Onion / Shallots; Kash.- Praan [UC]. An aggregate type onion with number of bulbs held together, commonly cultivated for its bulbs and its fistular leaves. Scape long, shining; spathes long beaked enclosing number of bulbils. Bulbs are used as condiment and to flavour various food preparations, however its use is restricted to local muslims only. Foliage is also cooked as vegetable (Fig. A 11).

Allium consanguineum Kunth. (Amaryllidaceae): Eng.- Wild Onion; Kash.- Ruhan Hakh. [W]. Perennial bulbous herb, on moist rocky cliffs at 2300-2500 m. Flower tepals yellow or pink. Bulbs cylindrical, outer coat coriaceous, striate. Leaves used as vegetable.

Allium griffithianum Boiss. (Amaryllidaceae) [W]. Perennial bulbous herb growing on Karewa lands, especially around saffron fields. Umbells hemi-spherical with light pink flowers. Leaves used as a condiment.

Allium humile Kunth. (Amaryllidaceae) [W]. Perennial bulbous herb, usually on moist alpine slopes at 2500-3000 m. Bulbs cylindrical with reticulate coat. Flower heads lax, tepals white. Leaves used as vegetable.

Allium jacquimontii Kunth. (Amaryllidaceae) [W]. A perennial bulbous herb, growing on some sub-alpine areas at 2000-2500 m. Bulbs solitary and oval. Leaves used as condiment and as herbal vegetable.

Allium macleanii Baker (Amaryllidaceae): [W]. Perennial bulbous herb, growing in sub-alpine grasslands, in moist and shaded situations. Umbells spherical, crowded with many purple flowers.

Allium nigrum L. (Amaryllidaceae): Eng.- Broad Leaved Leek: [W]. Biennial herb, usually on sub-alpine grasslands Bulbs asymmetrical, round umbels with whitish flowers. Leaves used as herbal vegetable.

Allium oreoprasum Schrenk. (Amaryllidaceae) [W]. Perennial bulbous herb, growing across cold-arid zone of Ladakh. Bulbs tufted, light -brown, reticulate and fibrous. Tepals pale-red with dark purple veins. Bulbs used to increase stamina.

Allium porum L. (Amaryllidaceae): Eng. – Leek; Hindi- Vilaiti Pyaze; Kashmiri – Viliati Rohan [UC/W]. A hardy biennial bulbous plant cultivated as well as found in wild state, however its cultivation is restricted to some areas only where it is cultivated along with garlic. Leaves distichous, sheath usually smooth, blade linear to long – lanceolate, 30-60 cm long and up to 4 cm wide; flowers light purple in globose umbels. Bulbs narrow and tunicate. Leek bulb is larger but mildly flavoured and is, therefore, used as a substitute to garlic.

Allium przewalskianum Regel. (syn. Allium stoliczkii Regel.) (Amaryllidaceae): Eng.- Wild Onion; Ladakhi – Skotse [W]. Pernnial bulbous herb growing in alpine ranges above 2500 m. Dried bulbs are given to women after delivery to boost energy. Fresh or dried bulbs are particularly used for preparation of a Ladakhi dish *–Thupka.*

Allium rosenbachianum Regel. (Amaryllidaceae) [W]. A perennial bulbous herb growing in alpine grasslands. Scape tall, with a spherical umbel of many reddish-purple flowers with long pedicels. Leaves used as herbal vegetable by tribals.

Allium roylei Stearn. (Amaryllidaceae) [W]. Perennial bulbous herb, growing on alpine forest slopes. Bulbs ovoid, reddish brown. Tepals reddish.Leaves used as herbal vegetable.

Allium rubellum M. Bieb. (Amaryllidaceae): Kash.-Wan Pran [W]. A wild bulbous herb growing around saffron fields of Kashmir. Flowers rosy pink. Its leaves are used as stimulant and as laxative (Fig. A 12).

Allium sativum L. (Amaryllidaceae): Eng. – Garlic; Hindi- Lehsun; Kash.- Ruhoun.; Ladakhi- Semsetz [UC]. Bulbous perennial herb. Leaves elongated, flat. The plant is grown for its compound bulbs. In villages the spathes of the plants are also sundried and cooked with fish. Cloves are smaller, enclosed within dark pinkish scales and are strongly flavoured. In Ladakh fresh cloves are taken in pasted form to relieve indigestion and gastric ailments.

Allium schoenoprasum L. (Amaryllidaceae): Eng.- Wild Chives; Hindi- Piazi [W] Perennial bulbous herb growing in humus rich soils on alpine slopes. Leaves and bulbs used as condiment/vegetable. Scapes hollow and tubular. Flowers pale purple. Bulbs slender, conical and grow in clusters.

Allium stipitatum Regel. (Amaryllidaceae): Eng.- Drumstick Allium [W]. Perennial herb, growing around sub-alpine grasslands, in shaded situations. Umbells spherical on tall spikes bearing lilac flowers. Leaves used medicinally.

Allium tuberosum Rottl. ex Spreng (Amaryllidaceae): Eng. – Chinese Chives; Hindi- Banga Ganda; Kash.- Wan Pran; Ladakhi- Rirgok [W]. Grows wild on hilly slopes at Pahalgam and Sonmarg at 2200-2500 m amsl. Leaves and inflorescences edible. It occasionally forms bulbs as it stores its reserve food in well developed rhizomes which look like tubers hence the name. Leaves narrow, grass like, spreading and drooping. Flavours garlic like and used in small quantities to season dishes. In Ladakh aerial portion of the plant is crushed and made into tablets which are used against constipation.

Allium umbelicatum Boiss. (Amaryllidaceae): Eng.- Wild Onion [W]. Prennial bulbous herb, growing as a weed in orchards of Jammu region. Umbells hemi-spherical with pink flowers. Bulbs oval. Leaves tubular, used as vegetable.

Allium victorialis L. (Amaryllidaceae): Kash.- Wan Pran [W]. Perennial bulbous herb, growing on moist shady cliffs, around alpine slopes. Bulbs cylindrical, coats reticulate, fibrous and have antifungal and anti -bacterial properties. Leaves used for diarrhoea.

Alnus nepalensis D. Don (Betulaceae): Eng.- Nepalese Alder; Hindi – Udis [W]. A large deciduous conical tree with dark green bark which becomes silvery grey in the open. Leaves are elliptic to ovate with rounded or shortly pointed apex, or with an undulate margin, finely hairy on the veins beneath when young. Plant is monoecious. Male catkins slender, female clusters cone like.

Alnus nitida (Spach.) Endl. (Betulaceae): Eng.- West-Himalayan Alder; Hindi- Kunis; Kash.- Champ Kul / Kanzal [W/UC]. A large deciduous conical tree with ovate to elliptic, and thin leaves, narrowed from about the middle to the apex, usually toothed. Monoecious flowers in catkins. Male catkins usually grouped in 5, female cones mostly solitary, axillary nutlet with a narrow leathery margin; bark smooth dark brown and shining, becoming rough and furrowed on old trunks. Twigs used for making baskets.

Aloysia triphylla (L' Herit.) Britton. (Verbenaceae): Eng.- Lemon Verbena [NI]. A bushy upright, deciduous half-hardy ornamental shrub. Leaves lanceolate, lemon scented, rough, bright green in whorls of 3, short stalked, acuminate, base cuneate, margins entire, glandular spots beneath. Flowers, pale lilac to white in slender panicles.

Althaea officinalis L. (Malvaceae): Eng.- Marsh Mallow; Hindi- Gul-khera; Kash.-Saze-posh/Saze-mool [W]. Annual or perennial herb commonly growing in gardens and wastelands. Stems simple 90- 120 cm; leaves soft, velvety, short petioled. Flowers pale coloured. The flowers have cooling and diuretic properties. Roots are astringent and demulcent. Roots/flowers/seeds are used for the treatment of cough and cold and as blood purifier. Also used by ladies for a healthy hair.

Althaea rosea L. (syn. Althaea rosea (L.) Cav.) (Malvaceae): Eng.- Hollyhock; Hindi- Gul-khera; Kash.- Saze-posh [UC/W]. An annual/ biennial herbaceous border plant. Leaves light green, rough and hairy. Stems rigid bearing single or double flowers in shades of pink from July to September. *Single Mixed* and *Double Triumph Mixed* are the two commonly cultivated varieties (Fig. A 13).

Amaranthus blitum var. oleraceus (L.) Costus (Amaranthaceae): Eng.- Purple Amaranth; Hindi –Chauli; Kash.- Bustanafroz /Lissa [W/ RC]. A tall vigorous cultivated herb, 90-150 cm tall. Leaves long petioled, elliptic or ovate- lanceolate; flowering spikes purplish- crimson, pale-green or dull white. Spikes in dense thyrses; seeds orbicular, compressed, greenish-white, testa crustaceous. In recent past the grains of the plant were toasted and popped or ground into flour for making bread, and taken especially after a fast. The parched and grounded seeds were also sometimes eaten with milk. Traditionally the amaranth grain was rated as a heating food by Kashmiris and mostly consumed after milling. Hindus used to eat it on their fast days. Kashmiri washermen used to extract an alkaline substance from the burnt ashes of the plant for dying woollen clothes.

Amaranthus caudatus *L.* **(Amaranthaceae): Eng.- Love Lies Bleeding / Pendant Amaranth [NI]**. A hardy annual famous for its plume like drooping flower racemes. Leaves ovoate, light green and entire. Flower heads of crimson colour appear from July to October. Variety *viridis* with pale green variegated foliage is commonly used in flower arrangements.

Amaranthus hybridus L. (Amaranthaceae): Eng.- Green Amaranth; Hindi-Ramdana; Kash. Ganhaar [W/RC]. A tall robust annual, wild as well grown as a grain yielding plant. Leaves purple or green, alternate; spikes slender, lax with dull green flowers; seeds orbicular and shining black. The tender shoots and leaves used as vegetable. The seed is popped for a local confectionary items. The grains are also parched, milled and dough patted into *chapattis*.

Amaranthus spinosus L. (Amaranthaceae): Eng.- Spiny Amaranth; Ladakhi- Swaktzetz [W]. Annual branched herb growing in Ladakh, mostly as a weed. Flowers green in axillary clusters in lower parts of the plant. Leaves cooked as vegetable.

Amaranthus tricolor L. (Amaranthaceae): Eng.- Josephs Coat; Hindi-Lal Sag [UC] A strong growing bushy plant, 2-3 ft tall grown for its attractive foliage. The ovate, pointed leaves are scarlet or crimson, overlaid with yellow, bronze and green. The racemes of flowers are dense erect and brush like, deep red in colour. Cultivated as border plant or pot plant for cool green house plant.

Amaranthus viridis L. (Amaranthaceae): Eng.- Green Amaranth; Hindi-Chauli; Kash.- Lissa [W]. An annual herb growing wild as a common weed in waste places, orchards and maize fields. Flowers green. Seeds dark-brown/ black, compressed. Tender leaves are commonly used as vegetable in early spring.

Amorpha fruticosa L. (Fabaceae): Eng.- False Indigo bush [UC]. A fast growing, spreading ornamental shrub. Leaves pinnate with oval or oblong leaflets. Orange or yellow anthered, purple-blue flowers are produced in narrow racemes. Pods usually with two seeds.

Anagallis arvensis L. (Primulaceae): Eng.- Scarlet Pimpernel; Hindi-Junge-muri; Kash.- Ansu/Chari saban [W]. Weak summer annual creeping herb, growing as a weed of cultivation and in waste lands. Leaves bright green, ovate, sessile, opposite. Flowers reddish. Plant used as antidote to different forms of poisoning.

Anagallis linifolia L. (Primulaceae): Eng.- Blue Pimpernel [UC]. An annual- sub-hardy ornamental herb grown for edging borders of sub-alpine gardens. Leaves are dark- green and narrowly lanceolate. Flowers are saucer shaped, gentian blue with red- purple underside.

Anaphalis nepalensis (Spr.) Hand.-Maz. (Asteraceae): Ladakhi- Kricheng [W]. Perennial herb of alpine pastures. Flowers white. Leaves ovate-lanceolate. Plant used for post -natal bath of ladies to relieve stiffened muscles.

Androsace mucronifolia Watt. (Primulaceae): Ladakhi- Singtezing [W].A tall, caespitose stoloniferous perennial herb on alpine slopes. Flowers rose purple in compact umbels. Leaves pale-green in dense rosettes. Powdered plant is taken for ringworm infection (Fig. A 14).

Androsace sarmentosa Wall. (Primulaceae): Eng.- Rock Jasmine; Ladakhi- Machetz [W]. Glandular pubescent perennial stoloniferous herb on alpine slopes. Leaves deep green in rosettes and elliptic- ablanceolate. Flowers bright pink with a yellow centre. In Ladakhi Amchi system of medicine powdered flowers, mixed with iron powder, are recommended for flatulence and acidity.

Androsace sempervivoides Jacq. ex Duby (Primulaceae):Eng.- Candelabra/ Sempervivum leaved rock Jasmine; Ladakhi- Singtez [W]. A small stoloniferous herb inhabiting open alpine slopes above 3000 m and forming large mats or compact cushions. Leaves densely rosulate, ciliate; scape solitary. Flowers pink with dark eyes (Fig. A 15).

Androsace septentrionalis L. (Primulaceae):Eng.- Northern Fairy Candelabra; Ladakhi- Singtez [W]. Perennial herb growing on dry alpine slopes. Basal leaves densely packed and candelabra. Stem lopped by tiny numerous white flowers. Ash of dried roots is applied on wounds as antiseptic.

Andropogon munroi Clark. (syn. Andropogon tristis Nees ex Hack) (Poaceae) [W]. Annual herb forming important part of alpine pastures in North western Himalaya. Racemes more than two.

Anemone coronaria L. (Ranunculaceae): Eng.- Poppy flowered Anemone [UC]. A hardy herbaceous tuberous rooted perennial grown as spring ornamental. Flowers cup-shaped, poppy like and leaves are multilobed and deep- green. Flowers produced in April and are white or in several shades of blue and red.

Anemone falconeri T. Thoms. (Ranunculaceae): Ladakhi- Pharenge [W]. Perennial herb, with a stout, fibrous rootstock; usually common in deciduous forests on soils covered with rich litter. Radical leaves numerous, long petioled and covered by long patent hairs. Flowers white. In Ladakhi system of medicine powdered plant is taken for cold, fever and sometimes for gastritis (Fig. A 16).

Anemone narcissiflora L. (Ranunculaceae): Eng.- Narcissus flowered Spring Anemone [W]. A tufted species bearing white or cream flowers, sometimes flushed pale purple in bud from May-June. Inflorescence 2–8 flowered. Growing on mountain alpine grasslands.

Anemone obtusiloba D. Don. (Ranunculaceae): Eng.- Wind Flower; Kash.- Kawe ashude [W]. Perennial tufted herb, with woody rootstock, common on moist open alpine slopes. Stem softly hairy, flowers white with bluish purple spots. Roots used to treat convulsions, also used as blistering agent (Fig. A 17).

Anemone rupicola Camb. (Ranunculaceae): Eng.- Rock Anemone; Ladakhi – Tscheska [W]. Perennial herb growing along alpine streams, stone cliffs and woods. Flowers showy and white. Powdered plant used as tonic and stimulant.

Anemone tetrasepala Royle. (Ranunculaceae): Eng. Four petalled Anemone [W]. A robust perennial with erect hairy stem. Basal leaves a heart or kidney shaped, deeply 5-lobed on long stalks. Stem leaves sessile. Flowers white with 4-5 obovate sepals, looking like petals. Grows on shady alpine slopes.

Angelica archangelica L. (Apiaceae): Eng.- Garden Angelica; Kash.- Pikhe Zuere [W] Perennial herb,up to 2 m tall, growing along stream banks,damp meadows and forest gullies across western Himalayas at 2500- 3800 m altitudes. Rootstock thick and branched. Flowers white in large umbels. Leaves, fruit and seeds aromatic. Fruit used in flavouring food articles. Root used in perfumery.

Angelica glauca Edgew. (Apiaceae): Eng.- Smooth Angelica; Hindi- Chora; Kash.- Gandrayani [W]. Perennial aromatic herb, usually among alpine scrubs on humus rich soils. Leaves bi-pinnate. Umbells of greenish white flowers. Root used against dysentery and constipation.

Anthemis cotula L. (Asteraceae): Eng.- Congress Weed /Stinking Chamamile; Kash.- Phakae gausse [NI]. Biennial herb growing as weed of grasslands, orchards as an invasive species. Leaves dissected aromatic with unpleasant smell. Flower heads white.

Anthriscus nemorosa (Bieb.) Spreng. (Apiaceae): Eng.- Chervil; Ladakhi- Sunakh [W]. Perennial herb, 1- 3m tall, common on ravines and road sides, mostly around Ladakh region. Leaves bi-pinnate. Flowers white. Inhaled smoke from the powdered plant is used for curing rheumatism. Paste of the plant used to cure inflammations.

Antirrhinum majus L. (Plantaginaceae): Eng.- Snap-dragon; Hindi – Sher-dahan [NI]. A herbaceous sub-shrub grown as valuable garden plants, having numerous varieties in different colours. Leaves ovate and glossy. Flowers are produced from June to October. Number of varieties with well shaped spikes of orange flushed- rose, pale yellow, white to dull white are cultivated in gardens and parks.

Aphragmus oxycarpus (Hook. f. & Thomson) Jafri (syn. Lignariella duthiei Naqshi (Brassicaceae) [W]. A dwarf herb, glabrous, purple. Radical leaves 4- 20, shortly petioled or sub-sessile; cauline leaves dislinet, oblong-elliptic, entire, sessile. Flowers white or pinkish. Fruits erect, linear, terete or sub-compressed. Seeds brown.

Apium graveolens L. (Apiaceae): Eng.- Garden Celery [NI]. Cultivated marginally for its leaves/seed. Leaves taken as salad, and as blood purifier. Seeds used as stimulant, carminative, nerve sedative and tonic

Aquilegia fragrans Benth. (syn. Aquilegia vulgaris subsp. pyreniaca (DC.) Hook. f. & Thomson. (Ranunculaceae): Eng.- Sweet scented Columbine; Kash.- Mameri [W]. Tall semi-glabrous perennial herb among alpine scrubs at 2000- 3500 m altitudes. Plants densely hairy with glands near the flowers. Flowers white and fragrant. Root extract used on inflammations (Fig. A 18).

Aquilegia nivalis (Baker) Falc. ex Jacks. (Ranunculaceae): Kash.- Mameri [W]. An erect perennial herb with woody root-stock on humus rich soils at altitudes of 3000 -4000 m. Leaves few and ternate, divided into 3-lobed leaflets. Flowers purple. Root used against inflammations (Fig. A 19).

Aquilegia pubiflora Wall. ex Royle (Ranunculaceae): Eng.- Columbine [W]. Erect perennial pubescent herb, in alpine slopes and pastures. Stems 30- 70 cm tall, pubescent. Radical leaves bi-ternate, leaflets sessile or with short petioles, trifid, hairy beneath. Flowers bluish-purple.

Arabidopsis pumila (Stepf.) Busch. (syn. Sisymbrium pumilum Stepf. ex Willd. (Brassicaceae): Eng.- Dwarf Rocket [W]. Annual herb, usually on mud wall tops and waste places. Flowers yellow. Seeds used for curing eye sores.

Arabidopsis thaliana (L.) Heynhold (syn. Sisymbrium thaliana (L.) Gray.) (Brassicaceae): Eng.- Thale Cress [W]. Sub-alpine annual herb, common in waste lands, roads and orchards. Flowers white. Seeds medicinal.

Arabidopsis wallichii (Hook. f. & Thomson) N Bush (syn. Sisymbrium wallichii Hook. f. & Thomson (Brassicaceae): Eng.- Thale Cress; Ladakhi- Pasaka [W]. A perennial herb on open marshy grassy slopes up to 3500 m altitudes. Stem simple or branched, hispidly hairy. Racemes up to 15 flowered. Flowers white or pale-pinkish. Radical leaves crowded, runcinately lyrate, terminal lobes large, cauline leaves few linear – lanceolate, pinnatifid or entire. In A*mchi* system powdered plant is used for treating measles and chicken pox.

Arabis kashmeriana Naqshi & Javed (Brassicaceae): Eng.- Rock Cress [W]. Low hardy annual, branched. Basal leaves oblong-ovate, glabrous, long petioled; cauline few, long petioled. Siliqua ascending. Inhabits meadow lands in alpine and sub-alpine areas. Seeds medicinal.

Araucaria columnaris (G. Forst.) Hook (syn. Araucaria excelsa (Lamb.) R. Br.) (Araucariaceae): Eng.- Araucaria / Island Pine [NI]. An evergreen dioecious coniferous tree with dark- green leaves, rigid and spine tipped, grown as an indoor pot plant. Tree attains 6 ft height and is pyramidal in outline with tiered whorls of branches. Used for landscaping.

Archangelica officinalis Hoff. var. himalaica Clarke (Apiaceae) [W]: An aromatic perennial alpine herb of Himalayas, inhabiting moist situations, mostly along streams. Flowers white.Rootstock used for intestinal problems, to promote digestion.

Arctium lappa L. (Asteraceae): Eng.- Beggers Buttons, Thorny Burr [W]. Tall biennial herb of shaded wastelands. Leaves long petioled and pubescent on lower surface. Heads purplish. Roots, leaves, fruits used in medicines- as blood purifying agents and as diuretic.

Arenaria foliosa Royle ex Edgew & Hook. f (Caryophyllaceae): Eng.- Sandwort; Ladakhi – Lingze [W]. A spreading, tufted perennial herb, inhabiting alpine grasslands above 3000 m. Flowers white, large. Plants in vegetative phases fed to sheep and goats. Leaves also used as vegetable in Ladakh.

Argemone mexicana L. (Papaveraceae): Prickly Poppy; Hindi- Shialkanta; Ladakhi- Gailshe [W]. A prickly annual herb naturalized in cold arid climate of Ladakh. Flowers yellow. In Ladakhi system of medicine an aqueous extract of the powdered leaves is used externally to treat eczema and eye diseases.

Arisaema jacquemontii Blume. (Araceae): Eng.- Jacquemont's Cobra Lily/ Cobra Plant; Hindi-Jangosh; Kash.- Hapat Gogij [W]. A perennial tuberous herb distributed in temperate Himalayas on scrubby alpine slopes. Bulbs sub-globose, 2-3 cm broad. Characteristically having white stripped spathe with long up-curved green or dark purple tail like tip. The tubers are used to treat chronic boils and skin eruptions. Also given to cattle for collics and to remove worms (Fig. A 20).

Arisaema flavum (Forsk.) Schott. (Araceae): Eng.-Yellow Cobra Lily; Hindi-Jangosh; Kash.- Hapat Gogij [W]. Perennial bulbous sub-alpine herb on scrubby slopes and humus rich soils. Leaves 2 or 1 pedate, leaflets 7-11, elliptic ovate. Flowers – spathe yellow green, purplish inside. Bulbs edible, also used for skin ailments.

Arisaema propinquum Schot. (syn. Arisaema wallichianum Hook. f.) (Araceae): Eng.- Snake Lily; Kash.- Hapat Gogij [W]. An erect perennial tuberous herb inhabiting open alpine slopes/rocky crevices of temperate Himalaya, at 3000- 4000m altitudes. Leaves with three leaflets which are rhombic- orbicular, shortly petioled. Spathe stripped with dark purple veins. Tubers used against skin eruptions.

Arisaema tortuosum (Wall.) Schott. (Araceae): Eng.- Whipcord Cobra Lily; Hindi- Bagh Jandhra; Kash.- Hapat Gogij [W]. A large clamp forming tuberous annual of Himalayan meadows. Spadix whip like, green. Fruits green maturing red. Crushed tubers used against poisoning and as anthelmintic; also used to treat colics in sheep and goats and to deworm them.

Arnebia benthami (Wall. ex G. Don) Jonhst. (syn. Macrotomia benthami Wall. ex. G. Don) DC.) (Boraginaceae): Eng.- Himalayan Arnebia; Hindi- Gaozaban; Kash.-Kahzaban /Gule- Kahzaban [W]. Perennial hairy herb, with thick rootstock, growing in W. Himalaya from Kashmir to Kumoan at 3000 to 4000 m altitudes. Stem densely covered with trichomes. Flowers purplish brown. Leaves and inflorescence of the plant used for the treatment of diseases of tongue, throat and other respiratory tract infections (Fig. A 21).

Arnebia euchroma (Royle.) Johnst. (Boraginaceae): Ladakhi- Deemok [W]. Erect perennial caespitose herb, on highland alpine slopes above 3500m altitudes. Flowers purple. Root used for cough and for profuse hair growth.

Arnebia guttata Bunge. (Boraginaceae): Ladakhi- Deemok [W]. Annual herb growing in Kashmir Himalaya above 3000m. Flowers heterostylous, in long coiled inflorescence with yellow petals. Leaves greyish-green and hairy. Powdered roots used to relieve kidney problems (Fig. A 22).

Artemisia absinthium L. (Asteraceae): Eng. -Wormwood; Hindi – Viliati Afsantin; Kash.- Tethwaen/Damer [W]. An aromatic, pubescent perennial herb distributed in sub- alpine Himalyas at 1700- 2100 m. Leaves bi- to tri-pinnate and spirally arranged. Flowering tops and leaves are source of a drug santonin which is used as vermifuge, stimulant, and anthelmintic and in curing digestive disorders. Plant has powerful influence on central nervous system. Dried plant parts used to protect garments from insects (Fig. A 23).

Artemisia biennis Willd. (Asteraceae): Ladakhi –Magrass [W]. A non-aromatic perennial high altitude herb of alpine ranges. Leaves alternate, hairless, pinnately lobed. Powdered seeds are taken with luke warm water regularly to decrease obesity.

Artemisia dracunculus L. (Asteraceae): Eng.- Tarragon; Ladakhi– Sache [W]. A perennial herb, up to 60 cm tall, growing in alpine grasslands and slopes. Flower heads small, round with yellow flowers. Leaves lanceolate, entire. Plant used for throat infections.

Artemisia indica Willd. (Asteraceae):Eng.- Mugwort; Hindi- Afsantin [W]. Perennial much branched herb/sub-shrub, commonly growing under forest shade in forest openings on humus rich soils at 1700 - 2200 m. Leaves shortly petioled, leaf blade abaxially densely gray, arachnoid tomentose. Flowers yellowish and used as a vermifuge and stimulant.

Artemisia japonica Thunb. (Asteraceae): Eng.- Japanese Wormwood; Hindi- Pamasi; Ladakhi – Zieo [W]. A perennial herb, 50-90 cm tall arising from a woody rootstock. Basal and lower stem leaves wedge shaped, coarsely toothed at tip; middle and upper leaves palmate or irregularly divided into linear segments. Flower heads numerous, yellow. Leaves used for skin infections.

Artemisia laciniata Willd. (Asteraceae): Eng.- Siberian Wormwood; Ladakhi- Khampa [W]. Perennial herb, common on cold-arid slopes of Ladakh. Leaves cauline, sparsely hairy. Corollas yellow. Powdered roots are rubbed as an antiseptic on cuts and wounds.

Artemisia macrocephala Jacq. ex Besser (Asteraceae): Ladakhi –Khampa [W]. Annual herb, pubescent, usually growing in waste areas, dry places at 1700 – 5500 m. Basal and middle stem leaves 3-5 sect, lobules narrowly linear, filiform; upper leaves entire or 3-lobed. Capitula shortly pedunculate, nodding. Leaves used for healing of cuts and wounds. Also used for treating foot infections in livestock. Ladakhis use its young shoots in the preparation of famous local alcoholic drink *Changue* (Fig. A 24).

Artemisia maritima L. (Asteraceae): Eng.- Sea Wormwood; Kash.- Murin /Seski; Ladakhi- Khanchuk [W]. Perennial herb growing in alpine vegetation above 3000m. Rootstock woody with 2-pinnatisect leaves. Leaves and stem covered with cottony hairs. The extract of the plant used to cure arthritis. Decoction of the leaves is also used to kill round worms.

Artemisia moorcroftiana Wall. (Asteraceae): Kash.- Jangli Tethwaen [W]. A sub-shrub growing wild in valleys rocky hills, sub-alpine meadows above 2000m. Rootstock horizontally creeping, woody. Lower and middle stem leaves oblong, petiolate, ovate or elliptic, with tomentose surfaces. Capitula sessile, flowers purplish. Leaves used for abdominal pain due to indigestion.

Artemisia roxburghiana Wall. ex Besser. (Asteraceae): Eng. Roxburg's Wormwood; Hindi- Afsantin; Kash.- Tethwaen; Ladakhi- Chuchok [W]. A perennial, robust medicinal herb, common on alpine slopes at 2800 – 3800m. Rootstock creeping with deeply dissected hairless leaves. The paste of fresh leaves used against itching and athlete foot.

Artemisia scoparia Waldst. & Kitam. (Asteraceae): Eng.- Red stemmed Wormwood; Kash.- Taethwaen [W]. Perennial hoary herb of alpine dry slopes. Leaves petioled, broadly ovate, 1-3 pinnatisect.Heads small. Leaves used against jaundice and infections of gall bladder.

Artemisia sieversiana Ehrh. (Asteraceae): Eng.- Worm Wood; Kash.- Taethwaen; Ladakhi – Khanatung. [W]. Perennial pubescent herb, usually on alpine slopes and waste lands. Leaves bi-pinnate. Flower heads purplish black. Powdered roots are taken with luke warm water to relieve bronchitis. Also fed to *Pashmina* goats for increasing wool production.

Artemisia tournefortiana Rchb. (Asteraceae): Eng. Worm Wood; Hindi- Afsantin; Kash.- Bazar bangae [W]. A tall annual medicinal herb usually on shaded wastelands at 1700- 2600 m. Heads greenish. Leaves pinnatisect, rachis serrate. Capitula many, erect, almost sessile, greenish and congested. Leaves and flower heads used as vermifuge and for healing broken and dislocated joints as anti-inflammatory. Leaves used in the preparation of local Ladakhi drink *Chhangue* (Fig. A 25).

Arthraxon lancifolius (Trin.) Hockst. (Poaceae): Eng. Beard Grass; Kash. Peelwa [W]. A common annual grass of eroded surfaces, waste places and eroded hills at 1700-1900 m. Inflorescence of 2-9 racemes. Flowers purplish. A valued fodder plant.

Arum jacquimontii Blume. (Araceae): [W]. A tuberous perennial herb, inhabiting forests and open alpine slopes. Leaf blade hastate, basal lobes acute or obtuse. Spathe green, limb lanceolate, acuminate. Berries red.

Arundinella nepalensis Trin. (Poaceae) [W]. Tufted perennial herb, arising from a stout hard rootstock. Culms tufted purplish green up to 97 cm tall. Culm nodes pubescent. Distributed on sloppy grasslands up to 2000 m.

Arundo donax L.(syn. Arundo sativa Lamk.; Donax arundinaceus P. Beauv.; Cynodon donax (L.) Raspail.) (Poaceae): Eng.- Giant Reed; Hindi- Narhal; Kash.-Nurgausse; Ladakhi- Dambu [W]. A giant reed inhabiting waste lands. Culms very strong, somewhat woody, 1.8-6.0 m tall, arising from horizontally knotty rootstocks. Leaves alternate with hairy tuft at base. Leaf blades long, flat, glabrous except on edges, regularly distichously spaced along the culm, auriculate at base, the throat of the sheath with a hairy tuft. Panicle erect, spikelets numerous and crowded, pointed at anthesis displaying the copious hairs. Hollow culms used for preparation of *hooka* pipes and for writing. Leaves have forage value.

Askellia flexuosa (Ledeb.) W A Weber (syn. Crepis flexuosa (Ledeb.) Benth ex Clarke. (Asteraceae): Ladakhi- Samboo [W]. A much branched glabrous annual, upto 1 m tall, of alpine meadow lands of Ladakh, above 3000 m altitudes, forming rounded tufts. Flower heads yellow. Latex of the plant used for constipation.

Asparagus filicinus Bach.-Ham. ex D. Don. (Asparagaceae): Eng.- Fern Asparagus; Hind- Musli [W]. Perennial diocious herb, usually on loose humus rich soils in forests. Cladodes in whorls. Rhizomes with fleshy thick swollen fasciculed roots. Flowers white. Berries deep green. Young shoots and roots used for arthritis.

Asparagus racemosus Willd. (Asparagaceae): Hindi- Shatavari; Kash.- Musli [W]. A common asparagus species growing in Himalayas in orchards and undisturbed soils, preferring gravely and rocky substrats. Phylloclads are needle like, shiny green. Flowers white on spiky stem. Fruit blackish purple globular berries. Roots used as tonic for ladies.

Asperugo procumbens L. (Boraginaceae): Eng.- Madwort [W]. A spreading annual herb,as a weed of orchards and scrubs. Flowers blue.Whole plant used for throat infections.

Asplenium adiantum- nigrum L. (Aspleniaceae): Eng.- Black Spleen wort; Kash.- Theire [W]. A fern growing on moist humus rich soils, on rocky crevices.Frondstriangularwithslightlyhairyandshinyrachis.Theextract of the plant used for treating chest problems and jaundice (Fig. A 26).

Asplenium bulbiferum L. (Aspleniaceae): Eng.- Mother Spleenwort; Kash.-Dadae [W]. An evergreen fern growing near banks of streams and moist shadedplaces.Eachfronddividedinto20-30leaflets.

Smallbulbilsappear on upper surface of many leaflets. Fronds eaten as vegetable (Fig. A 27).

Asplenium trichomanes L. (Aspleniaceae): Eng.- Maidenhair Spleenwort [W]. A small fern growing mostly in rocky habitats. Fronds tapering, simple divided into dark-green pinnae; rachis darker. Fronds used for dysentery.

Aster amellus L. (Asteraceae): Eng.- Michael mass Daizy [NI]. Cultivated in gardens and parks. Flowers lilac. Stem woody and flower from August to September. Several varieties with different flower colours are cultivated.

Aster diplostephioides (DC) Clark. (Asteraceae): Eng.- Wild Daizy; Kash.- Sahour [W]. Perennial, tall, villous herb, common on alpine pastures at 3000-3500 m. Flower heads solitary, white, pappus red, disc golden yellow. Plant has been used as a washing agent for washing clothes (Fig. A. 28).

Aster flaccidus Bunge (syn. **Aster tibeticus Hook. f.) (Asteraceae): Ladakhi-Utpalvanpo[W].**Perennialpubescentherb,20-30cmtall,ofalpinezoneof Ladakh. Stem slender. Heads solitary. Flowers ray blue. Powdered flowers are taken with luke warm waters to cure bronchial troubles.

Aster methodorus (Benth.) Govaerts. (syn. Brachyactis menthodora Benth.) (Asteraceae): Kash.- Sathi. [W]. A stout, tall, glandular pubescent herb of alpine areas. Lower leaves petioled, ovate – lanceolate, serrate. Pappus pale brown. Aerial parts used for arthritis and rheumatic pain.

Aster thomsonii Clarke. (Asteraceae): Eng.- Alpine Aster /Thomson's Aster; Kash.- Khatai [W]. A hardy herbaceous perennial, commonly growing wild in alpine and sub-alpine grasslands. Leaves are grey-green, spathulate. Flowers are light blue, to pale-pink and white with orange-yellow centre. Flowering from June to July.

Astracantha strobilifera (Benth.) Podlech. (syn. Astragalus strobiliferus Benth.) (Fabaceae): Eng.- Liquer Rice; Hindi - Binge; Kash.- Gagar kound / Gagar Kan [W]. A perennial woody herb growing on alpine meadows, distributed in western Himalayas from 2400-3900 m. Corollas yellow. Stem is a source of gum used in confectionery and for glazing. The seeds are given for colic and also for leprosy.

Astragalus falconeri Bunge. (Fabaceae): Eng.- Milk Vetch; Ladakhi-Munsear [W]. A low spiny, prostrate, perennial herb endemic to alpine desert lands of Ladakh. Stipules foliacious, pilose. Corolla pink-purple. Powdered seeds used to cure whooping cough.

Astragalus grahamianus Benth. (syn. Astragalus cicerifolius Fisch.) (Fabaceae): Eng.- Tragacanth; Hindi- Garmezu; Kash. – Gagar-kound [W]. A low very spiny shrub of alpine/sub-alpine grass lands, with spreading branches. Leaves small, pinnately compound, with 8-14 leaflets, ovate-blunt with a fine point. Stipules large. Flowers yellow, calyx sparsely hairy, with linear lobes; spines needle – like, slender. Pods, with silky adpressed hairs. Underground parts are chewed or taken in powdered form to stop vomiting during long journeys and for toothaches (Fig. A 29).

Astragalus nivalis Kar & Kir. (Fabaceae): Eng.- Tragacanth; Ladakhi – Cherchuk [W]. Perennial herb, common on alpine meadows. Stem appressly white, pubescent. Leaf imparipinnately compound. Decoction of the plant used for intestinal disorders.

Astragalus peduncularis Royle. (Fabaceae): Ladakhi- Sheatz [W]. A low perennial herb of Ladakh highlands. Stem covered with hairs. Leaves sub-sessile. Petals greenish-white. Powdered roots used as diuretic and to reduce obesity.

Astragalus rhizanthus Benth. (Fabaceae): Ladakhi- Kretseng [W]. A low perennial herb of alpine zone. Leaves imparipinnately compound. Corolla yellow. A cold extract of the powdered roots is taken as heart stimulant (Fig. A 30).

Astragalus tibetanus Bunge. (Fabaceae): Ladakhi- Magress [W]. Perennial sub-erect pubescent herb on rocky slopes of cold-arid region of Ladakh. Stem with black and white hairs. Extract of the whole plant is used against gout.

Astragalus zanskarensis Bunge. (Fabaceae): Ladakhi- Dianko [W]. Perennial under-shrub on alpine ranges of Ladakh. Calyx densely silky. Powdered seeds are taken with milk as an anthelmintic.

Astrophyllum capricorne L. (Mniaceae): Eng.- Goats Horn Cactus [NI]. A globular cactus grown for its dull green bodies having sharply crested ribs bearing depressed areoles from which brown-black spines arise. Flowers are daisy like, yellowish with red centre.

Athanasia linifolia Burm. f. (syn. Tanacetum longifolium Wall. ex DC.) (Asteraceae): Eng.- Long leaved Tansy; Ladakhi- Yakzae [W]. A silky villous perennial aromatic herb growing on alpine slopes between 3300 – 5000 m altitudes. Heads yellow. Roots used as incense (Fig. A 31).

Atriplex hortensis L. (Amaranthaceae): Eng.- Mountain Spinach /Orach; Hindi- Chakwat; Kash.-Waste hakh [RC]. An annual stout herb cultivated as spring vegetable. Two variants of the plant are recognised in early spring, one with straw green and other with purple coloured foliage. The leaves and soft shoots of the both the types are used as vegetable however, the type with coloured leaves is much sought after. During cooking house made vinegar/tamarind paste is sometimes added to the preparation for making it tasty and digestive.

Atropa acuminata Royle ex Lindley (Solanaceae): Eng.- Indian Belladona; Hindi- Sagangur; Kash.- Brande [W]. A perennial glabrous branched herb occurring on shady alpine slopes from 2400- 3300 m. Leaves petioled, acuminate; flowers greenish yellow. Berries spherical and black. Leaves and roots are the source of a drug- atropine used in the treatment of asthma, whooping cough. Root decoction is used to treat ulcers, gout and rheumatic inflammations. Berries are highly poisonous and suppress glandular secretions (Fig. A.32).

Atropa belladonna L. (Solanaceae): Eng.- Deadly Nightshade; Hindi- Sag-angur, Angurshefa / Beladon [NI]. A perennial herb, being cultivated by some medicinal growers. Flowers bell shaped, with green tinges. Berries ripening shiny- black. Roots are used as sedative, stimulant and antispasmodic. Atropine drug obtained from the plant is used for ophthalmic use (Fig. A 33).

Aucuba japonica Thunb. (syn. Aucuba japonica var. variegata Dombrain.) (Garryaceae): Eng.- Spotted Laurel [NI]. A medium-sized evergreen shade-loving dioecious shrub, forming dense rounded bushes with elliptic to ovate, glossy, mid green leaves blotched with yellow/white dots/ streaks. Flowers small, reddish-purple. The male plants with conspicuous creamy white anthers. Female plants produce bright red berries. Used for outdoors as well as indoor planting for its ornamental foliage.

Avena fatua L. (Poaceae): Eng.- Wild Oats; Kash.- Hoone Vishqa [W]. Annual herb commonly growing in most of the grasslands and also as a weed of cultivation. Stem erect, hollow bearing nodding panicles of spikelets. Relished by cattle and used as hay.

Avena sativa L. (Poaceae): Eng.- Oats; Hindi- Jai; Kashmiri-- Oats [WC]. A sub-erect bi annual herb cultivated for forage purposes. Roots shallow and adventitious. Culms with 3-5, nodes, hollow, leaves long and narrow, apex drooping, ligules well developed and membranous. Inflorescence much branched panicle; the spiklets occur at the end of branches and pedicellate; each spikelet consists of two or more flowers enclosed by two large glumes; flowers bisexual; caryopsis spindle shaped firmly adhering to the lemma and palea.

Fodder oats, after having been introduced, are popularly grown in the valley of Kashmir, as a *rabi* crop exclusively for forage purposes. Besides green forage oat hay or straw is also made from it and fed to cattle during winter. Recently released and popularly grown fodder oat varieties include:

1. **Sabzaar:** Released and developed by Sher-e Kashmir University of Agricultural Sciences & Technology,Kashmir (SKUAST Kashmir) in 2001 for valley plains of Kashmir and high altitude areas of Jammu Division. Plants tall; leaves deep green;grains bold, amber coloured. Matures in 225-230 days. Moderately resistant to loose smut and *Helminthosporium* leaf spot and major insect pests. Average forage yield is 350q ha -1. Harvestable for forage purposes 170 days after sowing, under valley conditions.
2. **Shalimar Fodder Oats -1(SKO-20):** Variety developed and released by SKUAST of Kashmir at state level during 2010. It is a cross between EC -13178 and *Sabzaar* followed by selection in successive generations. The variety has higher green fodder yield potential of 400-450 q/ha, with a yield superiority of 15% over *Sabzaar* and 35% over national check *Kent.* Harvestable for forage purposes after second fortnight of April. Fodder richer in proteins. Very well suitable for cultivation in temperate areas of Jammu & Kashmir.
3. **Shalimar Fodder Oats -2 (SKO- 90):** Released at national level for temperate and mid-hill regions of J & K and Himachal Pradesh. Green fodder yield ranges between 380-400 q/ha; dry matter yield ranges between 81-85 q/ha. Seed yield potential is 25-27 q/ha. (Fig. A 34).
4. **Shalimar Fodder Oats -3 (SKO- 96):** Variety released at national level for temperate and mid-hill region of the country. Average seed yield is 26 q/hac. Fodder yield is about 400 q/ha. (Fig. A 35).
5. **Shalimar Fodder Oats -4 (SKO- 108):** A state released fodder oats genotype bred through selection from a cross between Sabzaar and EC-96539. Days to 50% flowering ranges from 171-180 days. Plants are medium tall with larger leaf size. Fodder yield ranges between 390-400 q/hac.

Fig. A1. Abelia triflora

Fig. A2. Acantholimon lycopodioides

Fig. A3. Achillea millefolium

Fig. A4. Aconitum chasmanthum

Fig. A5. Acorus calamus

Fig. A6. Actaea spicata

Fig. A7. Adonis chrysocyathus

Fig. A8. Aesculus indica

Fig. A9. Ajuga integrifolia

Fig. A10. Allium carolinianum

Fig. A11. Allium cepa var viviparum

Fig. A12. Allium rubellum

Fig. A13. Altheae rosea

Fig. A14. Androsace mucronifolia

Fig. A15. Androsace sempervivoides

Fig. A16. Anemone falconeri

Fig. A17. Anemone obtusiloba

Fig. A18. Aquilegia fragrans

Fig. A19. Aquilegia nivalis

Fig. A20. Arisaema jacquemontii

Fig. A21. Arnebia benthami

Fig. A24. Artemisia macrocephala

Fig. A22. Arnebia guttata

Fig. A25. Artemisia tournefortiana

Fig. A23. Artemisia absinthium

Fig. A26. Asplenium adiantum-nigrum

Fig. A27. Asplenium bulbiferum

Fig. A28. Aster diplostephioides

Fig. A29. Astragalus grahamianus

Fig. A30. Astragalus rhizanthus

Fig. A31. Athanasia linifolia

Fig. A32. Atropa acuminata

Fig. A33. Atropa belladonna

Fig. A34. Avena sativa (var. Shalimar Fodder Oats-2

Fig. A35. Avena sativa var. Shalimar Fodder Oats-3

B

Barbarea vulgaris var. taurica (DC) Hook. f. & Anders. (syn. Barbarea vulgaris R. Br.) (Brassicaceae): [W]. A herb of damp soils inhabiting water bodies. Flowers yellow. Leaves used for healing of frost bites.

Bassia scoparia (L.) A J Scott. (syn. **Chenopodium scoparia L.; Kochia scoparia (L.) Schrad.) (Amaranthaceae): Eng.- Goose foot; Kash.- Paunjabe Lautchije [W].** An annual bushy herb, growing wild in orchards. The plant is also planted as a border plant in vegetable fields and later used as a broom after leaf fall. In villages, its leaves are cooked as vegetable along with the leaves of local tomato variety. Seeds used to regulate hypertension and obesity.

Bellevalia ciliata (Cirillo) T. Nees. (syn. Hyacinthus azureus Baker / Hyacinthus ciliatus (Cirillo) T. Nees (Asparagaceae) [NI]. A bulbous perennial ornamental herb grown as border ornamental. Leaves strap shaded and dark green. Spikes sparsely compact bearing bell shaped, light blue flowers. Flowering February to March. Also seen growing under moist shady places around grasslands and gardens.

Bellis perennis L. (Asteraceae): Eng.- Lawn Daisy /Bruisewort [W]. Perennial herb inhabiting lawns of gardens, especially colonizing under chinar trees. Plant has creeping rhizome, bearing rosette of small round leaves and short stalked flower heads with white ray florets and yellow central disc (Fig. B 1).

Berberis aristata DC. (Berberidaceae): Eng.- Indian Barberry; Hindi– Dar-hald; Kash.- Daunde–laeder/ Rass-ashude; Ladakhi– Kraring [W]. A semi-evergreen spiny shrub with attractively pendulous twigs. Leaves grouped 3-6 in clusters, usually elliptic and often spineless, green above, whitish or yellow-green beneath. Flowers bright yellow, often tinged red, grouped in long racemes. Fruits fusiform, dark red, finally blue purple. Root powder used as anti-bacterial and anti-inflammatory and for rheumatic pain (Fig. B 2).

Berberis jaeschkeana Sch. (Berberidaceae): **Eng.- Jaeschke's Barberry; Kash.- Kawe-dach [W]**. An armed deciduous shrub of alpine slopes endemic to Kashmir Himalaya. Leaves oblong-elliptic, usually 2-5 spinose serrulate at margins. Flowers yellow; fruit ovoid, red. Root used for rheumatic pain.

Berberis kashmirana Ahrendt. (Berberidaceae): Eng.- Barberry; Hindi- Dar-hald; Kash.- Kawe-dach [W]. Shrub of sub-alpine zone, endemic to Kashmir. Leaves sub-sessile. Roots used medicinally.

Berberis lycium Royle (Berberidaceae): Eng.- Barberry; Hindi- Dar-hald/ Zari- shak;; Kash.- Kawe-dach; Ladakhi -Kraring [W]. Semi –evergreen armed shrub, common on dry exposed slopes at 1700-2500 m. Flowers in May June. Stems white, leaves oblanceolate, light green above, bluish green beneath. Flowers golden yellow, in elongated racemes. Fruit elliptic, black and persistent at maturity. Extract of leaves and twigs is used to treat opthalmia, piles and as laxative. Kashmiris have been using its fruit/root as anti-bacterial, anti-inflammatory and to fight Cholera (Fig. B 3).

Berberis nervosa Pursh (syn. Mahonia nervosa (Pursh.) Nutt. (Berberidaceae): Eng.- Holly Grape [NI]. A dwarf, suckering shrub with pinnate glossy dark-green leaves which often turns red in December; bears yellow flowers in long racemes, from mid to late April followed by spherical blue black berries.

Berberis orthobotrys Bien. ex Aiteh (Berberidaceae): Eng.- Common Barberry; Hindi- Dar-hald [W]. A medium-sized shrub of vigorous, upright habit, growing on alpine slopes. Twigs pale-brown or yellowish, armed with spines. Leaves obovate, cuneate, slender petioles, pale beneath. Flowers yellow, racemes-umbellate; berries oblong or ellipsoid, red, coated with greyish-white bloom.

Berberis pachyacantha Bien. ex Koehne. (Berberidaceae) [W]. A deciduous shrub with glabrous dark-red to pale twigs, with spines. Leaves obovate to elliptic-oblong, prominently reticulate, cuneate into a short slender petiole; margin with numerous finely thorn close set, irregular serrations. Flowers yellow, borne on pendulous racemes; berries epruinose, oblong-ovoid red in colour and persistent.

Berberis piperiana (Abrams.) Mc Minn (syn. Mahonia piperiana Abrams.) (Berberidaceae): Eng.- Oregon Grape [NI]. An erect ornamental shrub. Leaves pinnate glossy green above and papillose beneath. Flowers yellow in erect racemes at the tips from early to late April; followed by spherical blue-black berries.

Berberis thunbergii DC (syn. Berberis thunbergii var. atropurpurea Chenault; Berberis nepalensis Spreng.) (Berberidaceae):Eng.- Common Barberry; Hindi- Kashmal; Kash.- Chouka Phal [W]. A dense, rounded, deciduous wild shrub with dark red -purple or purplish bronze foliage, turning red in autumn. Umbel like racemes with, red tinged yellow flowers, produced along the branches; fruit ellipsoid, glossy and cherry red.

Berberis ulicina Hook f. & Thomson. (Berberidaceae): Eng.- Gorse Barberry; Ladakhi- Sinshingnama / Shinnar [W]. Deciduous armed shrub on scrubby alpine slopes of Ladakh region at altitudes of 3000-3700 m. Flowers orange yellow. Berries globose, blackish, 3-5 seeded. Powdered fruits are taken for ringworm infections. In Ladakh its twigs are used as fuel (Fig. B 4).

Bergenia pacumbis (Buch. –Ham ex D. Don) Wu & Pan (syn. Berginia himalaica Boriss; Bergenia ligulata Engl.) (Saxifragaceae): Eng. – Rock foil; Kash.- Pahand / Palpout /Zakhmi hayat [W]. A perennial succulent herb found on rocky substrate in alpine and sub-alpine areas of Kashmir. Rootstock rhizomatous, woody and underground. Basal leaves large and with bristly margins. Flowers rose to purple on erect panicles. Decoction of roots used to cure fevers, coughs, diarrhoea and wounds (Fig. B 5).

Bergenia stracheyi (Hook. f. & Thoms.) Engl. (syn. Saxifraga stracheyi Hook &. f. & Thoms.) (Saxifragaceae): Eng.- Bergenia; Kash.- Ghee-pathi, Silphari; Ladakhi- Gatils [W]. A hardy herbaceous perennial herb, growing wild in alpine pasture lands, on rocky surfaces, at 3100-3500 m, and also grown as a pot plant. Leaves leathery, ovate and dark-green. Sprays of pale pink to white flowers are produced from March to April. Root paste used for ulcers, cuts and swellings.

Betula utilis D.Don (Betulaceae): Eng.- Himalayan Birch; Hindi – Bhojpatra; Kash.- Burza [W]. A moderate sized tree, growing in alpine zone of Kashmir Himalaya, as last tree line, above 3000m altitudes forming pure stands or mixed with coniferous elements. Leaves ovate, fine-pointed, irregularly saw-toothed, finely hairy when young, that turn yellow in autumn. Male catkins reddish, appearing mostly on bare branches. Fruit a catkin, cylindrical. The thin papery bark is shiny, reddish white or white with horizontal lenticels which peels off in very thin pale, almost transparent horizontal strips. Up to first quarter of 20th century the thin papery bark of the tree, known locally as *burza*, was of great importance to Kashmiris in being used as paper; and being moisture resistant the under coverings of roofs all better houses and shrines were formed from birch bark overlaid with thick earthen layer. Many old manuscripts are still preserved on birch bark. Bark was also medicinal in being useful in bronchitis and convulsions. Plant had a social significance as its bark was used to perform the ceremony of lustration after a mother was bathed on the sixth day after child birth in which wisps of birch bark were lightened and waved over the heads of the mother and the child and later extinguished in a pot of water.

Bidens biternata (Lour.) Merr. & Sherff. (Asteraceae): Kash.- Kound Chour [W]. Annual herb, usually in wastelands, among scrubs. Leaves cauline, softly tomentose on both sides, divided pinnately into lealets. Capitula radiate, yellow. Whole plant used in medicinal preparations.

Bidens tetraspinosa Kak & Javeid (Asteraceae): Kash.- Kound Chour [W]. Annual diffuse herb, usually in wastelands and ditches. Lower leaves entire, upper linear, acute, irregularly serrate. Flowers yellow. Whole plant used medicinally.

Biebersteinia emodi Jaubert & Spach. (Biebersteiniaceae):Ladakhi-Drakspose [W]. Annual herb growing on mountainous rocky surfaces of Ladakh near glaciers. Plant parts used to cure wounds, ulcers, cuts and peptic ulcers.

Biebersteinia odora Stephan. ex Fisch. (Biebersteiniaceae): Eng.- Fragrant Yellow Geranium; Ladakhi- Shekairpa [W]. A woody perennial glandular and hairy herb growing across Ladakh and Kashmir regions. Flowers yellow, borne in racemes. An extract of the powdered aerial parts used to treat ophthalmic diseases (Fig. B 6).

Boletus edulis Bull. (Boletaceae): Eng.- Porcino; Kash.- Jungle haedre. An edible fungus growing in deciduous and coniferous forests of Kashmir. Cap large, brown, up to 30 cm in dia.; stipe is stout white or yellowish in colour with reticulations.

Bothriochloa bladhii (Retz.) Blake (syn. Bothriochloa intermedia (R. Br.) A. Camus) (Poaceae): Eng.- Australian Blue Grass/Blue Grass [W]. A perennial wild grass occurring on level sites in shades or on open grasslands in alpine and sub-alpine forest areas.

Bothriochloa ischaemum (L.) Keng. (syn. Andropogon ischaemum L.) (Poaceae): Eng.- Purple Grass/ Plains Blue-stem Grass; Kash.- Touje gasse; Ladakhi- Jamak [W]. A perennial tufted grass species of North- western Himalayan grasslands having fodder value. Culms 1-1.5 m long, slender, with a terminal spike. Also used for hay making. Dried and jointed culms used as toothpicks.

Brachyactis pubescens (Candolle) Aitch. ex Clarke (Asteraceae): Ladakhi-Tcha [W]. An erect, stout annual on alpine slopes. Stem densely leafy, brownish yellow, stipitate, glandular. Leaves also densely glandular. Powdered flowering heads are taken with cold milk to expel intestinal worms.

Brachyactis umbbrosa Benth. (Asteraceae): Kash.- Sathi [W]. Annual herb of alpine grasslands. Aerial parts used against arthritis, rheumatism.

Brachypodium sylvaticum (Huds). P. Beauv. (Poaceae): Eng.- Bunch Grass [W]. A tall tufted perennial bunch grass of alpine pastures, with hairy nodes and without rhizomes. Culms pillose, spikelets long. Leaf blades yellow-green and drooping.

Brassica cretica Lam. (syn. Brassica oleracea var. botrytis L.) (Brassicaceae): Eng.- Cauliflower; Hindi- Phool Ghobi; Kash. - Phool gobhi [WC]. A low herb, cultivated for the terminal head or curds. Stem stout and short. Leaves long- oblong, elliptic, ascending; flowers abortive on closely crowded head with short fleshy stocks, forming a dense terminal head overlapped by leaves. Siliqua thin and long. Lately introduced and popular varieties cultivated include *Pusa Katki, Improved Japanese, Pusa Snow Ball, K-1, Snow Crown* and *Silver.*

Brassica nigra (L.) Koch. (Brassicaceae): Eng. Black Mustard; Hindi – Rai; Kash.–Ausoore [RC]. A much branched annual herb, 1.0- 2.1 m or more in height; lower leaves lyrate- pinnatisect with 1-3 pairs of lateral lobes and a much larger terminal lobe, hispid on both surfaces; upper leaves linear – oblong, entire or sinuate; flowers yellow in corymbose racemes; seeds very small, angled, minutely pitted. Black mustard is grown marginally in Kashmir and used as a condiment in pickles and curries.

Brassica oleracea L. (syn. Brassica oleracea var. acephala DC) (Brassicaceae): Eng.- Kale; Hindi – Kadam Sagh; Kashmiri – Haakh [WC]. A biennial, usually grown as an annual in Kashmir. Stem robust, branched, 60-80 cm in height. Leaves large or oval, on long petioles with uneven margins. Flowers milky white/light yellow on long elongated inflorescences. Kales and collards have been among the most popular cultivated leafy vegetables in Kashmir and are locally known as *Haakhs*. Due to prolonged association the plant has become a part of Kashmiri culture and forms an integral part of diet of a Kashimiri.

Since the leafy Brassicas i,e kales have received much attention in Kashmir than other brassicas, therefore, continuous selection for desirable traits and subsequent breeding has eventually resulted in numerous botanical forms and cultivars. The leaves of most of the types are succulent, nutritious with high digestibility and are cultivated only for production of leaves and thus pinched regularly for fresh leaves. The plants of summer types are detopped in Autumn and left for winter which later produce a fresh crop of young leaves in Spring, used as fast green vegetable of the season. Based on plant type, season of cultivation, use and cooking qualities some popular types of kales grown in Kashmir are:

- **G M Dari / Jumadari Haakh**: The plant has swollen stem bearing a tuft of leaves which are large with wavy margins. G.M. Dar has been a progressive vegetable growers of Srinagar area and this kale type was introduced and popularized by him, hence named after its initial cultivator.
- **Dale Haakh**: A kale type grown on the floating gardens of Dal Lake. Leaves very crispy with fleshy petioles, having good cooking quality.
- **Krame Haakh:** A hard kale type grown in all the vegetable growing areas of Kashmir as a *kharif* crop. Stem is swollen, light green/purple in colour; stem leaves broad with shallowly lobed margins and light greenish or dark pink petioles. Swollen stems and leaves are cooked and also used in the preparation of pickles in rural areas.
- **Gaume Haakh:** Mostly grown in villages in their kitchen gardens as *kharif* vegetable crop. Plant sturdy, frost resistant, bearing large leaves of light green or light purplish colour. Stems swollen and used as vegetable along with leaves. Large quantities also sundried and then gardlanded for subsequent use during winter.
- **Kawdari Haakh:** This kale type is named after a vegetable growing areas of Srinagar namely, Kawdara. Its leaves are tender, with almost entire margins and deep bluish green tinge. It is cultivated mostly as a *rabi* crop in Dal Lake area of Srinagar.
- **Haunze Haakh:** The cultivation of this cultivar was initially by boat dwelling people called *Hanjis*, living on the banks of river Jhelum and Dal Lake, hence its name. It is still cultivated as a *rabi* season crop mostly by vegetable growers of Dal Lake. Stem is slightly swollen, producing a tuft of large leaves.
- **Baktoor Haakh:** Named after Baktoor locality of Gurez where it is commonly grown, however its cultivation has spread to other areas of Gurez and Tilial. Leaves are tender, smaller than other Kashmiri kale types, light green and less fibrous. It is cultivated as *kharif* crop (Fig. B 7).

Brassica oleracea L. (syn. Brassica oleracea. var. capitata L.) (Brassicaceae): Eng.- Cabbage; Hindi- Band Gobhi; Kash.- Band Gobhi [WC]. A small, low biennial herb. Stem stout, simple. Leaves glaucous radical leaves cauline large, often 30 cm across, smooth and compact in a dense globose, greenish white head. Cabbages are grown for enlarged edible terminal head used in salads and as vegetable. Among the lately introduced and high yielding cabbage varieties under cultivation include *Golden Acre, Pride of India, Sindhu selection, Krishna* and *Pusa Drum* Head.

Brassica oleracea L. (syn. Brassica oleracea var. gonglodes L.) (Brassicaceae): Eng. Khol Rabi; Hindi- Ganth Gobhi; Kash.-Munje Haakh [WC]. Commonly cultivated local and introduced Khol rabi varieties include:

- **Kashmiri Munje Hakh:** A low,glabrous biennial. Thickening of stem above cotyledons results in the formation of tuber or knob. Tubers -5-10 cm in dia, green or purplish with large leaf scars on the surface. Leaves thin, 20-25 cm long oval to oval-oblong, irregularly dentate with long selender petiole. Flowers whitish yellow, in long racemes; siliqua 5-10 with usually very short beak.
- **White Vienna:** Introduced and widely cultivated. Leaves green, knobs pale green; flesh greenish white. Sowing time March- April, transplanting by May to June.
- **Purple Vienna:** Introduced and widely cultivated. Leaves dark green with purplish veins and midribs and petioles. Knobs purplish, flesh greenish white. Sowing time March- April. Transplanting by May to June.

Brassica oleracea L. (syn. Brassica oleracea var. kashmeriana Naqshi & Javeid) (Brassicaceae): Eng.- Curly Cale; Hindi -Sagh; Kash.- Khanyari Haakh [WC]. A very popular indigenous Kale type cultivated throughout year for its crisp and tender leaves which are characteristically crumpled and deep green. Stems long, erect and stout, less branched, bearing a plume of short petioled and globosed leaves. It is the hardest of Cole crops and can withstand drought and even sub-zero temperatures. The morphotype has its origin in Khanyar locality of Srinagar city, hence its name (Fig. B 8).

Brassica rapa L. (syn. Brassica campestris L.; Brassica rapa var. campestris (L.) Peterm. (Brassicaceae); Eng.- Brown Sarsun; Kashmiri- Tilgogul **[WC].** A biennial plant grown for its seeds which yield oil. Flowers yellow. Varieties under cultivation include:

- **Kashmiri Sarson:** A local sarson type, under cultivation. Plant is tall, herbaceous, annual, 1-1.5 m tall; stem glaucous with leaves of two types-the lower ones known as the stem leaves and the upper arriving on the axis of the inflorescence known as floral leaves. The stem leaves are amplexicaul, while floral leaves are lanceolate. Inflorescence is long corymbose raceme; flowers light yellowish and self incompatible; seeds are of light brown colour. The traditional cultivar has small to medium sized seeds, and yields low (6 to 8 q ha-1) with an oil bearing capacity of about 30 % only.

- **KOS -1:** A brown sarson variety grown in many oilseed producing belts of the Kashmir valley. It was developed by scientists of the State Department of Agriculture and released for general cultivation in the valley during 1978. The variety yields 10-12 q ha-1, matures within 225-230 days and fits well in oilseed-rice cropping sequence. The plants are erect with non- hairy foliage, bright yellow flowers and dark brown seeds.
- **Gulchin (KS-101):** A sarson variety developed by SKUAST-K, and released for cultivation in the Kashmir valley during 1995. The plants are erect up to 110 cm high and bears bright yellow flowers and dark-green, non-hairy leaves. Seeds are bold, dark brown with oil content of 40%. The crop matures within 230-235 days and yields 12-14 q ha-1. The variety is being cultivated in all traditional oil seed producing belts of the valley mostly under rice- oilseed cropping sequence (Fig. B 9).
- **Shalimar Brown Sarson-1:** A lately released variety developed and released by SKUAST Kashmir during 2010. The variety has been synthesized by crossing Gulchin and NRCV -16 followed by several cycles of selection and adaptation. It is highly suitable for cultivation in Kashmir region. Posseses disease resistance, high yield potential as well as high oil content (42%), exhibiting a significant yield superiority of 16.7% over the best check *Gulchin.*

Brassica rapa L. (syn. Brassica campestris subsp. rapifera (Metz.) Sinskaya) (Brassicaceae): Eng. -Turnip; Hindi- Shalgam; Kash.- Gogij [WC]. A green leaved distinctly hairy biennial herb, with a swollen fleshy napiform tap root, usually white fleshed. Leaves prominent lyrate or interrupted below, green hairy or bristly; radical leaves soft, prickly. Flowers bright white maroon in corymbose racemes; silique with slender beaks; seeds blackish or reddish brown.Young plants, with nut sized roots, are woven into garlands for sundrying. The dried foliage called *Gogji patre* is later cooked as vegetable. Thick, swollen roots of the plant are also sliced for sun drying. Dried slices called *Gogji Aure* are cooked with fish/meat during winters. Common types of turnips grown in Kashmir include:

- **Islamabadi Gogij:** Cultivated mostly in Distt. Anantnag because of local preference. Its shelf life is longer than white type and the plant is hardy and cold tolerant. The size of the root is larger than other types. Sown from September to October.

- **Dal Gogij:** The cultivar is usually preferred to in Distt. Srinagar for table purpose as well as a vegetable. Roots napiform, discoid, skin white, light pink around the condensed stem disc. Cultivated by the vegetable growers of Dal Lake and by other vegetable growers in and around Srinagar city. Roots after being cut into slices are garlanded for sun- drying.
- **Gaume Goguje:** Roots discoid with characteristic greenish area around the stem disc. Mostly cultivated by villagers as fodder as well as vegetable. Large quantities of the plant are sundried by villagers when their roots are of walnut size, along with foliage, for use during winter.
- **Nageen -1:** A turnip variety developed by SKUAST-K and released for general cultivation in 2002. Roots large, compact and pure white. Flesh white, fine grained and moderately sweet. Yielding 350-400 q ha -1.

The other lately introduced and commercial turnip cultivars include *Purple Top* and *Golden Ball.*

Braya humilis (C A Mey.) B.L. Rob. (syn. Torularia humilis (C.A Mey) O E Schulz.) (Brassicaceae): Ladakhi- Erekarpo [W]. Hoary, diffuse perennial, hairy herb on alpine slopes. Flowers pale white. Powdered roots are administered with honey, as carminative and stimulant in dyspepsia.

Bromus japonicus Thunb. (syn. Bromus barobalianus G. Sing) (Poaceae) [W]. Annual herb, softly hairy all over. Leaves linear. Panicle spreading with 6-14 branches, densely softly hairy. Spikelets drooping, compressed and solitary, 6-12 flowered. Glumes and lemma hairy and anthers larger. Grows along mountain slopes overlying Harwan.

Bromus ramosus Huds. (Poaceae): Eng.- Rescue Grass/ Hairy brome [W]. A perennial herbaceous bunch grass, up to 1-2 m tall, of alpine meadowlands, in glades, in high altitude forests and on shady places on alpine slopes. Leaves long, usually drooping, wide, finely hairy.

Bromus scoparius L. (Poaceae) [W]. Annual tall grass of Himalayan grasslands and moist fields. Culms erect, geniculately ascending. Leaf surface pilose, hairy on both sides. It is grazed by cattle.

Bromus tectorum L. (Poaceae): Eng.- Drooping Brome /Cheat Grass [W]. Annual bunch grass flowering in early summer, often found growing gregariously in waste places and has a good fodder value.

Buddleja crispa Benth. (Scrophulariaceae): Eng.- Butterfly Bush; Hindi – Niwarpati [NI]. A medium-sized to large densely woolly-haired shrub. Leaves variable lanceolate to ovate, with an enlarged base, conspicuously toothed, silvery grey when young. Grown as an ornamental for its fragrant purple flowers.

Buddleja davidii Fanch. (Scrophulariaceae): Eng.- Butterfly Bush; Hindi –Nimda [NI]. A medium sized shrub. Leaves ovate - lanceolate, acuminate, rounded or tapering at base, short petioled, fine toothed, dark green above, densely clothed beneath with white tomentum. Flowers in terminal umbels; corolla 3-4 times as long as the calyx, lilac, throat orange coloured, fragrant. Commonly grown vars. include: *Epire blue* (Rich violet- blue with orange eye); *Fascination* (Wide, panicles of vivid lilac pink) and *White Profusion* (Large panicles of pure white, yellow-eyed flowers).

Buglossoides arvensis (L.) I.M. Johnst.(syn. Lithospermum arvense L.) (Boraginaceae): Eng.- Field Gromwell/Bastard Alkanet [W]. Annual herb, usually a weed of cultivation, also on dry scrubby slopes. Leaves alternate,sessile, hairy on both sides. Flowers white. Roots and seeds used as sedative and for treating ulcers and wounds.

Bunium persicum (Boiss) Fedtsch. (Apiaceae): English –Earth Nut /Black Caraway; Hindi – Kala Zeera; Kashmiri- Gunyun/Zeure [W]. A much branched, perennial tuberous herb, 15-75 cm high, growing wild in alpine and sub- alpine grasslands and foot hills of Kashmir especially in the areas like Gurez, Tilale, Khrew, Paddar, Kishtawar, Chari sharif at altitudes ranging from 1800 -3,300 m amsl. Plants have thick tuberous roots, bi to tri -pinnate compound leaves with linear segments. Flowers are small white and borne in dense compound umbels. The light black mericarps are aromatic which are used as spice in the same way as caraway for seasoning dishes and drinks. Seeds diuretic, anti- dysenteric, carminative and used for flatulent colic and other human digestive problems. Tubers also eaten as vegetable (Fig. B 10).

Bupleurum jucundum Kurz. (syn. Bupleurum jucundum var. cashmiricum Clarke. (Apiaceae): [W]. Perennial herb, endemic to Kashmir, usually common along savana slopes. Stem fistular. Lower leaves lanceolate,sessile to sub-sessile. Fruit oblong, ridges slightly winged, furrows 3 -vittate. Flowers yellow. Roots used against liver problems.Traditionaly used in treating fevers, pain and inflammations associated with influenza and common cold.

Bupleurum longicaule Wall. ex DC (syn. Bupleurum longicaule var. himalayense (Klotsch.) Clark.; Bupleurum himalayense Klotzsch. (Apiaceae): Ladakhi- Sah-Kuchlas [W]. Perennial prostrate herb on alpine dry slopes. Flowers yellow,glabrous. Dried extract used for renal colic, hepatitis and gastrointestinal disorders (Fig. B 11).

Bupleurum thomsoni Clarke. (Apiaceae): Eng.- Thomson's Thorowax [W]. Erect herb of alpine slopes. Leaves strict, lanceolate, broad at base, stem clasping, heart shaped. Flowers yellow. Fruits used medicinally.

Bupleurum marginatum Wall. ex DC (syn. Bupleurum falcatum var. marginatum (Wall. ex DC) Clarke) (Apiaceae): Eng.- Sickle Hare's Ear; Kash.- Billuiren; Ladakhi- Zeera karpo [W]. Perennial herb, usually common on dry scrubby and sunny savanah slopes. Umbells yellow. Extract of the root used as an anti-inflammatory and also as diaphoretic. In Ladakh root used for liver diseases.

Buxus sempervirens L. (Buxaceae): Eng.- Common Box Tree; Hindi-Chikri [W/UC]. A wide spreading ornamental shrub or small tree with slightly winged branches. Leaves oval to oblong lanceolate, emerginate at apex, shining dark green above and pale beneath; flowers pale green with yellow anthers are honey scented. Two vars. grown include *Arhorescens* (Large shrub or occasionally a small tree with medium to large, dark green leaves) and *suffruticosa* (compact, slow growing dwarf with small leaves).

Buxus wallichiana Baill. (Buxaceae): Eng.- Box Tree; Hindi- Chikri; Kash.-Chikre [W]. An open shrub of alpine vegetation, with narrowly ovate-lance-shaped, glossy bright green leaves, notched at the tips; flower pale green in dense, axillary clusters. The wood of the plant is specially used for making musical instruments, small boxes and combs (Fig. B 12).

Fig. B1. Bellis perennis

Fig. B4. Berberis ulicina

Fig. B2. Berberis aristata

Fig. B5. Bergenia pacumbis

Fig. B3. Berberis lycium

Fig. B6. Biebersteinia odora

Fig. B7. Brassica oleracea (var. Baktoor Haakh)

Fig. B10. Bunium persicum

Fig. B8. Brassica oleracea (var. Khanyari Haakh)

Fig. B11. Bupleurum longicaule

Fig. B9. Brassica rapa (var. Gulchin)

Fig. B12. Buxus wallichiana

C

Cajanus cajan (L) Millsp. (Fabaceae): Eng.- Pigeon Pea / Cajan Pea; Hindi – Arhar; Kash.- Lokut Chana [RC]. An annual erect herb, stem woody, leaves trifoliate. Flowers yellowish diffuse red, borne in axillary racemes. Pods flattish with fawn brown coloured seeds having light yellow cotyledons. The pulse has been lately introduced and is under marginal cultivation.

Calamagrostis pseudophragmites (Haller f.) Koelar (Poaceae): Ladakhi- Jamak [W]. Perennial tall herb on open alpine slopes, near moist situations. Spikelets in the overhanging panicles are violet coloured. Plant serves as an excellent fodder to cattle in fresh as well as in dried forms.

Calendula officinalis L. (Asteraceae): Eng.- Pot Marigold; Hindi- Zer-gul; Kash.- Hamesh Bahar [NI]. An annual herbaceous ornamental producing daisy like orange to yellow flower heads from May to first frost. Grown in all gardens, parks as a spring flowering plant. Flowers used against burns.

Callianthemum pimpinelloides (D.Don. ex Royle) Hook. f. & Thomson (syn. Callianthemum cashmerianum Cambess (Ranunculaceae) [W]. A densely tufted glabrous herb of alpine meadowlands. Leaves 2-3 pinnate, segments orbicular, acute. Flowers pale with orange nectar.

Caltha palustris var. alba (Camb.) Hook.f. & Thoms. (Ranunculaceae): Eng.- Marsh Marigold; Kash.- Jungli [W]. Perennial alpine/sub-alpine herb along mountain streams and wetlands. Leaves dark-green, fleshy, borne in a rosette. Root stock densely fibrous. Flowers white. Fresh plants taken to relieve constipation (Fig. C 1).

Camellia japonica L. (syn. Thea japonica (L). Baill.) (Theaceae): Eng.- Garden Camellia / Rose of Winter [NI]. An upright to spreading evergreen shrub grown for its single and double pink reddish coloured flowers which resemble rose. Leaves alternate, dark-green above and pale –green below.

Campanula cashmeriana Royle. (Campanulaceae): Eng.- Kashmiri Bellflower; Ladakhi- Kratsenagpe; Kash.- Chaeri Hakh [W]. An attractive perennial herb, with branched stem, grows on crevices of rocky cliffs across alpine areas. Flowers blue, bell shaped. Stem often tufted, hairy. Leaves oblong, sessile, tomentose on both sides. Powdered seeds mixed with butter given as appetizer, against indigestion and post vomiting nausea (Fig. C2).

Campanula latifolia L. (Campanulaceae): Eng.- Giant Bellflower [W]. Perennial herb on shaded forest slopes. Stem unbranched. Leaves stalked, broadly ovate, with cordate base. Flowers bluish purple, bell shaped. Fruit a hairy nodding capsule. Plant has ornamental importance.

Campanula pallida Wall (syn. Campanula colorata var. tibetica (Hook. f.& Thomson) H Hara (Campanulaceae): Eng.- Himalayan Bellflower /Tibitan Bell Flower [W]. A much branched, hairy stemmed herb of exposed slopes; endemic to Kashmir and Ladakh. Flowers bell shaped, light purplish-blue; sepals curved outwards. Leaves stalkless, narrow elliptic, softly hairy, margins toothed. Whole plant used medicinally.

Campsis grandiflora (Thunb.) K. Schum. (syn. Bignonia grandiflora Thunb.) (Bignoniaceae):Eng.- Chinese Trumpet- Creeper [NI]. An attractive, deciduous, ornamental climber with pinnate dark green leaves. Flowers funnel-shaped in loose, terminal, large panicles. Corolla tube flared, funnel form, tangerine red, inside deep yellow; fruit an elongated capsule.

Campsis radicans L. Seem. (syn. Bignonia radicans L.) (Bignoniaceae): Eng.- Trumpet Vine; Hindi- Latkania [NI]. A vigorous climber grown for its graceful foliage and large showy flowers. Terminal cymes of 4-12 slender, tubular-trumpet shaped orange to red flowers are borne on current year's wood.

Campsis x tagliabuana (Vis.) Rehder. (Bignoniaceae): Eng.- Chinese Trumpet vine [NI]. A variable hybrid climbing shrub with pinnate dark green leaves. Flowers are in loose panicles, bearing trumpet shaped orange-red flowers. Fruit an elongated capsule beaked at apex.

Canna hybrida Hort. various cultivars (Cannaceae): Eng.- Cana. [NI]. A large tender herbaceous ornamental cultivated in green houses and as border plants. Leaves are large, green to brown or purple up to 60 cm long and 30 cm wide. The flowers are orange, red or rich yellow with gladiolus resemblance. Multiplication is through fleshy rhizomes.

Cannabis sativa L. (Cannabaceae): Eng.- Hemp; Hindi- Ganja, Bhang; Kash.- Bhangi [W]. A strong- smelling annual, growing as naturalized weed on river banks, waste places and orchards. Also cultivated for its narcotic resin. Stem slender, angular, grooved; leaves 7-20 cm long, palmately 3-11 partite, stacked, leaflets lanceolate and serrate. Plants diocious; male flowers yellowish in short dense cymes; female flowers light green, solitary in the axils of small membranous bracts. The drugs–*bhung* / *charas* are derived from the flowers and leaves and the resinous matter obtained from them. *Bhang* – is the dried and finely powdered/solidified mature leaves and flower shoots of the male and female plants. *Charas* is obtained from leaves and flowering tops of female plants which remain coated with resinous exudates in the month of September-October. For its extraction the foliage of the live plants is gently rubbed between bare hands and in the process a thick sticky dark greenish- substance gets accumulated on hands which is later peeled off forcefully from the hands and later pressed into small balls. These balls make commercial *charas* for sale and smoking.

The stem of the plant also furnishes an excellent fibre from which matting and strong and durable ropes are made. Besides the fibre is also used to make wooden boats water proof. In villages the stems of the plants are used for wattled fences around home gardens.

Capparis spinosa L. (Capparaceae): Eng.- Caper bush / Flinder's Rose; Hindi- Kabra; Ladakhi- Kabra [W]. A small prostrate shrub of alpine slopes. Leaves simple, leathery, shining with thorny stipules. Flowers pinkish- white. Fruit fleshy with red pulp. Flower buds and immature fruits used as vegetable and condiment. Decoction of leaves used against hyperacidity.

Capsella bursa- pastoris (L.) Medik. (Brassicaceae): Eng.- Shepherds Purse; Kash.-Krale mound. [W]. An annual herb of common occurrence in grasslands and as a common weed of cultivation. Pods triangular and purse like. Leaves of the plants cooked as vegetable in villages. Seeds used as cardiac stimulant. In Ladakh fresh leaves are boiled and rolled into balls which are carried in polythene bags during long journeys.

Capsella thomsonii Hooker (Brassicaceae): Eng.- Shepherds Purse; Ladakhi –Koezeat [W]. Annual herb of alpine slopes above 3000m altitudes.Flowers white. Leaves cooked as vegetable in Ladakh.

Capsicum annuum var. longum (DC) Sendtn. (Solanaceae): Eng.- Red Pepper; Hindi –Lal mirch; Kashmiri- Marchwangan [WC]. An annual herb, 60-90 cm tall, cultivated for its fruit. Leaves oblong, glabrous and dark green, flowers solitary, pure white. Berries green, 10 to 20 cm long, tapering towards apices, maturing into blood red pendent fruits. Fruits are extremely pungent and used both as green and ripe as spice and condiment. Cultivated varieties include:

- **Kashmiri Red Chillies:** Fruits long, very pungent with appealing blood red colour and good keeping qualities. Yield ranges from 70-90 and 15-20 q ha-1 respectively for green and dry red chillies (Fig. C 3).
- **Nishat -1**:A red chilli variety developed by SKUAST-K and released for cultivation in the temperate areas of state. Fruits long, tapering towards apex, uniformly deep red.

Capsicum annuum var. frutescens (L.) Kuntze. (Solanaceae): Eng.- Bell Pepper; Hindi – Shimla Mirch; Kash. - Shimla Marchaewangen [UC]. An annual herb grown for its fruit which are used as vegetable. Cultivated varieties include:

- **Shalimar Capsicum Hybrid -1:** A single cross hybrid developed and released by SKUAST-K through cross between two lines –SH-SP-2 and SH-SP-461. It has large sized dark green fruits. Its average yield is 446 q/ha. Matures in 55-60 days.
- **Shalimar Capsicum Hybrid –2:** A single cross capsicum hybrid between two lines –SH-SP-2 and SH-SP-11 and released by SKUAST-K during 2010. It has large sized dark green fruits with. Fruit yield of 417 q /ha. Maturity is within 55-65 days with longer period of availability.
- **California Wonder:** A European Bell Pepper variety introduced for cultivation. Fruits are large dark- green with thick flesh.

Many other pepper varieties from neighbouring states are also under cultivation whose fruit colours varies from red, yellow and brownish green. Red types are however cultivated commercially for seed production purposes only.

Caragana brevispina Benth. (Fabaceae): Kash.- Baure Kounde [W]. A medium-sized spreading armed shrub of open and dry sub-alpine habitats with arching branches and pink-tinged, downy young shoots. Leaves pinnate. Flowers yellow and attractive. Dried plants, bearing long spines, are used for fencing of orchards (Fig. C 4). Leaves taken for aching joints.

Caragana micropylla Lam. (Fabaceae): Eng.- Little leaf Peashrub [W]. A spreading, deciduous sub-alpine shrub of medium height. Young twigs silky pubescent, leaflets oval to obovate, emerginate, grey green; silky flowers bright-yellow, usually in pairs, short stalked, calyx cylindrical, short, toothed.

Caragana pygmaea (L.) DC. (Fabaceae): Eng.- Pygmy Peashrub; Ladakhi-Tamma/ Brema [W]. A high altitude fodder plant growing wild in Ladakh up to an altitude of 4500 m. It is a spiny deciduous bush, and flourishes extensively on sandy stretches in Rupshu, Leh and also act as sand binder. Leaves which are sessile, compound with linear lanceolate leaflets, have good fodder value. Flowers light yellow. It is used to treat sinusitis, gastric ulcers and as blood purifier (Fig. C 5).

Cardamine hirsuta L. (Brassicaceae): Eng.- Hairy Bitter Cress; Kash.- Haagsuiner [W]. Annual or biennial herb, in moist places at 1700-2600 m. Stem green and wiry. Flowers white. Decoction of the leaves used against rheumatism and epilepsy.

Cardamine impatiens L. (Brassicaceae): Eng.- Narrowleaf Bitter Cress; Kash.-Haagsuiner [W]. Slender biennial herb on damp soils in sub -alpine areas. Stem erect and glabrous. Leaves pinnate. Flowers white. The extract of the plant is used as stimulant and diuretic.

Carduus edelbergii Rech. F. (Asteraceae): Eng.- Edelberg's Cotton Thistle [W].A tall stout thistle inhabiting wastelands and orchards. Stem grooved, with waxy wings.Leaves lobed, thorny. Flower heads borne singly, pink.Whole plant medicinal.

Carduus nutans L. (Asteraceae): Eng.- Musk Thistle; Kash.- Chaeri Toumul; Ladakhi- Jangcher [W]. Biennial, stout, armed thistle, in wastelands and open dry places. Stem and leaves bear sharp spins. Flower heads with red-purple flowers. Seeds have wind dispersal. Fresh leaves and roots are chewed to initiate vomiting in case of indigestion. Decoction of roots used as blood purifier (Fig. C 6).

Carex kashmirensis C B Clarke (Cyperaceae) [W]. Perennial rhizomatous herb, endemic to Kashmir Himalaya, inhabiting shaded damp places in forests. Flowers greenish. Rhizomes used for joint problems.

Carex maubertiana Boott. (Cyperaceae): Ladakhi- Jangshar [W]. Perennial herb under forest shade, generally near wet places. Rhizomes thick and stout. Culms tufted, obtusely trigonous. Extract of the plant is used against rheumatism.

Carex nivalis Boot. (Cyperaceae): Ladakhi– Karcheot [W]. Perennial tufted, rhizomatous herb, common on open forest grasslands. Culms 20-40 cm long,trigonous, clothed at base. Spikes dark purple-red. Paste of powdered leaves is applied as antiseptic on wounds.

Carex nubigena D Don. ex Tilloch & Taylor (Cyperaceae): Kash.- Fical [W]. Perennial herb, usually on shaded damp places in forests. Culms tufted, trigonous. Flowers greenish. Leaves used for treating open wounds.

Carissa spinarum L. (Apocyanaceae): Eng.- Bush Plum; Hindi- Karanda [W]. A bushy evergreen ornamental shrub grown in hedges for its fragrant flowers. Leaves glossy green, opposite. Flowers white, star shaped. Wood has been used for making spoons and similar cutlery and combs.

Carpesium abrotanoides L. (Asteraceae): Kash.- Woliangil [W]. Annual herb, usually in shaded wastelands, also in forest in ravines. Flower heads many, sessile, spicately arranged and yellow. Used as an insecticide in folk medicine to treat bruises.

Carthamus lanatus L. (Asteraceae): **Eng.- Saffron Thistle** [W]. A glandular and spiny herb growing on karewa lands of Kashmir. Heads straw-yellow. Plant is sudorific, febrifuge and anthelimintic.The plant is a wild relative of cultivated safflower (Fig. C 7).

Carum carvi L. (Apiaceae): Eng.- Caraway; Hindi- Zeera; Kash.- Zeure; Ladakhi-Khonat [W]. Annual herb grown for its aromatic seeds. Flowers white in compound umbels. Powdered seeds are used to treat abdominal disorders as carminative and digestive.

Carum roxburghianum (DC) Kurz. (Apiaceae): Eng.- Wild Celery; Hindi-Karfuss Kash.- Pekhae zuere [W]. An erect branched aromatic annual inhabiting woodlands. Stem longitudinally triped. Leaves double compound, ultimate segments all linear. Flowers in compound umbels with white or pink petals. Fruits ovoid, shining and yellow. Used for seasoning of dishes.

Carya illinoinesis (Wang) Koch. (syn. Carya pecan (Marshall) Engl. & Graeb.) (Juglandaceae): Eng.- Pecan /Pecanut; Kash.- Pista magaz [W]. A large and valuable nut-bearing deciduous tree, with ornamental, furrowed grey bark; pinnate, mid-green leaves, nuts thick shelled edible when ripe. The tree grows from 1650 to 1850 m in forests and protected areas. Fruit a drupe with a single stone.

Carya ovata (Mill.) K. Koch (Juglandaceae): Eng.- Pecan / Shagbark Hickory [W]. The most valuable nut-producing, broadly conical tree with ornamental peeling, grey to brown bark. The pinnate, mid green leaves, have usually 5 long-pointed leaflets. Leaves turn golden yellow in November. Fruit a drupe, globose, edible nut within a bony shell, contained in thick husk.

Cassia angustifolia M. Vahl.; Hindi – Burge sana (Fabaceae) [NI]. An ornamental shrub 0.5 to 1 m tall; stem pale green, erect with long spreading branches bearing 4 – 5 pairs of leaves. Flowers yellow in a raceme.

Cassiope fastigiata (Wall.) D Don (Ericaceae): Eng.- Himalayan Heather [W]. A much branched tufted shrub of Kashmir Himalaya. Many stems remain clustered together. Flowers are pendulous,white, like bells.

Castanea sativa Mill. (syn. Castanea vesca Gaertn.) (Fagaceae): Eng. – Sweet Chest nut; Kash.- Panjaeb Gour [RC]. A vigorous, broadly columnar tree, upto 30m tall, with spirally furrowed bark. Leaves oblong, toothed, glossy, tomentose beneath. Male flower catkin yellowish- green, long, being extremely ornamental; female flowers bears nut fruits in October. Nuts are enclosed by spiny protective husks. Seeds are edible.

Catalpa bignonioides Walt. (Bignoniaceae): Eng.- Indian Bean Tree / Cigar Tree [W/UC]. A medium sized, spreading deciduous tree. Leaves broadly ovate, entire, heart shaped and mid green. White flowers, marked with yellow and purple-brown spots are borne in upright panicles. Fruit thin bean like pods. Seeds with papery wings. Grown as an ornamental cum avenue tree in gardens and road sides.

Catalpa speciosa (Warder ex Barney) Warder ex. Engelm. (Bignoniaceae): Eng.- Northern Catalpa [W/UC]. A tall spreading tree, with broad ovate, glossy dark green leaves being densely hairy underneath. Large white trumpet shaped flowers marked with yellow and purple are sparsely borne in upright panicles. Pods slender bearing seeds with papery wings.

Cedrus deodara (Roxb. ex D Don.) G. Don (syn. Pinus deodara Roxb.; Abies deodara Lindley) (Pinaceae): Eng. - Himalayan Cedar; Hindi- Deodar; Kash.- Rayil [W/UC] A large conical, coniferous tree with spreading branches and drooping, branchlets and leading shoots. Bark smooth with vertical fissures. Leaves either solitary or in dense clusters, dark green, rigid, leathery, three-sided, sharp pointed, produced in whorls of 20-30; male cones cylindrical; female cone ovoid cylindrical, erect. Young cones bluish-purple. Its wood used for making bridges, house building, furniture works and boat making. Also cultivated as an ornamental.The plant is sacred to Hindus as they consider it as an abode of *Devtas* (Fig. C 8).

Celastrus paniculatus Willd. (Celastraceae): Eng.- Staff Vine; Hindi- Mal-kangni [W/UC]. A large climber with corky bark, young shoots marked with lenticels. Leaves variable in size and shape, glossy green above turning clear yellow in Autumn. Panicles terminal, large, drooping, branching into compound cymes. Flowers pale-green, calyx-lobes rounded, toothed, petals oblong, capsule globose bright yellow when ripe.

Celosia argentea L. (Amaranthaceae): Eng.- Cocks Comb; Hindi- Morsikha; Kash.- Moval [RC]. An annual cultivated herb, planted as a border plant, in vegetable fields, for obtaining flowering heads. The plant has greenish-pink leaves and purple coloured inflorescence which is the source of a deep pink dye used in colouring of dishes. Leaves are slightly sour in taste and used as vegetable, especially cooked with fish (Fig. C 9).

Celosia argentea var. cristata (L.) Kuntze. (Amaranthaceae): Eng.- Cocks Comb; Hindi- Safed Murga; Kash.- Mavali posh [NI]. An annual hardy herb planted for its showy crested flower heads and foliage. Leaves lanceolate appearing crowded from dwarf stem. Heads crested, red, orange, or yellow produced from July to September. *Celosia argentea* L. var *pyramidalis,* having feathery flower plume is often grown as pot plant.

Celtis australis L. (syn. Celtis alpine Royle.) (Cannabaceae): Eng.- Hackberry / Lote Tree; Kash.-Bremij [W]. A spreading, deciduous tree with elliptic, dark green leaves. Fruits pea sized, yellowish at maturity and edible. The tree grown for its shade and wood, is mostly seen growing around sacred groves, road sides and grave yards.

Centaurea depressa M.Bieb. (Asteraceae): Eng.- Low Cornflower; Ladakhi-Vashaka [W]. Erect branched herb of Ladakh Himalaya; stem cottony; lower leaves petioled, pinnatifid, upper sessile, irregularly toothed. Heads pale. Whole plant used for cough, chest pain, head-aches, and abdominal distress. Fruit used as blood purifier.

Centaurea iberica Trev. ex Spreng. (Asteraceae): Eng. –Iberian Star Thistle; Hindi- Harshaf; Kash.-*Kretch.* [W]. A perennial herb growing wild, especially in moist situations on pastured slopes. Flower heads purple. Seeds plumed. Leaves of the plant are cooked as herbal vegetable in fresh form or after sun drying (Fig. C 10).

Centella asiatica (L.) Urban (syn. Hydrocotyle asiatica L.) (Apiaceae): Eng.-Indian Pennywort; Hindi- Brahmi [W]. A prostrate herb rooting at inter-nodes, occurring in moist soils. Runners of the plant used to poison fish and wild life. Tincture of the plant used for treating skin disorders.

Cephalanthera longifolia (L.) Fritsch. (Orchidaceae):Eng.- Narrow leaved Helleborine [W]. A rhizomatous herbaceous erect perennial with glabrous multiple stems.The leaves are dark-green, long and narrowly tapering. The inflorescence is a lax, 5-20 flowered spike bearing white, orange lipped bell shaped flowers. Plant inhabits damp woodlands, near forest edges and rocky slopes.

Cerastium cerastoides (L.) Britton. (Caryophyllaceae): Eng.- Mountain Chickweedi; Ladakhi- Spengyan karpo [W]. A rhizomatous mat forming perennial with creeping stems inhabiting alpine snow beds near glaciers above 4500m.Leaves sessile, somewhat succulent. Inflorescene lax, 1-3 flowered cymes. Flowers white. In Ladakh plant used for renal colic and headaches.

Cerasus spachiana f. spachiana (syn. Prunus subhirtella Miq. var. pendula (Sieb. & Maxim.) Y.Tanaka (Rosaceae): [NI]. A large ornamental tree, with a thick trunk. Leaves ovate, acuminate. Semi double rich pink flowers are produced in great profusion during March to early April.

Ceratocephalus falcatus (L.) Pers. (Ranunculaceae): Kash.-Kinial [W]. Annual herb, along roadsides and open slopes. Flowers yellow. Powder of the plant used against ringworm infections.

Ceratophyllum demersum L.(Ceratophyllaceae): Eng.- Horn Wort; Kash.-Hill [W]. A submerged hydrophyte of water bodies of Kashmir. Dried leaves and twigs used as adultrant of salt tea.

Cercis siliquastrum L. var. boilant (Fabaceae): Eng.- Judas Tree. [NI]. A graceful small sized deciduous ornamental tree, very attractive when in full bloom. The trunk is stout and twisted, bark rough, generally branching low and forming a broad crown. Leaves simple entire, alternate pure green and glabrous. The most distinguishing feature is its habit of producing rosy lilac flowers all over the wood, from joints, the trunk and the principal branches. Fruit a pod, dark brown in colour.

Chaenomeles japonica (Thunb.) Lindl. ex Spach. (Rosaceae): Eng.- Japanese Quince [W/UC]. A dwarf, thorny high shrub, with warty pubescent young growth. Leaves nearly orbicular, roundish - oval to obovate and glabrous. Flowers orange scarlet. Fruit nearly globular, yellow with red blush.

Chaenomeles speciosa (Sweet.) Nakai (syn. Chaenomeles lagenaria (Loisel.) Koidz.) (Rosaceae): Eng.- Maules quince [NI]. A vigorous, wide spreading shrub with tangled, spiny branches and broad ovate, glabrous leaves. Flowers borne in clusters scarlet to crimson in colour. Fruits yellowish green with several deep furrows, aromatic. Three commonly cultivated vars. are *moerloobi* –(pink/white flower in clusters), *simmonii*- (blood- red, semi double flowers) and *umblicata* (deep salmon pink flowers).

Chaerophyllum acuminatum Lindl. (Apiaceae): Eng.- Turnip rooted Chervil; Ladakhi- Beerananrpo [W]. Perennial, pubescent herb, usually on forest openings and alpine slopes. Leaves 1-2 pinnate.

Flowers white. Powdered fruits after mixing with honey used for cold and other respiratory infections.

Chaerophyllum villosum Wall. (Apiaceae): Eng.- Rouge Chervil; Kash.- Kow kund. [W] Perennial much branched herb,usually in forest openings at 1700- 1800 m altitudes. Flowers in June- August, white. Leaves finely divided, larger.

Chamaemelum nobile (L.) All. (syn. Anthemis nobilis L.) (Asteraceae): Eng.- Common Chamomile [NI]. Herbaceous perennial mat forming ornamental with dissected aromatic leaves. Produces daisy like flowers from June to August.

Chenopodium album L. (Amaranthaceae): Eng.- Pigweed; Hindi- Bathua; Kashmiri - Wastahakh; Ladakhi- Janchikarpo [RC]. An annual herb grown for its foliage. Two chenopod types having purple and dull green leaves are grown in Kashmir. Foliage of 4 to 5 leaved seedlings is used as vegetable in early spring. In *Amchi* system of medicine the extract of the leaves is given for gastric and liver troubles.

Chenopodium ambrosiodes L. (Amaranthaceae):Kash.- Jangli javend [W]. An annual herb of alpine grasslands. Seeds used for abdominal pain.

Chenopodium foliosum Asch. (syn. Chenopodium blitum Hook.f.) (Amaranthaceae): Eng.- Leafy Goosefoot /Pig weed; Hindi- Bathua; Kash.- Wan Palak [W]. An annual herb growing in -alpine grasslands of Kashmir Himalayas up to 3500 m amsl. Leaves fleshy and used as vegetable. Fruits red berry like glomerules, used to promote apetite and for treatment of stomach and liver diseases (Fig. C 11).

Chenopodium murale L. (Amaranthaceae): Eng.- Nettle leaved Goosefoot; Ladakhi- Janchikarpo [W]. Annual shade loving herb upto 70 cm tall, growing as a weed. Stem erect, leaves triangular, toothed. Whole plant is dried and stored to be used as vegetable in winter. Fresh leaves also used as vegetable.

Chenopodium scoparia L. (Amaranthaceae): Eng.- Goose foot; Hindi- Bathua; Kash.- Latchije [W]. Annual herb,usually along pathways and cultivated fields.Flowers June – August, green. After leaf fall plant used as a broom, especially in villages.

Chesneya cuneata (Benth.) Ali (Fabaceae): Eng.- Wedgeleaf Chesneya; Ladakhi- Beegamboo [W]. A tufted perennial herb. Tap root thick and woody.Leaf imparipinnately compound. Flowers purple. The paste of powdered seeds is used as antiseptic on open wounds.

Chimonanthus praecox (L.) Link. (Calycanthaceae):Eng.- Winter Sweet [NI]. A vigorous, broadly upright, deciduous shrub, with entire, lance shaped glossy mid-green leaves, arranged in opposite pairs. Pendent fragrant, sulphur-yellow flowers, stained brown or purple inside, produced on the leafless branches from late December to February. Fruit completely enclosed by the ovoid or pyriform receptacle.

Chimonanthus praecox (L.) Link var. grandiflorus (Calycanthaceae) [NI] An ornamental shrub. Flowers long deeper yellow, conspicuously stained with maroon, produced on the leafless branches from mid December to mid February. Fruit completely enclosed by the ovoid or pyriform receptacle.

Christolea crassifolia Camb. (Brassicaceae):Eng.- Afghani Christolea; Ladakhi- Krascheotz, Kharla [W]. A leafy perennial herb endemic to high altitude desert lands of Ladakh. Stem sub-erect, densely villous. Flowers white with purple base. Fresh leaves used as vegetable. Powdered roots are taken as purgative (Fig. C 12).

Chrysanthemum griffithii Clarke. (Asteraceae): Ladakhi- Chackerkeh [W]. A small shrub,up to 50 cm tall,with several erect to ascending green arachnoid hairy branches arising from woody rootstock. Leaves hairy, bi or tri-pinnatisect divided into small segments. Capitula solitary on apex of branches, whitish. Plant inhabits rocky crevices, gravely beds of alpine zone above 3000 m altitudes. Powdered plant used as poultice to relieve joint pain.

Chrysanthemum pyrethroides (Kar. & Kir.) Fedt. (Asteraceae): Eng.- Himalayan Chrsanthemum [W]. Dwarf, densely woolly perennial of dry alpine slopes at 3300 – 4800 m altitudes, common in Ladakh. Cauline leaves 2-3 pinnatisect; peduncles long,selender, 1-2 leaved. Involucral bracts ovate. Ray-florets pinkish with yellow central disc. Whole plant medicinal.

Chrysanthemum maximum L. (Asteraceae): Eng.- Maxa Daisy; Kash.- Gule Dawood [NI]. A rhizomatous, herbaceous green house perennial autumn flowering ornamental grown as a border cum pot plants in green houses or outdoors. Leaves dark green, toothed; flower heads borne on strong stems, white with yellow disc florets. Numerous varieties with different shade of cream-white light yellow and pink are grown in gardens and parks.

Chrysopogon fulvus (Spreng) Chiov. (syn. Pollinia fulva Spreng.; Andropogon montanus Koen. ex Trin.) (Poaceae): Hindi –Goria.; Kashmiri- Baeren [W]. A variable perennial grass growing in tufts over hilly slopes and grasslands up to 2000 m amsl. Racemes bearing triad of spikelets. Esteemed as a fodder grass; it can also be made into hay.

Chrysopogon gryllus (L.) Trin. (syn. Andropogon gryllus; Chrysopogon echinulatus (Nees) W. Wats.) (Poaceae): Eng. – Bunch Grass; Hindi– Salum [W]. A perennial tall tufted grass found growing in sub-alpine grasslands up to 2500 m altitude under open sunny and drier locations. Flowers purplish. Panicles with branches tipped by a raceme.

Cicer arietinum L. (Fabaceae): Eng.- Chick Pea; Hindi- Chana; Kashmiri- Boda Chana [RC]. Herbacious annual with diffuse branching. Leaves pinnately compound with small leaflets. Flowers axilary and greenish white. Fruit a turgid pod with 1 to 2 seeds which are light fawn in colour with thick yellowish white cotyledons.

Cicer microphyllum Benth. (Fabaceae): Ladakhi – Sari [W]. A high altitude cold adapted cicer species of Trans-Himalayan zone of Ladakh. Branched glandular –pubescent stem. Rachis ending in a tedril. Flowers royal purple or white. Pods beaked many seeded. In Ladakh aerial portions of the plant used for tongue infections (Fig. C 13).

Cicer songaricum DC (Fabaceae): Eng.- Dzungarian Chick pea; Ladakhi- Sari [W] An annual herb growing in Alpine meadows of Ladakh. Flowers violet- blue. Pods pubescent, beaked. Grains eaten in parts of Ladakh. Dried plants and seed fed to animals during winter to give them warmth and energy.

Cichorium intybus L. (Asteraceae): Eng.- Chicory/Blue Dandelion; Hindi- Kasni dashti; Kash.- Kasnil Hande [W]. A perennial, hispid herb commonly occurring in meadows and wastelands in Kashmir Himalaya. Flower heads light blue. Roots and flowers are demulcent and with cooling properties. A decoction of seeds is given in treating menstruation and bilious disorders. Dried and powered roots are blended with coffee (Fig. C 14).

Cicuta virosa L. (Apiaceae): Eng.- Cowbane / Northern water Hemlock [W] A perennial herb (1 – 2 m tall) of stream banks and wet lands. Stem smooth, branching, swollen at base, purple stripped. Leaves alternate, tri-pinnate, coarsely toothed. Flowers small, white in umbel shaped umbrellas. Stem used in convulsions.

Cimicifuga foetida L. (Ranunculaceae): Eng.- Bugbane; Kash.-Schurer [W]. Stout perennial herb on humus rich soils. Leaves ternately pinnate. Flowers yellowish white. Extract of the roots is used for rheumatic pain.

Cirsium arvense (L.) Scop. (Asteraceae): Eng.- Creeping Thistle; Ladakhi – Jangcher [W]. A rhizomatous perennial herb of sub-alpine / alpine grasslands, forming extensive colonies. Stem green, smooth and glabrous. Leaves spiny, lobed. Heads pink-purple. Aerial parts used to induce vomiting.

Cirsium falconeri (Hook.f.) Petrak. (Asteraceae): Eng.- Falconer's Thistle; Ladakhi- Jangcher; Kash.- Boban [W]. Perennial spiny herb of Kashmir and Ladakh regions. Flower heads cream coloured. Leaves with lobed margins carrying pale spines. Fresh leaves are chewed to induce vomiting in case of indigestion.

Cirsium wallichii D.C. (syn. Cirsium nepalensis DC)(Asteraceae): Ladakhi-Tsejangchar Tagar [W]. Biennial spiny herb, usually in waste and alpine lands. Stem villous. Flowers creamy white /light pink, densely woolly. Powdered roots dissolved in water are used as cooling drink. Extract of the shoots is used to expel intestinal worms.

Citrus trifoliata L. (syn. Poncirus trifoliata (L.) Rafin) (Rutaceae): Eng.- Trifoliate Orange [NI]. A rounded, bushy, deciduous shrub or small tree with rigid green shoots armed with very sharp spines. Alternate, 3-palmate leaves, comprise 3 obovate to elliptic, crenate, somewhat leathery dark-green leaflets turning yellow in October. Produce solitary, saucer-shaped, fragrant white flowers in mid April; followed by a dull yellow-green, finely pubescent, 3-5 cm thick non-edible lemons.

Citrullus vulgaris Schrad. (Cucurbitaceae): Eng.- Watermelon; Kash.- Hendewaend [UC]. Annual trailing viny herb grown for its large edible fruits.

Clematis alpina subsp. sibirica (L.) O.Ketze (Ranunculaceae): Eng.- Alpine Clematisi; Ladakhi- Bisho [W]. A woody climber, common on scrubby slopes of cold-arid mountain slopes of Ladakh Himalaya. Powdered roots are taken as laxative.

Clematis jackmanii Moore (Ranunculaceae) [NI]. A superb large-flowered hybrid climber. Leaves simple to trifoliate, dark green above, glabrous, lighter beneath, lightly pubescent. Flowers usually in groups borne at the ends of the current year's wood, violet purple, flat spreading. *Nelly koster* and *Nellymoser* are the two commonly grown varieties differing in flower colour.

Clematis grata Wall. (Ranunculaceae): Eng.- Virgin,s Bower; Kash.- Agar -ranth [W/UC]. A strong growing climber with deeply furrowed stems; leaves with usually 5 leaflets, ovate-Ianceolate long pointed, strongly toothed or lobed, hairy beneath. Flowers in small cream-coloured clusters, fragrant, with spreading petals. Fruit indehiscent, pubescent achene.

Clematis montana Buch-Ham. ex DC. (Ranunculaceae): Eng.- Himalayan Clematis [W/UC]. A very vigorous deciduous climber, with trifoliate leaves; leaflets short stalked, ovate, usually deeply dentate. Flowers white, grouped 1-5 in axillary clusters, slightly fragrant. Achenes flat, hairless.

Clematis orientalis var. obtusifolia Hook. f. & Thomson. (Ranunculaceae): Eng.- Oriental Clematis; Ladakhi – Yamongnakpo [W]. Deciduous woody climber, growing in scrubby, rocky areas in forests. Flowers creamish. Powdered roots and shoots are used to cure stomach troubles. Also grown as ornamental to cover walls.

Clerodendrum indicum (L.) Kuntze. (Lamiaceae): Eng.- Turk's Turban [NI] An tall shrub grown as an ornamental. Root used for healing wounds, digestive problems and against cough and cold.

Clinopodium umbrosum (M. Bieb.) Kuntze. (syn. Calamintha umbrosa Fisch. Et. Mey.) (Lamiaceae): Eng.- Shady Calamint [W]. Perennial herb,usually in forest shade, damp places at 1700-2500 m. Flowers pink. Rhizomes used as cardiac tonic.

Clinopodium vulgare L. (syn. Calamintha clinopodium Benth.) (Lamiaceae): Eng.- Wild Basil [W]. Perennial rhizomatous herb, usually growing among scrubs,on dry scrubby and savannah slopes at 1700-2200 m. Flowers purple. Leaves opposite, hairy, ovate or lanceolate. Rhizomes used as tonic.

Clitocybe nuda (Bull) Bigelow Sm (Tricholomataceae): Eng.- Wood Blewit; Kash.- Saeki-Boube [W] An edible mushroom inhabiting coniferous and dicidous woodlands on decaying leaf litter. Stem stout, enlarged at base. Flesh soft purplish or whitish caps convex.

Codonopsis clematidea (Schrenk.) Clark. (Campanulaceae): Eng.- Asian Bellflower; Ladakhi- Brukutung [W]. A low upright climber plant on rocky slopes,with several stems. Flowers blue, showy. Seed used for swelling due to sprain.

Codonopsis ovata Benth. (Campanulaceae): Eng.- Kashmir Bonnet Bellflower; Kash.-Kishgur; Ladakhi- Ludut [W]. Perennial trailing herb, among scrubs in alpine pastures at 3000-3800 m. Leaves ovate, hairy. Flowers dirty blue, bell shaped and nodding. The paste of roots and leaves is used in treating bruises and wounds. Tap roots used as famine food (Fig. C 15).

Codonopsis rotundifolia Benth. (Campanulaceae): Ladakhi – Kaeremapo [W] Twining herb on open alpine slopes above 3000 m. Roots carrot like. Flowers yellow-green with purple reticulate markings. Powdered roots used as poultice to stop cutaneous eruptions.

Colchicum luteum Baker. (Colchicaceae): Eng.- Meadow Crocus/ Yellow Colchicum; Hindi- Suranjan-e-talkh /Hiran-totiya; Kash.- Vir-keom [W]. A perennial herb of meadows/ open slopes of temperate Himalaya, and flowering in early spring. Corm ovoid. Perianth golden yellow, funnel shaped. Extracts of bulb used to relieve pain of gout and rheumatic inflammations, diseases of liver and spleen; also applied to old piles and wounds to relieve pain (Fig. C 16).

Colutea arborescens L. (Fabaceae): Eng.- Bladder Senna [NI]. An ornamental shrub with pinnate, pale green leaves. Produces 3-8 yellow flowers in racemes from early to late May, followed by green, and translucent seed pods.

Colutea media Willd. (Fabaceae): Eng.- Bladder Senna [NI]. A bushy ornamental shrub with pinnate bluish green leaves. From early to late May, it bears orange-brown flowers sometimes flushed yellow in the centre in racemes. Seed pods, long, greenish-brown at first then turn translucent.

Comastoma pedunculata (Rottb.)Toyok. (syn. Gentiana tenella Rottb.) (Gentianaceae): Kash.-Gule Maidan; Ladakhi- Narpowanglucktsetz [W]. Annual herb of alpine pastures. Flowers bluish purple. Decoction of the plant used for renal infections. In Ladakhi system of medicine powdered plant,mixed with vegetable oil, is used to cure frost bites.

Conium maculatum L. (Apiaceae): Eng.- Devils Beard/ Poison Parsely; Hindi- Mohra kuche; Kash.- Fauke baudyane [W]. A herbaceous biennial inhabiting roadsides, waste places and disturbed soils, having become an invasive species, forming extensive stands. Stem smooth, hollow, streaked purple. Leaves bi- to tri-pinnate and emit an unpleasant smell. Flowers white, clustered in compound umbels.Threpeutically used as sporofic.

Consolida ajacis (L.) Schur. (syn. Delphinium ajacis L.) (Ranunculacea): Eng.- Rocket Larkspur [NI]. A hardy herbaceous annual with spurred leaves grown in gardens mostly as border plants. Branches upright, and sparce with finely cut fern like leaves. Racemes are loose and bearing blue violet flowers.

Consolida incana (E.D. Clarke) Munz. (syn. Delphinium incanum Clarke.). (Ranunculaceae): Eng.- Foothill Larkspur; Kash.- Kari pate [W]. Perennial herb, growing on forest slopes on rich scrubby soils. Flowers blue. Used as a veterinary medicine to kill ticks (Fig. C 17).

Consolida regalis Gray. (syn. Delphinium consolida L.): Eng.- Forking Larkspur. [NI]. Erect growing, well branched ornamental bearing long racemes having blue, purple, red, pink to white flowers. Leaves finely cut into segments. The species is represented by many strains and varieties. The other cultivated varieties belonging to other species like *Delphinium cardinate, D. grandiflorum* and *D. nudicaule* are also cultivated in gardens.

Convolvulus arvensis L. (Convolvulaceae): Eng.- Field Bindweed; Kash.- Thoure; Ladakhi- Tiktikma [W]. Annual herb commonly growing in grasslands and orchards or dry open slopes, as a weed. Flowers rosy-white. Leaves spirally arranged. Serves an an excellent fodder for milking cows. In Ladakh powdered leaves are used as blood coagulant on fresh wounds.

Coprinus comatus (Mull) Pers. (Agaricaecae): Eng.- Shaggymane; Kash.- Doda-Kashaer [W]. A common fungus growing on lawns, gravel roads and waste areas. Caps white and covered with shaggy scales. Flesh white commonly cooked soon after collection.

Coriaria nepalensis Wallich (Coriariaceae): Eng.- Tanner's Tree; Hindi- Mokkola [NI]. A small to medium sized deciduous, hairless shrub with 4-sided arching reddish-brown branches. Leaves, ovate to oblong, distinctly 3 veined, glabrous, entire, base somewhat cordate and nearly stalkless. Flowers conspicuously 3 veined borne on the previous years wood in cylindrical racemes, greenish-yellow in colour. Stamens and styles red or purplish, conspicuous and protruding. Fruit black, encircled by 5 large purple fleshy persistent petals. Fruits have been used as famine food.

Coriandrum sativum L. (Apiaceae): Eng. – Coriander; Hindi- Dania; Kash.- Dhaniwal [WC]. Annual aromatic herb cultivated widely for foliage and spicy fruits; white or pinkish purple flowers borne on compound terminal umbels. The lower leaves are broad with crenately lobed margins while the upper ones are narrow, finely cut with linear lobes. The fruits are globular and ribbed, yellow brown, in colour. The stems, leaves and fruits have a pleasant aromatic odour. The entire plant, when young is used in preparing *chatneys* and sauces and the leaves are commonly used for flavouring curries and soups. The fruits are employed as a condiment the preparation of curry powder and for seasoning of dishes.

Coreopsis drummondii (D Don.) Torr & Gray (Asteraceae): Eng.- Golden Wave [NI] Herbaceous annual, upright, with deep cut leaves and colourful daisy like flowers. Flowers bright yellow with a deep purple central disc and a red-brown blotch at the base of each ray floret. Flower from July to September. Single colour varieties having yellow, crimson and crimson scarlet flower are cultivated in gardens as border plants.

Cornus sanguinea L. (Cornaceae): Eng.- Common Dogwood [NI] A deciduous ornamental shrub with dark greenish-brown branches and twigs. Leaves opposite, oblong with entire margins. Berries called dog berries.

Cortaderia selloana (Schult. & Schult. f.) Aschers. & Graebn (syn. Arundo selloana Schult. & Schult. f.) (Poaceae): Eng.- Pampas Grass [NI]. Perennial ornamental grass forming large clumps; leaves long and narrow, rough margined, pistillate panicles white, silky-hairy, plumy and fluffy; staminate panicles with naked spikelets; spiklets 2-3 flowered, with slender-awned lemmas.

Cortia depressa (Don.) Norman (syn. Daucus depressus Spreng) (Apiaceae): Eng.- Prostrate Cortia [W]. A low growing,usually stemless alpine plant endemic to Kashmir Himalaya; with many spreading and radiating long reddish primary rays giving secondary umbels. Leaves double compound. Flowers white to dark red.Whole plant medicinal.

Corydalis adiantifolia Hook. f. Thomson (Papaveraceae) [W]. A high altitude herb with woody rootstock, growing at 4000 – 5000 m altitudes, preferring dry arid stony and gravely slopes. Leaves fast green, fleshy. Radical leaves pinnate, pinnae often kidney shaped. Flowers yellow in terminal lax clusters. Extract of herb used to treat skin ailments in Ladakh.

Corydalis diphylla Wall. (Papaveraceae) [W]. A perennial herb of temperate alpine, sub-alpine grasslands. Stem simple. Tubers irregularly shaped. Leaves opposite to whorled, 2-3 ternately cut. Flowers purple with dark tips.

Corydalis falconeri Hook.f. & Thoms. (Papaveraceae): Ladakhi – Ralchat; Kash.- Aet Neel [W]. Perennial herb growing in scrubs at 2500-3000 m. Rootstock slender, tuberous. Flowers July-August. Flowers yellow. Roots used for treating skin affections. In Ladakh the shoots are used to treat fevers and colds. In Kashmir the shoots and leaves were used as hair tonic to render hair thick, fragrant, soft and long (Fig. C 18).

Corydalis govaniana Wall. (Papaveraceae): Hindi- Bhut Kesi; Eng.- Govan's Corydalis; Kash.- Rhus ashude; Ladakhi –Ral chat Nakpo [W]. Erect tufted perennial herb of W. Himalayas, growing in moist situations. Leaves 2-pinnatisect. Flowers yellow in dense clusters. Roots and seeds are taken as carminative and for reducing inflammations (Fig. C 19).

Corydalis kashmeriana Royle. (Papaveraceae):Eng.- Kashmir Corydalis; Kash.- Kari pate [W]. Perennial herb usually among moist open alpine slopes. Cauline leaves oblong-lanceolate, sub-sessile. Flowers lax, sky blue with purple tips(Fig. C 20).

Corydalis meiofolia Wall. (Papaveraceae): Ladakhi- Stogzil [W]. Herb used as hepatic tonic and stomachic.

Corydalis ramosa Wall ex Hook. f. Thoms. (Papaveraceae): [W]. A herb of open sub-alpine,alpine Himalayan slopes.Leaf segments linear, acute; racemes dense; flowers yellow. Whole plant medicinal.

Corydalis thyrsiflora Prain. (Papaveraceae): Eng.- Thyrse Corydalis [W]. A glabrous erect much branched perennial inhabiting alpine rocky crevices. Leaves glaucous, three times cut into oblong to linear, usually pointed segments, pinnules petioled. Racemes dense. Flowers yellow, borne in thyrse or dense clusters.

Corylus avellana L. (Betulaceae): Eng.- Filbert / Cob Nut; Hindi- Findak [RC] An upright deciduous shrub, 3-8m tall, occasionally tree with glandular pubescent branches. Leaves rounded-broad ovate, abruptly acuminate, doubly serrate to weakly lobed. Pendent yellow catkins appear from late February to early March. Nut roughly spherical to oval with a pale scar at base.

Corylus colurna L. (Betulaceae): Eng.- Turkish Hazel nut; Hindi- Fundak; Kash.- Virin [W]. Deciduous tree, often forming stands with *Ulmus wallichiana* subsp. *xanthoderma* in coniferous forests at 2200-2800 m. A conical tree with broadly oval, shallowly lobed, long double serrate, dark-green leaves, turning yellow in October. Pendant reddish brown catkins are borne from late February to early March. Fruit 4-6 together in ball-like clusters, nut very thick shelled and edible.

Cosmos bipinnatus Cav. (Asteraceae): Eng.- Cosmos [NI]. An annual hardy free flowering ornamental grown from its dahlia like flowers which are white, crimson, rose or pink and appear from August to September.

Cotinus coggygria Scop. (syn. Rhus cotinus L.) (Anacardiaceae): Eng.- Wig Tree [W] A shrub of alpine slopes.Leaves used for dying clothes.

Cotoneaster aitchinsonii C K Schneider (Rosaceae): Kash.- Linu [W]. Tall semi -evergreen dwarf spreading shrub, common on moist slopes and in forests at 1700-2400 m. Flowers white. Branches used for making agricultural implements and baskets.

Cotoneaster bacillaris var. affinis (Lindl.) Hook.f. (syn. Cotoneaster rosea Edgew.) (Rosaceae): Eng.- Open fruited Cotoneaster; Kash.-Lenu [W]. A large, spreading, deciduous shrub, on humus rich soils at 2100-

2500 m. Branches arching with variable lanceolate, ovate or obovate leaves. Flowers in clusters. The blue-black bloomy fruits, are borne in large clusters. Stout branches are used as walking sticks and for handling of axes.Tender shoots used for basket making (Fig. C 21).

Cotoneaster horizontalis Decne. (Rosaceae) [NI]: A low growing shrub of spreading habit, with branches of characteristic herring bone pattern. Leaves broadly elliptic, glossy, dark-green which turn red in November. The shrub bears pink-tinged white flowers, singly or in pairs, followed by bright red spherical fruit. Planted on garden slopes and rocky surfaces.

Cotoneaster frigidus Wall. ex Lindl. (Rosaceae): [W] A deciduous shrub of sub-alpine Himalaya. Fruits given to weak patients.Twigs used for making baskets.

Cotoneaster microphyllus Wallich ex Lindley (Rosaceae):Eng.- Littleleaf Cotoneaster {NI]. A compact, prostrate, evergreen robust shrub, with rigid much branched stems. Leaves ovate to elliptic glossy, dark green, bristly-hairy beneath. Bears usually solitary white flowers, in late April to early May followed by almost spherical dark red fruit. Wood used as fuel under Ladakh conditions.

Cotoneaster racemiflora (Desf.) C. Koch. (syn. Cotoneaster nummularia Fisch. & Mey) (Rosaceae) [W]. A deciduous, semi-evergreen shrub growing on dry slopes and forest openings at 1700-2400 m; characterized by its tall, slender, arching branches with small leaves which are rounded to broadly elliptic. Flower clusters very short, with 2- 3 white flowers. Fruit brick- red and showy. Bark and leaves medicinal.

Cotula anthemoides L. (Asteraceae): Eng.- Buton Weed; Kash.-Thoule Bouble [W] Decumbent, pale-green,glabrous annual herb, growing in gardens and orchards as weed. Flower heads yellow, solitary. Extract of the plant used as tonic, carminative and for rheumatism (Fig. C 22).

Cousinia thomsonii Clarke. (Asteraceae): Ladakhi- Kreatising [W]. A tall herb, 30-60 cm tall, biennial growing in gravelly places at 3700- 4000 m altitudes. Stems greenish white, erect. Leaves leathery; capitula many with purple-pink flowers. An aqueous extract of the shoots is drunk as diuretic. Leaves and spines edible and taken as wild food.

Crambe cordifolia var. kotschyana (Boiss.) Jafri (Brassicaceae): Eng.- Colewort [W]. Clump forming tall perennial herb, on humus rich soils, usually below rocky cliffs at 2000- 2500 m. Leaves dark-green and kidney shaped. Flowers white. Roots edible. Plant used as a cure for itching.

Crataegus laevigata (Poir.) Dc. (Rosaceae): Eng.-English Hawthorn [NI]. A rounded, thorny, deciduous small tree with ovate, shallowly 3 to 5 lobed, obtusely serrate, glossy, mid-green leaves. Corymbs of 8-10 pink flowers, are produced in May followed by spherical to ovoid red fruit which are showy and ornamental.

Crataegus monogyna Jacq. (Rosaceae):Eng.- Hawthorn; Kash.– Rhinga [W]. A deciduous tree 5-14 m tall, common on open slopes and in forests at 1700-2500 m. Flowers white in lax flat-topped clusters. Fruit red globular and fleshy. Used as home remedy for dropsy, hypertension and sore throat. Also used in making herbal tea.

Crataegus rhipidophylla Gand. (syn. Crataegus oxyacantha L.) (Rosaceae): Kash.- Ringae [W]. A small deciduous tree of woodlands, scrubs, foothills, and waste places of Kashmir. Leaves deeply cut into two-third width, blade broadly ovate or rhombic in outline. Flowers white, in flat topped clusters. Fruit red, globular and fleshy. Fruit used in treating gastrointestinal problems, as nerve tonic and heart problems. Also used in making herbal tea (Fig. C 23).

Crataegus songarica G.Koch (Rosaceae):Eng.- Hawthorn [NI]. A deciduous ornamental tree with greyish branches. Leaves deeply cut to lobes with coarse acute teeth; blade broadly ovate rhombic in outline. Flowers long stalked. Fruits purplish black, globular. Leaves have been used for hepatoprotective effects.

Cremanthodium arnicoides (DC ex Royle) R D Good (Asteraceae): Eng.- Himalayan Daisy; Ladakhi- Reskusemar [W]. Endemic to Kashmir Himalaya at 3300 – 4800 m altitudes, inhabiting gravelly mountainous lands. Stem solitary, erect. Basal leaves petiolate, middle leaves shortly petioled while distal stem leaves sessile. Flower heads nodding, yellow with dark central disc. Used for peptic ulcers, dysentery and liver diseases (Fig. C 24).

Cremanthodium decaisnei Clarke (Asteraceae): Eng.- Decaisne's Cremanthodium; Ladakhi- Rashkun [W]. Low, weak, glabrous perennial herb, 10-20 cm tall, growing on open alpine slopes. Flower heads solitary, yellow. Whole plant used as vegetable particularly by shepherds in alpine pastures.

Cremanthodium retusum (DC.) R D Good (Asteraceae): Ladakhi- Richkut [W] A rhizomatous herb of alpine Himalaya, stems stout up to 90 cm tall. Radical leaves reniform, retuse, acutely dentate. Capitula radiate, broadly campanulate, ray florets yellow. Stem used for joint pains.

Crepidifolium tenuifolium (Willd.) Sennikov. (syn. Youngia tenuifolia (Willd.) Babc & Stebbins. (Asteraceae): Ladakhi- Chumthang [W].

A perennial herb. Flower heads yellow. Extract of the leaves is used as mouth wash and to treat bleeding gums. Also taken for diarrhoea and dysentery.

Crepis dachhigamensis G. Sing (Asteraceae): Eng.- Hawk,s Beard [W]. Annual herb inhabiting Harwan slopes of Kashmir at 1900 m altitude. Radical leaves less cut, cauline leaves also present. Receptacle naked. Achenes stout, longer with rigid involucral bracts.

Crepis flexousa (DC.) Benth. (Asteraceae):Eng.- Tangled Hawk's Beard; Ladakhi- Samboo [W]. A much branched perennial herb, forming rounded tufts with a tangle of slender, rigid nearly leafless forking branches ending in small yellow flower heads. The plant is prominently growing in Ladakh Himalaya at 3300 – 4200 m. Latex of the plant used for constipation (Fig. C 25).

Crocus sativus L. (Iridaceae): Eng.- Crocus; Hindi- Zaffron; Kash.- *Kounge* **[RC].** A small cormed, autumn flowering perennial, cultivated for its large, scented flowers whose trifid orange-red stigmas form the saffron of commerce. Corms flat, up to 25 mm in diameter; leaves solid, 6-10 as tall as fls., very narrow, ciliate- edged; parianth tube little exerted, oblong and obtuse, bright lilac, throat pubescent, anther yellow, longer than filaments, stigma trifid and orange- red coloured. At harvest time the whole flower is hand picked and the three long, red orange stigmas are separated from flowers and dried and form the commercial saffron. In Kashmir saffron has been traditionally used as a condiment to colour various dishes and delicacies, and also in the preparation of famous hot drink known as *Kashmiri Zaffrani Kehwa*. Crocus also features prominently in ceremonies connected with religious and social functions of Kashmiris. Hindus value the plant as sacred and use its pigment to mark their foreheads. They also offer the flowers to their Gods and Goddesses during worship and bath their famous idols in saffron coloured water. The coffin of dead body is often sprinkled with saffron dye. Saffron has also been used to dye the shawls and other costly garments (Fig. C 26).

Crocus sieberi (L.) Gay. (Iridaceae): Eng.- Snow Crocus [NI]. A bulbous perennial grown as spring flowering garden plant. Flowers are showy, pale mauve, yellow at base appearing in March.

Crocus speciosus M Bieb. (Iridaceae): Eng.- Garden Crocus [NI]. A bulbous perennial producing bright lilac blue flowers with yellow anthers and scarlet stigmas. Flowers in October. Grown as border plants in garden and home gardens.

Crucihimalaya himalaica (Edgew.) Al-shahbaz, O Kane & Price (syn. Arabidopsis himalaica (Edgew) O E Schulz) (Brassicaceae): Eng.- Himalayan Rock-Cress [W] A biennial/ perennial erect pubescent, much branched herb inhabiting open sub-alpine slopes at 2400- 3000m. Branches ascending. Leaves coarsely toothed, amplexicaule. Racemes bracteate upto apex. Flowers lilac. Siliqua needle like (Fig. C 27).

Crucihimalaya wallichii (Hook. f. Thom.) Al- shahbaz, O Kane & Rice (syn. Arabidopsis wallichii (Hook. f. & Thomson) Busch. (Brassicaceae): Eng.- Rock Cress [W]. Annual erect biennial herb, occurring on dry rocky slopes at 2500-4000 m. Stems erect, few branched, tomentose with finely branched stellate trichomes. Flowers purple, white or tinged red. Seeds brown. Powder of the dried plant is used against measles, and chicken pox (Fig. C 28).

Cryptomeria japonica (Thunb. ex L.f.) D Don (Cupressaceae): Eng.- Japanese Cedar [NI]. A large fast-growing tree of broadly columnar habit with reddish shedding bark and spreading or decurved branches. Leaves mid to deep green, curved awl-shaped, 4-angled with thickened bases, arranged spirally on the twigs. Female cones brown, globular, composed of 20-30 overlapping cone- scales each with 4-6 spiny tips. The commonly grown cv. *Elegans Compacta* is a slow growing conical shrub with soft glossy, dark green juvenile leaves turning bronze in winter.

Cucumis melo L. (Cucurbitaceae): Eng. –Muskmelon; Hindi- - Kharbuza; Kash.- Kharbuze [UC]. A creeping, hispid annual cultivated for its fruit which are eaten as dessert.

Cucumis sativus L. (Cucurbitaceae): Eng. –Cucumber; Hindi – Khira; Kashmiri- Laure [WC]. A trailing or climbing annual, cultivated for its edible fruits which are usually used as salad vegetable. Commonly cultivated varieties include:

- **Kashmiri Cucumber:** A local cucumber cultivar having become adapted to agro-ecological conditions of the valley. A climbing annual, bearing elongated, thick, cylindrical fruits of varying sizes. The fruits mostly have bitter ends, dark green in colour at edible stage and assume a yellow- greenish colour at maturity. Fruits consumed fresh as salad. Plant yields 60-80 q ha-1.
- **Japanese Long Green:** An introduced variety with long and cylinderical light green fruits. It is the most commonly cultivated commercial cucumber cultivar.
- **Shalimar Cucumber Hybrid -1:** A single cross cucumber hybrid between two lines –SH-K-1 and SH-K-11 having very high yield potential

of 653q/ha. It matures in 49 days and is tolerant to leaf spot, powdery mildew, downy mildew and mosaic. Fruit are crispy and available for longer duration. The hybrid has been developed by SKUAST-K during 2010 and released for cultivation in temperate areas.

• **Shalimar Cucumber Hybrid-2**

A single cross cucumber Hybrid developed by SKUAST-K between two lines –SH-K-1 and SH-K-12 with a yield potential of 615 q/ha. It matures in 51 days and is tolerant to angular leaf spot, powdery mildew, downy mildew and mosaic. Fruits are crisp and remain available for longer durations.

Cucurbita maxima Duch. (Cucurbitaceae): Eng. – Red Gourd; Kash.- Paurim Aule. [WC]. A stout hispid trailing herb. Leaves shallowly lobed with petioles covered with prickly hairs. The fruits large (upto 35 kg) have hard, deeply furrowed peduncles without any enlargement at the point of attachment. The fruits have pale, scentless flesh, and are cooked fresh. The fruit is also used for making pickles and even sundried after slicing, for subsequent use during winters. *Arka Surya* and *Arka Chandan* are the recently introduced pumpkin varieties under cultivation all over the valley. Besides fruit, male flower buds dipped in a flour paste of local water chest-nut/ basin, are fried and served fresh.

Cucurbita moschata Duchesne (Cucurbitaceae): Eng. –Skuash /Crook-neck Pumpkin; Hindi – Mitha Kadu; Kahmiri – Dogre Alae [UC]. An annual trailing herb with large leaves. The fruits are large (2 to 5 kg) ridged, and with ridged peduncles, enlarged at the point of attachment and matures by Oct-Nov. The fruits having long shelf life are often stored for subsequent use in winter months. The fruits are cooked and served often with curds.

Cucurbita pepo L. (Cucurbitaceae): Eng. Pumpkin; Hindi- Vilaiti Kadu; Kashmiri- Squash Alae [WC]. A trailing herb with long, hairy/ stem, bearing large flaccid leaves. Fruits oval/elongated, spotted with yellowish flesh. Yields 150-200 q ha-1. It is the earliest type of cucurbit fruit available in the valley during June/July months. The commonly cultivated variety is *Australian Green.*

Cupressus cashmeriana Royle. ex Carriere (Cupressaceae): Eng.- Kashmiri Cyperuss / Weeping Cyperuss [W/UC]. A medium sized elegant tree, growing in coniferous forests at 1800- 3000 m, with wide-spreading branches and drooping branchlets, pendulous; and lanceolate mucronate sage-green leaves with a spreading apex. Grown for hedging and decoration.

Cupressus lusitanica Miller (syn. Cupressus glauca Lamk.) (Cupressaceae): Eng.- Mexican Cyperuss [NI]. A conical to columnar, coniferous tree with rich brown, peeling bark; bears scarcely scented glandless grey green leaves, spherical, glaucous blue thin shiny brown female cones with conical prickles.

Cupressus macrocarpa Hartw. (Cupressaceae): Eng.- Montery Cyperuss [NI] A fast growing tree of medium to large size, conical or broadly columnar in habit. Leaves closely appressed, swollen towards the tip, obtuse, dark green. Cones 2.5 - 3.5 cm across, with 8-12 scales, with short ridge-like obtuse boss on buck. Grown as an ornamental.

Cupressus sempervirens L. (Cupressaceae): Eng.- Italian Cyperuss; Hindi – Sarwa; Kash.- Shamshad [NI]. A narrowly conical or columnar to broadly spreading coniferous ornamental tree with horizontal branches which bear dense sprays of grey-green or dark green, glandless leaves with rounded, abruptly pointed tips; spherical to ovoid, prickly brown female cones.

Cupressus torulosa D. Don (Cupressaceae): Eng.- Himalayan Cyperuss; Hindi- Devidiar; Kash.- Shamshad / Sarwa [W/UC]. A large evergreen tree with a pyramidal crown, whorled spreading branches and drooping branchlets, bark peeling off in long strips. Leaves dark green, opposite, ovate-triangular and with white margins, closely overlapping. Cones loosely clustered, bluish, globular, dark brown with a violet bloom.

Cuscuta capitata Roxb. (Convolvulaceae): Ladakhi- Hande- thapa [W] A parasitic climber of high altitude areas. Stem thin, brown to violet, usually not interlaced, making a few spirals around host stem. Leaves oblong-elliptic, obtuse. Flowers in cyme clusters. Whole plant used for fevers and melancholia.

Cuscuta europaea L. (syn. Cuscuta brevistyla A Br. ex Rich.) (Convolvulaceae): Eng.- Dodder; Kash.- Wazul Kukli-poute [W]

Parasitic twinner, annual, usually seen parasitic on *Isodon rugosus, Origanum normale, Capsicum annuumm,* etc.Used for relieving joint pain.

Cuscuta reflexa Roxb. (Convolvulaceae): Eng. -Dodder; Hindi– Akash-bel; Kash.- Kukli pout [W]. A wide spread parasitic twinner, found from temperate to alpine regions at 1700-2900 m. Leaves twining, thin, yellowish. Flowers small bell shaped. Plant used in the treatment of joint pain, inflammations, arthritis, diseases of eye and heart and in bilious disorders (Fig. C 29).

Cyanus depressus (M. Bieb.) Sojak.(syn. Centaurea depressa M. Bieb. (Asteraceae): Eng.- Low Cornflower; Ladakhi- Vashaka [W]. Erect, branched herb. Stem cottony. Heads pale. Fruits are taken for the treatment of cough, chest pain, fevers, headache and abdominal disorders.

Cydonia oblonga Mill. (syn. Cydonia vulgaris Pers.; Pyrus cydonia L.) (Rosaceae): Eng.- Quince; Kash.- Bomtsunt [WC]. A deciduous, low spreading type tree. Flowers in the first week of April. Bearing is heavy and annual. Flowers white to pale rose appearing with the main foliage. Fruit golden yellow, fragrant, fuzzy, pyriform, and aromatic and with good keeping qualities. Two types are quince are grown differing in taste-one being sour is known as *Chok bamtsunt* while another being sweet is known as *Modur bamtsunt*. Fruits are cooked fresh and also sundried after slicing for use in winters.

Cymbopogon distans (Nees. ex Steud) Watson. (Poaceae): Eng.- Ginger Grass [W]. A perennial, caespitose grass species of alpine slopes of Himalayas characterized by aromatic lemon scent in its tissues. Used as flavouring for cooking.

Cynanchum arnottianum (Wight) Wight (Apocyanaceae): [W]. Perennial climbing herb, on open humus rich soils. Flowers purplish. Dried leaf powder of the plant used against muggets infecting wounds of cattle, horses and sheep.

Cynara scolymus L. (Asteracae): Eng.- Globe Artichoke; Hindi- Hathichuk [NI] A perennial herb grown as vegetable as well as ornamental for its attractive floral heads. Leaves glaucous, green and deeply lobed.

Cynodon dactylon (L.) Pers. (Panicum dactylon L.; Digitaria dactylon (L.) Scop.) (Poaceae): Eng.- Bermuda Grass; Hindi- Doob; Kashmiri -Dramun Kache; Ladakhi- Jamak [W]. A perennial grass. Culms prostrate, creeping and roots at nodes, forming wide mats on the soil across all grasslands. Inflorescence a whorl of 2-6 spikes radiating from the top of slender peduncle. An esteemed lawn grass relished by all grazing animals.

Cynoglossum lanceolatum Forssk. (Boraginaceae): Eng.- Lanceleaf Forget-me not [W]. Biennial sub- alpine erect herb, usually in deciduous forests. Stem and branches covered with rigid white hairs. Flowers white with blue scales. Fruit consists of 4 small nutlets covered in sticky hairs. Roots used for eye ailments. Leaves used for treating wounds.

Cynoglossum wallichii var. glochidiatum (Wall. ex Benth.) Kazmi (syn. Cynoglossum glochidiatum Wall. ex Benth.)(Boraginaceae): Eng.- Prickly Hound's Tongue; Kash.- Binjeer; Ladakhi- Krishkeot [W]. Erect biennial herb of W- Himalayas, usually under forest shade. Stem branched, hirsute. Flowers with bluish scales. Troublesome weed as its burs stick to wool of sheep and other animals. In Ladakhi system of medicine seeds are eaten for potency and fertility.

Cynoglossum zeylanicum (Vahl.) Thunb. ex Lehm. (Boraginaceae): Eng.- Indian Hound's Tongue. [W]. An erect branched sub-alpine herb. Leaves elliptic- lanceolate, margins ciliate. Flowers bluish white in terminal racemes. Root and leaves used medicinally.

Cycas revoluta Thunb. (Cycadaceae): Eng.- Sago- Palm [NI]. A robust-stemmed cycad, erect at first but gradually reclining with age, and suckering and branching when mature; arching leaves, sickle shaped, glossy, dark green leaflets. The male inflorescence pine apple-scented, female inflorescence producing ovoid yellow fruit. Grown as ornamental in gardens.

Cypripedium cordigerum D. Don (Orchidaceae): Eng.- Orchid of N.W. Himalaya / The heart shaped Cyprepedium / Lady,s Slipper Orchid [W]. A perennial terrestrial herb, up to 60 cm tall, often in colonies, in alpine meadows. Rhizomes creeping. Leaves sub-orbicular to elliptic, acute to acuminate, margins ciliate with glandular hairs. Inflorescence one flowered. Flowers white with purple spots.

Cyperus rotundus L. (Cyperaceae): Eng.- Nut Grass; Hindi- Motha [W]. Perennial sedge,usually in pastures, along stream sides, cultivated areas. Flowers chocolate brown. Powdered rhizomes of the plant used as perfume and in treating cutaneous diseases. Tubers resemble nuts. Flowering stems have a triangular cross section.

Cyperus serotinus Rottb. (Cyperaceae): Eng.- Guinea Rush [W]. Perennial herb, usually along rice fields and shallow water bodies.Flowers yellow green. Used as fodder for milking cows.

Cytisus scoparius (L.) Link. (Fabaceae): Eng.- Common Broom [NI]. An upright, deciduous ornamental shrub, 1-3 m tall, with erect slender branches. Leaves short petioled, trifoliate, often simple at the branch tips, oblanceolate. Flowers usually axillary, large, golden-yellow. Pods narrow oblong, brownish black, villous along margins.

Fig. C1. Caltha palustris

Fig. C2. Campanula cashmeriana

Fig. C3. Capsicum annuum (var. Kashmiri - Red Chillies

Fig. C4. Caragana brevispina

Fig. C5. Caragana pygmaea

Fig. C6. Carduus nutans

Fig. C7. Carthamus lanatus

Fig. C10. Centaurea iberica

Fig. C8. Cedrus deodara

Fig. C11. Chenopodium foliosum

Fig. C9. Celosia argentea

Fig. C12. Christolea crassifolia

Fig. C13. Cicer microphyllum

Fig. C14. Cichorium intybus

Fig. C15. Codonopsis ovata

Fig. C16. Colchicum luteum

Fig. C17. Consolida incana

Fig. C18. Corydalis falconeri

Fig. C19. Corydalis govaniana

Fig. C20. Corydalis kashmeriana

Fig. C21. Cotoneaster bacillaris

Fig. C22. Cotula anthemoides

Fig. C23. Crataegus rhipidophylla

Fig. C24. Cremanthodium arnicoides

Fig. C25. Crepis flexousa

Fig. C26. Crocus sativus

Fig. C27.Crucihimalaya himalaica

Fig. C28. Crucihimalaya wallichii

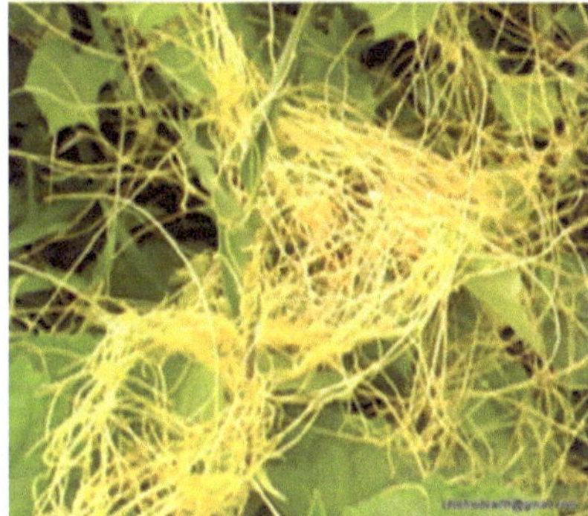

Fig. C29. Cuscuta reflexa

D

Dactylis glomerata L (syn. Bromus glomeratus (L.) Scop. (Poaceae): Eng.- Orchard Grass; Kash.- Dramun [W]. Perennial bunch grass,usually in dense shaded forest areas. A very important pasture and hay grass of temperate grasslands. Culms glabrous, erect. Inflorescences are erect panicles bearing spikelets in dense one sided clusters.

Dactylorhiza hatagirea (D. Don) Soo. (Orchidaceae): Eng.- Himalayan Marsh Orchid; Kash.- Salem panja; Ladakhi- Wangluk [W]. Perennial herb of temperate Himalaya with distinctly palmately lobed root tubers. Leaves oblong-lanceolate, clothing to stem, flowers rosy purple. The extract of the underground parts, in milk, is used as tonic/ aphrodisiac (Fig. D 1).

Dactylorhiza incarnata (L.) Soo (syn. Orchis latifolia L.) (Orchidaceae): Eng.- – Early Marsh Orchid; Kash.- Barf-ke Gughtli / Salem panja [W]. A perennial tuberous herb of wet meadows of Himalayas. Stem thick, hollow. Perianth irregular, pink/dark-purple. Leaves alternate, sessile. Tubers used as aphrodisiac and as tonic.

Dahlia variabilis Cav. (Asteracae): Hindi– Gulechin [NI]. Half hardy, tuberous perennials, grown as garden decoration and as cut flowers, as mixed border plants or as bedding dahlias. Leaves pinnate and deep- green. Most of the modern dahlias, with widely ranging forms and colours, are under cultivation for garden decoration representing varieties and mutants from all three Mexican species namely *D pinnate, D. coccinea* and *D. rosea* which have probably originated from *D. variabilies*. Cultivated varieties belong both to bedding and border dahlias and include *Single flowered, Anemone flowered, Collerette, Paeony flowered; Decorative, Pompon cactus* and *Semi-cactus*.

Dalbergia sissoo DC. (Fabaceae): Eng.- –Indian Rosewood; Hindi- Shisham [NI] A deciduous tree distributed in the foothills of Jammu zone. Introduced as ornamental. Leaves leathery, alternate, pinnately compound. Flowers whitish to pink, fragrant. Bark thin, longitudinally furrowed.

Daphne mucronata Royle. (Thymelaeaceae):Eng.- Kashmir Daphne; Ladakhi- Kutlal [W]. Erect much-branched evergreen wiry shrub, with narrowly inverted- lance shaped grey leathery leaves. Flowers densely woolly haired, creamy-white or yellowish and fragrant. Fruit ellipsoid, reddish-orange enclosed in sepals. Leaves used for treating boils (Fig. D 2).

Daphne papyracea Wall. ex G Don (Thymelaeaceae):Eng.- Kashmir Daphne; Ladakhi- Kutlal [W]. Evergreen hrub of Kashmir forest shrubries. Flowers yellowish. Roots used medicinally.

Daphne oleoides Schreb. (Thymelaeaceae): Kash.- Gandh Lenu /Vethru [W]. A slow growing evergreen upright shrub, with grey pubescent branches and oblanceolate leaves. Produces terminal clusters of 3-6, usually scented, sometimes pink, tinged, creamy white flowers followed by downy, fleshy orange fruit. Leaves used for treating boils and tumors.

Dasiphora fruticosa (L.) Rydb. (syn. Potentilla fruticosa L.) (Rosaceae): Eng.- Shrubby Cinquefoil; Kash.- Van-chai [W]. A hardy deciduous flowering upright to sprawling shrub inhabiting high altitude mountain slopes on moisture retentive soils. Plant is densely leafy, the leaves are pubescent and divided into 5-7 leaflets. The flowers are produced terminally on the stems and are yellow, butter cup shaped. Fruit a cluster of achenes covered with hairs. Leaves and flowers used against high fevers.

Datisca cannabina L. (Datiscaceae): Eng.- False Hemp; Hindi- Akalber; Kash.- Waft tung, Aqal bir [W]. A bushy herb, up to 1.5 m tall, distributed in alpine meadows. Leaves coarsely serrate. Flowers diocious, yellow, sub-sessile. Roots and bark used for colouring wool and cotton. Also used against fevers and rheumatic pain.

Datura stramonium L. (Solanaceae): Eng.- Jimson Weed /Thorn Apple; Hindi- Dhatura; Kash.- Datoure [W]. A foul smelling, under shrub occurring in road sides and wastelands from Kashmir to Sikkim up to 2500 m altitude. Flowers funnel shaped, white. Fruit a capsule with slender spines. Dried leaves and seeds are used in the treatment of boils, inflammations, painful piles and biliousness. Plant is a source of scopolamine which is used as pre-anaesthetic in surgery.

Daucus carota L. (Apiaceae): Eng.- Birds Nest; Kash.- Jangli Gazer [W]. A wild biennial herb growing on fallow lands, apple orchards and grasslands. Plant a bristly, biennial herb, with twice/thrice pinnate leaves and flattish umbels with small white flowers. Fruits are oblong, dorsaly flattened with minute bristles. Roots slender, elongated, woody and tapering with strong aromatic odour and with acrid, disagreeable billerish taste. Seeds are aromatic, with a pungent bitterish taste and used for urinary infections.

Daucus carota subsp. sativus (Hoffm.) Arcang. (Apiaceae): Eng.- Carrot; Hindi- Gajer; Kash. - Gazer [WC]. An annual or biennial herb, grown for its edible roots, stem erect, much branched, 40- 150 cm high arising from a thick fleshy tap root. Leaves pinnately compound. Flowers white, borne on umbels. Fruits oblong, with bristly hairs along ribs. Following are the commonly cultivated carrot varieties.

- **Black Carrot**: Roots are dark violet, tapering toward lower end and with light yellowish/light voilet flesh.
- **Nantes:** An exotic carrot cultivar now cultivated by commercial growers for table and vegetable purposes. It is being consumed in fresh and after cooking especially by urban population because of its tenderness, orange flesh and in having minimum core.
- **Chamman:** A carrot variety developed by SKUAST-K and released for cultivation during 2001 for general cultivation under temperate conditions. High yielding and with early maturity. Roots orange coloured, long, cylindrical, semi blunt and resistant to cracking and forking.

Delphinium brunonianum Royle (Ranunculaceae): Eng- Musk Larkspur; Ladakhi - Chargopoz [W]. A high altitude perennial herb inhabiting mountain slopes of Ladakh above 4000m. Plant has strong musky smell. Leaves petiolate, round, alternate and palmately lobed. Flowers blue, cup shaped. Tepals strongly veined and have slender white hairs. Juice of the leaves is used in treatment of throat complaints and for getting rid of head lice.

Delphinium cashmirianum Royle. (Ranunculaceae): Eng.- Kashmiri Larkspur; Kash.- Kaeri paete; Ladakhi – Chargosposz [W]. A perennial herb on alpine slopes above 3000 m altitudes. Leaves palmately 3-7 lobed. Flowers woolly haired, conspicuously veined bluish-purple. Roots used medicinally. Root also used in the preparation of *Changue* - a local alcoholic drink of Ladakh (Fig. D 3).

Delphinium denudatum Wall. ex Hook. f. & Thoms. (Ranunculaceae): Kash.- Jadwara /Mori [W]. Tall perennial herb of Kashmir Himalaya, on dry scrubby slopes and forest openings above 3000 m altitudes. Leaves with round blade, cut into 3-5 narrow lobes. Flowers blue. Tubers chewed to cure toothache. Also used as tonic and stimulant (Fig. D 4).

Delphinium incanum Clarke (Ranunculaceae): Eng.- Foothill Larkspur [W]. A perennial herb, growing on forest slopes on rich scrubby soils. Flowers blue. Used as veterinary medicine to kill animal ticks.

Delphinium roylei Munz. (Ranunculaceae): Kash.- Kare paet; Ladakhi- Mamer /Mori [W]. A perennial endemic herb 60-100 cm tall, stem simple or with few branches. Leaves palmately multipartite. Flowers blue borne on branched racemes. Root used for liver infections in cattle. Also used for tooth aches and rheumatic pains. Plant occurs in alpine grasslands at 2300- 2800 m altitudes.

Delphinium viscosum Hook. f. & Thomson. (Ranunculaceae): Ladakhi- Bilamonokh [W]. Erect decumbent herb of Ladakh. Stem densely hairy. Flowers hairy purplish-blue.The paste of fresh leaves and shoots is applied on swollen joints to relieve pain and swelling.

Delphinium vestitum Wall ex Royle. (Ranunculaceae): Eng.- Clothed Delphinium; Ladakhi- Bilamonokh [W]. Hairy perennial herb of open alpine slopes and shrubberies at 2700- 4000m altitudes, restricted to Kashmir Himalaya and North Pakistan. Flowers bluish, hairy, veined borne on long spike. Leaves rounded, deeply 5 lobed. Leaves and stem used for cuts, piles and tooth aches. Also used for intestinal troubles and ulcers.

Descurainia sophia (L.) Webb. & Prantl. (Brassicaceae): Eng.- Tansy Mustard; Ladakhi- Poshtenarpo; Kash.- Chari Laschije [W]. Annual pubescent herb, upto 80 cm tall, on alpine slopes. Leaves alternate, bluish-green, double/ triple pinnately lobed. Stem haired. Flowers yellow or pale. Powdered seeds in milk used to treat measles, chronic bronchitis and as expectorant.

Desmodium elegans D C (syn. Desmodium tiliaefolium G Don.) (Fabaceae): Eng.- Elegant Tick Clover; Kash.- Hin, Chamara, Chamkat [W]. Deciduous sub-alpine densely pubescent shrub on forest slopes. Flowers bluish borne in a panicle. Pods segmented into one seeded sections, covered with hooked bristles. Roots medicinal.

Desmostachya bipinnata (L.) Stapf. (syn. Desmotachya cynosuroides (Retz.) Stapf ex Massey (Poaceae): Eng.- Halfa Grass [W]. A perennia tall, rhizomatous, tufted grass of waste places and orchards. Culms rigid and herbaceous, with glabrous nodes. Inflorescence erect, spike like panicles having above 100 spikes per panicle.The plant has a religious and social significance in Hindus as its leaves and shoots were used in marriage and religious functions. A girdle made out of plant used to be put around the neck of a bride during marriage; while a few leaves were twisted into a ring and worn in the fourth finger of right hand of a bridegroom.

Deutzia corymbosa R Br. ex G Don (Hydrangeaceae) [W]. A deciduous shrub of vigorous habit, with bright brown bark, peeling into rolls inhabiting open habitats. Leaves ovate, finely toothed. Flowers white, crowded into a corymb having charming hawthorn like scent. Aerial parts used medicinally.

Deutzia macrantha Hook. f. Thomson (Hydrangeaceae) [NI]. A vigorous, upright ornamental shrub with arching branches. Leaves ovate-oblong, finely serrate, rough and with scattered stellate pubescence, mostly 10-15 rayed hairs. Flowers white double in short dense paniculate clusters, calyx-teeth prevailingly persistent on fruit.

Deutzia rehderiana C K Schneid: (Hydrangeaceae) [W/UC]. A compact, rounded, bushy ornamental shrub; branches with brown bark usually peeling. Leaves ovate-oblong to ovate-lanceolate, finely serrate, with scattered mostly 4-6 rayed stellate hairs. Flowers pinkish outside, often fading white, in short paniculate clusters.

Deutzia scabra Thunb. (syn. Deutzia crenata Sieb. & Zucc.) (Hdrangeaceae): Eng.- Fuzzy Deutzia [NI]. A strong ornamental shrub, with many erect or ascending branches and brownish-bark. Leaves ovate to oval-oblong, dark green, rounded base, crenate-dentate, scabrid-pubescent on both sides with stellate hairs, the hairs beneath smaller and more (10-15)-rayed. Flowers white or bluish, in loose usually somewhat compound racemes. Two popularly grown varieties include var. *candidissima* – (Flowers pure white, double) and *plena rehd* – (Flowers rosy-purple).

Dianthus allwoodii (Caryophyllaceae): Eng.- Modern Pink/ Divine Flower; Hindi- Gule marjan [NI]. A hybrid between *Old Pink* and *Carnation*. Fast growing and produce many blooms during the season from June to October. Many varieties of the species are cultivated for their varied coloured, attractive and fragrant flowers.

Dianthus anatolicus Boiss. (Caryophyllaceae): Eng.- Anatolian Pink; Ladakhi- Isat-zingma [W]. Densely tufted perennial herb on rocky alpine slopes above 2500 m altitudes. Flowers rosy. Extract of the dried flowers is used against malaria and epidemic fevers (Fig. D 5).

Dianthus barbatus L. (Caryophyllaceae): Eng.- Sweet William. [NI]. A hardy biennial forming tufted mats with narrowly lanceolate leaves generally grown as border plant. Bears densely packed flattened heads with flowers of various colours from white to red or marked with other colours.

Dianthus caryophyllus *L.***(Caryophyllaceae):Eng.- Carnation; Hindi-Gule merjan [NI].** A border decorative plant having typical grey-green foliage. Flowers dull purple, with sweet clove fragrance.

Dianthus chinensis *L.* **(Caryophyllaceae): Eng.- Chinese Pink**. [NI]. Some cultivated and short lived pinks are grown in gardens for their showy flowers. Leaves are mild- green. Flowers are self coloured or in mixtures with or without intricate markings.

Dictamnus albus L. (Rutaceae): Eng.- Gas Plant / Burning Bush; Kash.- Tuenale [W] An aromatic under-shrub of Himalayan woodlands above 2500 m altitude, clothed with pustular glands throughout. Leaves leathery. Leaflets ovate, acute, serrulate and sessile. Flowers in pyramidal spike bearing pale purple to pink coloured flowers. Root poultice used for rheumatic pain and skin eruptions. Decoction of leaves used as post-natal analgesic bath.

Digitalis lanata Ehrh. (Plantaginaceae): **Eng.- Grecian Foxglove; Hindi-Salab Punja; Kash.- Lanter [W]**. A biennial herb found in sub-alpine regions of Kashmir Himalaya. Flowers in spikes, brownish with white lips. Leaves elongated and woolly. Flowers tubular, bell shaped and creamy white. Leaves are a source of glycoside digoxin used as cardiac stimulant and tonic (Fig. D 6).

Digitalis purpurea L. (syn. Digitalis tomentosa Hoffmanns & Link) (Plantaginaceae): Eng.- Foxglove [W]. A herbaceous perennial growing wild in alpine forest slopes and also grown in home gardens for its showy flowering spikes. The oblong–lanceolate leaves form a roseta from which arise long flowering spikes. Flowers tubular, pendant. Variants with deep pink, light pink and milky spotted flowers grow abundantly on Gulmarg and Khelanmarg slopes.

Digitaria abludens (Roem. & Schult.) Veldkamp. (Poaceae) [W]. Annual herb, of fodder value, growing along damp situations. Culms decumbent, 30-60 cm long without nodal roots. Leaf blade surface scaberulous, leaf blade apex attenuate, filiform. Inflorescence composed of many racemes.

Digitaria cruciata (Nees) A Camus (Poaceae): Eng.- Finger Grass /Crab Grass; Kash. - Srikanth. [W]. A common forage grass in open grasslands of Himalayas. Culms decumbent at base, 50- 130 cm long and rooting from lower nodes. Culm nodes pubescent. Leaf sheaths glabrous on surface to pillose. Used in religious ceremonies by local Hindus (Fig. D 7).

Digitaria sanguinalis (L.) Scop. (syn. Paspalum sanguinale (L.) Lamk.) (Poaceae) [W]. Annual herb, a widely distributed fodder grass of

N-Western Himalayan grasslands. Spikelets hairy, leaves linear. Flowers green. Lemma with spiny outgrowths. Plant usually grows along borders of rice fields. Flowers green. Plant has been used as famine food.

Dioscorea deltoidea Wall ex Griseb. (Dioscoreaceae).: Eng.- Wild Yam; Hindi – Gaithi; Kash.- Kritz/ Knis [W]. A perennial, rhizomatous, climbing herb distributed throughout N-W Himalaya at 1700-2800 m. Leaves cordate, clothed on stem. Rhizomes are used against rheumatic and opthalamic disorders. Steriods from the plant are also potential source for manufacturing contraceptive pills. Its rhizomes have been used as detergent to cleanse wool. Seeds taken for deworming.

Diospyros kaki L. f (syn. Diospyros chinensis Blume; Diospyros roxburghii Carr., D. Schitse Bunge) (Ebenaceae): Eng.- Japanese Persimmon; Hindi- Halwa Tendu [NI]. A round-crowned deciduous tree with brownish branches. Leaves ovate elliptic to obovate, deep green which turns yellow to orange red and purple in autumn. Flowers yellowish-white. Fruit slender conical, usually ribbed at base with thin orange yellow to reddish skin and orange coloured pulp surrounding the elliptic flattened seeds.

Diospyros virginiana L. (Ebenaceae): Eng.- American Persimmon; Kash.- Amlok. [NI]. A round-headed tree with spreading branches. Leaves oval to ovate, acuminate, cuneate, rounded or sub-cordate at base shining deep green. Flowers greenish- yellow. Fruit globose, pale orange cheeks, with oblong flattened seeds. Grown for fruit as well as ornamental.

Dipsacus mitis D Don. (Caprifoliaceae): Eng.— Himalayan Teasel; Hindi-Burash; Kash.- Wopal Haakh. [W]. A tall, robust, perennial herb growing on humus rich soils on alpine/sub-alpine grassy slopes of the valley at 1800- 3100 m. Leaves are sour in taste and cooked as vegetable in fresh as well as in sundried forms. Leaves opposite. Flower heads creamish white (Fig. D 8).

Doronicum kamaonense (DC.) Alv. Fern (syn. Doronicum roylei DC) (Asteraceae): Kash.- Duedaranh [W]. Perennial herb, in forest shade, usually along streams. Heads yellow. Extract of the roots used to cure high altitude sickness (Fig. D 9).

Draba nuda (Bel.) Al- shehbaz & M. Koch. (syn. Drabopsis brevisiliqua Naqshi & Javeid) (Brassicaceae): [W]. A scapose herb inhabiting hillsides, rocky slopes above 2000m. Stem erect, simple, rarely branched. Basal leaves sessile or sub-sessile, pubescent with stalked, stellate trichomes along margins. Flowers yellowish or yellowish white. Fruits linear, valves prominently veined and glabrous. Seeds medicinal.

Dracocephalum heterophyllum Benth. (Lamiaceae): Eng.- White Dragonhead; Ladakhi – Jimthiglae [W]. Perennial aromatic herb on disturbed soils, above 3000 m.Verticillasters 4-8 flowered. Stem densely retrose pubescent. Leaves leathery, stalked, oblong-ovate, toothed. Flowers bluish purple. An extract of the aerial portion of the plant used as eye disinfectant.

Dryopteris filix-mas (L.) Schott. (Dryopteridaceae): Eng.- Buckle Fern/ Male Fern; Ladakhi- Kakoi [W]. A hardy plant with deep- green, lanceolate fronds that are bipinnate. Frond stalks covered with orange brown scales. Grows wild in moist and shady places in alpine, sub-alpine forests and also as a potted plant. Used for worm infestations.

Dryopteris thelypteris (L.) A Gray (syn. Lastrea thelypteris Desv. (Dryopteridaceae): Eng.- Maiden Fern [W]. Long creeping perennial fern, usually along drains, damp places at 1700 -2500 m altitudes. Fronds deciduous, monocarpic, upto 75 cm high. Also grown as ornamental.

Dryopteris velata (Kunze.) Kuntze. (Dryopteridaceae): Eng.- Rigid Buckler Fern. [W]. An ornamental wild fern characterized by narrowly triangular, bi-pinnate fronds and grown as pot plants in home gardens.

Dysphania ambrosiodes (L.) Mosyakin & Clemants (syn. Chenopodium ambrosioides L. (Amaranthaceae): Eng.-Mexican Tea / Wormseed; Kash.- Jangli javend [W]. A short lived perennial herb 45- 90 cm tall growing in sub-alpine grasslands and forest margins. Plant irregularly branched, leaves oblong-lanceolate upto 10 cm long. Flowers small, green, produced in a branched panicle at the apex of the stem. Seeds used for abdominal pain. Leaves edible (Fig. D 10).

Dysphania botrys (L.) Mosyakin & Clemants (syn. Chenopodium botrys L.(Amaranthaceae): Eng.- Feathered Geranium; Ladakhi- Vastuli [W].Annual herb, usually along pathways and cultivated fields at 1700-2400 m altitudes. Flowers green. Leaves and root stock used for asthma (Fig. D 11).

Fig. D1. Dactylorhiza hatagirea

Fig. D4. Delphinium denudatum

Fig. D2. Daphne mucronata

Fig. D5. Dianthus anatolicus

Fig. D3. Delphinium cashmerianum

Fig. D6. Digitalis lanata

Fig. D7. Digitaria cruciata

Fig. D8. Dipsacus mitis

Fig. D9. Doronicum kamaonense

Fig. D10. Dysphania ambrosiodes

Fig. D11. Dysphania botrys

E

Echeveria derenbergii DC (Crussulaceae): Eng.- Painted Lady [NI]. A succulent house plant and green house perennial grown for its succulent coloured leaves which are arranged in rosettes. The orange coloured flowers appear in July on raised stem branches.

Echinochloa colona (L.) Link. (syn. Panicum colonum L.; Milium colonum (L.) Moench.; Echinochloa zonalis (Guss.) Parl. (Poaceae):Eng.- Jungle Rice / Shama Millet; Hindi –Sawank; Kash.- Hama [W]. A grass of damp soils where it grows in profusion. In vegetative phase it closely resembles rice plant hence it is considered as obnoxious pest of rice fields. It is greedily grazed by all herbivores.

Echinochloa crusgalli (L.) P Beauv. (syn. Panicum crusgalli L.; Milium crusgalli (L.) Moench.; Penisetum crusgalli (L.) Baung.; Echinochloa hispidula (Retz.) Nees ex Royle.) (Poaceae): Eng.-Barnyard Grass; Hindi- Sawank; Kash.- Hama [W]. A polymorphous, tufted annual grass, growing along roadsides, fields and as a weed of cultivation in rice fields. Culms erect branched at base. Used as fodder.

Echinops cornigerus D C. (Asteraceae): Eng.- Blue Globe Thistle; Ladakhi – Zanchar [W]. A perennial herb growing in alpine ranges above 3000 m. Leaves silvery green, divided pinnately into segments, each having spines. Inflorescence spherical, with pale- blue flower heads with long spine-tipped involucral bracts. Seeds used as tonic. Powdered leaves are used to cure yellownishness of eyes.

Elaeagnus angustifolia L. (syn. Elaegnus hortensis Bieb.) (Elaeagnaceae): Eng.- Russian Olive; Hindi – Shiulik [NI]. A large deciduous thorny tree with spreading red-tinted, spiny branches with silvery scales when young. Leaves lanceolate to oblong, dull green above, with silvery scales beneath. Flowers 1-3, axillary on the lower branch tips; calyx tubular, yellow inside, silvery outside, fragrant; fruit oblong, yellow with silvery-scales, mealy, sweet and edible. Grown as ornamental.

Elaeagnus parvifolia Wall ex Royle (syn. Elaegnus latifolia L.) (Elaeagnaceae): Eng.- – Autumn Olive / Bustard Oleaster; Hindi –

Ghiwain; Kash.-Gaun [W] A small spiny tree /silvery shrub distributed in temperate Himalayas.Leaves covered densely by small scales. Flowers aromatic. Fruits ovoid, succulent and edible.

Elaeagnus rhamnoides (L.) A. Nelson (syn. Hippophae rhamnoides L. (Elaeagnaceae): Eng.- Seabuckthorn; Hindi- Dhurchak; Ladakhi- Tsermang/Sastar [W]. A deciduous woody and thorny shrub of Ladakh generally used for hedging because of its spiny branches. The plant is dioecious in nature bearing male and female flowers on separate plants. Flowers in June and fruits ripen by mid September. Flowers insignificant, yellow, appearing before leaves; berries numerous, bright orange yellow or scarlet, persistent on the branches. It is economically importance for its berries as well its foliage which is highly palatable to cattle. Fruits edible and their juice, being rich source of vitamin C, is given to patients with lung complaints. Because of high vitamin and other essential nutrients it has assumed high commercial importance as it is used for making various ethnic foods and drinks (Fig. E 1).

Elaeaganus umbellata Thunb. (Elaeagnaceae): Eng.- Himalayan Oleaster; Kash.- Vishinear [W]. A wide-spreading, deciduous shrub with brown scaly, spiny shoots. Growing in ravines in sub-alpine vegetations at 1700-2200 m. Leaves elliptic to ovate-oblong, wavy-margined, silvery at first, maturing to bright green above. Flowers yellowish-white, clustered on small side branches, fragrant; fruit globose, silvery brown at first, red at maturity, edible. Flowers are astringent, used in pulmonary infections (Fig. E 2).

Elsholtzia densa Benth. (Lamiaceae): Ladakhi –Erzeotz / Cherukpa [W] Erect herb, 20-60 cm tall growing along alpine meadows above 2500 m altitudes. Stem pubescent, much branched from base. Leaf blade lanceolate with serrate margins. Spikes cylindrical, corolla purplish. Its young leaves are cooked with meat to impart flavour. Paste of the leaves is used as antiseptic.

Elsholtzia eriostachya (Benth.) Benth. (Lamiaceae): Ladakhi- Phloling [W]. Erect annual herb of hilly grasslands above 2500 m. Stems purple –red, puberulent. Verticellasters many flowered. Calyx campanulate, corolla yellow. Decoction of leaves is used to cure gastric troubles and renal pain.

Elymus caninus (L.) L. (syn. Agropyron caninum (L.) P Beauv.) (Poaceae): Eng.- Bearded Wheat Grass / Bearded Couch Grass [W]. A grass species of Kashmir alpine pastures growing above 2500 m. Plants perennial, caespitose with lax leaves.

Elymus dahuricus Griseb. (Poaceae): Ladakhi- Zesariatz [W]. Perennial herb of Ladakh alpine meadows. The plant has high fodder value in being nuitritive and is fed to horses and ponies after days work.

Elymus dentatus (Hook). f.) Tzvelev (syn. Agropyron dentatum Hook. f.) (Poaceae): Eng.- Twitch Grass [W]. A tufted perennial with creeping rhizomes. Culms 25-80 cm tall, erect or geniculately ascending. Lemmas awnless; leaf sheaths and blades glabrous. An endemic grass species of Kashmir Himalaya.

Elymus himalayanus (Nevski) Tzvelev (syn. Agropyron himalayanum (Nevski) Melderis): (Poaceae) [W]. Commonly growing grass species in Lidder valley of Kashmir.

Elymus semicostatus (Nees ex Steud) Melderis (syn. Agropyron semicostatum Nees ex Steud) (Poaceae) [W]. Perennial herb usually among scrubs on drier slopes. Widely distributed in grasslands of Kashmir. Leaves scaberulous, hairless and smooth.

Ephedra gerardiana Wall ex Stapf. (Ephedraceae): Eng- Ephedrine; Hindi – Khanda; Kash. Asmani booti; Ladakhi - Sephat, Cheldumb [W]. A small shrub, 30- 40 cm tall distributed in temperate and cold arid areas of Himalaya. Stem base woody giving rise to closely places, jointed branches. Fruit ovoid, fleshy and red succulent. Extract of glumes and twigs is used in treating liver and blood diseases, irregular mensuration asthama and cardiac stimulant in *Amchi* system of medicine of Ladakh (Fig. E 3).

Ephedra regeliana Florin. (Ephedraceae): Eng.- Joint Pine; Ladakhi- Tsecheldumb [W]. A small gymnospermous shrub of alpine slopes. Branches striate. Berry globose, red. Decoction of aerial portion is used to cure blood and liver diseases.

Epilobium angustifolium L. (Onagraceae): Eng.- Fire Weed; Kash.- Tulli woen; Ladakhi –Utpal wambo/Chung [W]. Tall, erect herbaceous perennial of sub-alpine grasslands. Leaves entire, lanceolate, pinnately veined. Flowers rose purple. Leaves used for abdominal pain and intestinal disorders.

Epilobium hirsutum L. (syn. Epilobium himalense Royle) (Onagraceae): Eng.- Hairy Willowherb [W]. Tall perennial softly hairy herb of grasslands. Leaves opposite, coarsely toothed. Flowers bight pink. Whole plant used medicinally.

Epilobium latifolium L. (syn. Chamerion latifolium (L.) Sweet.) (Onagraceae): Eng.- –Dwarf Fire Weed; Ladakhi- Utpalwambo [W]. Perennial herb of Ladakh highlands. Leaves 2- 10 cm long, lance shaped, pointed at tips and waxy. Flower deep pink, nodding. Decoction of plant is used as stomachic and for renal pain (Fig. E 4).

Epilobium laxum Royle (Onagraceae): Eng.- Lax Willowherb; Ladakhi –Mushtengste [W]. A perennial, often clumped alpine erect herb, usually in damp places along rocky streams, above 2500 m altitudes. Leaves ovate to acuminate, distinctly toothed. Flowers pink slightly nodding, borne at the end of branches. Extract of the dried plant used as drug for restoring vitality and vigour in old men.

Epipactis royleana Lindl. (Orchidaceae) [W]. A rhizomatous perennial orchid of Western Himalayas, growing on damp grassy slopes, especially near streams. Flowers rosy pink /greenish with red veining.

Equisetum arvense L. (Equisetaceae): Eng.- Common Horsetail; Kash.- Sehat-bund [W]. A herbaceous perennial plant, inhabiting damp forest grounds.The sterile stems are 30- 90 cm tall, with jointed segments.The fertile stems are succulent textured and offwhite. It is traditionally used for treating wounds and for kidney problems (Fig. E 5).

Equisetum ramosissimum Desf. (Equisetaceae): Eng.- Branched Horse tail [W] A perennial herb growing near dry places, forest openings at 1800-2500m altitudes. Stem hollow and ridged. Leaves greatly reduced and arranged in whorls at nodes. Spores produced in cone like strobili. Used as a medicinal for dropsy, dyspepsia and urinary infections.

Eragrostis polytricha Nees. (syn. Eragrostis minor Host.) (Poaceae):Eng.- Weeping Love Grass; Kashmiri- Love- gausse [W]. Annual herb, usually along roadsides, pathways and also common grass of North-western Himalayan grasslands. Flowers purplish.

Eremopoa persica (Trin.) Roshev. (Poaceae) [W]. A gregarious grass species growing over considerable areas in high altitude areas around Zanskar in Ladakh.

Eremurus himalaicus Baker. (Xanthorrhoeaceae): Eng.- Himalayan Desert Candle; Kash.- Walun [W]. Perennial herb,upto 100 cm tall, inhabiting humus rich soils. Roots fleshy.Leaves strap like strips, glabrous, margins scabrid.Scap long, raceme dense; flowers white. Capsules globose; seeds small black, narrowly winged. Leaves used as vegetable.

Erigeron bonariensis L. (Asteraceae): Eng.- Summer Startwort; Kash.- Shesherda [W]. An annual herb growing on waste lands, and orchards. Leaves covered with hairs and blue-green. Heads with white ray florets and yellowish disc florets. Young shoots used for burning sensation.

Erigeron canadensis L. (Asteraceae):Eng.—Canada Fleabane; Kash.- Poat, Shala lout [W]. Annual herb along wastelands and orchards. Plant extract used as diuretic, astringent and in intestinal disorders.

Erigeron multiradiatus (Lindl. ex D C) Benth & Hook.f (Asteraceae): Eng.- Himalayan Fleabane [W]. Erect sparsely pubescent herb of open slopes. Leaves inversely ovate, coarsely toothed. Flower heads solitary, with central yellow disc and dark purple ray florets.Whole plant used against inflammations.

Erigeron poncinsii (Franch.) Botsch. **(syn Psychrogeton andryoloides (DC) Novopokr ex Krasch.): (Asteraceae): Eng.- Ladakh Fleabane; Ladakhi-Burtsetzen [W].** A densely, grey woolly haired tufted perennial with a stout woody rootstock, commonly growing on mountain slopes of Ladakh. Flower heads solitary borne on leafless stalks. Ray florets light purple, disc florets yellow. Used as fuel.

Erigeron speciosus (Lindl.) DC. (Asteraceae): Eng.- Fleabane [NI]. A leafy stemed plant bearing terminal cluster of purple, daisy like flowers with arrow ray florets. Many varieties, having light to deep pink flowers with yellow central disc, are grown in gardens.

Erigeron umbrosus (Kar. & Kir.) Boiss (syn. Brachyactis umbrosa (Kar. & Kir.) Benth. (Asteraceae): Kash.- Sathi [W]. Annual herb of temperate grasslands. Aerial portions of the plant used for arthritis and rheumatism (Fig. E 6).

Eriobotrya japonica (Thunb.) Lindl. (Rosaceae): **Eng.- Loquat; Hindi-Loquat [NI]** A small round-headed tree with rusty-tomentose branches. Leaves stiff, obovate to elliptic-oblong, acute at both ends, with distinct venation, deep green above, brownish tomentose beneath. Flowers white, appearing in dry-bracted rusty pubescent terminal panicles in late autumn or early winter.

Eritrichium fruiticosum Klotz (Boraginaceae) [W]. Erect perennial or ascending herb of open alpine slopes; stems many from the base.Leaves lanceolate, upper sessile,lower short petioled. Flowers light blue.Whole plant medicinal.

Erodium cicutarium (L) L'Herit (Geraniaceae): Eng.- Filaree [W]. Annual herb,as a weed of cultivation, fairly common in open disturbed scrubby slopes. Flowers pink.Whole plant medicinal.

Erodium tibetanum Edgew. & Hook.f. (Geraniaceae): Eng.- Stork's Bill; Ladakhi-Zema [W]. Annual herb of alpine meadows of Ladakh. Flowers pink. Seeds are taken to relieve constipation and flatulence. Fresh plants are fed to pasmina goats to increase wool production.

Erophila verna (L.) D C (Brassicaceae): Eng.- Common Whitlow Grass [W]. Small annual herb common on disturbed slopes. Leaves in a basal rossete. Flowers white –corolla white, sepals reddish. Fruit many seeded silicula. Whole plant medicinal.

Eryngium caeruleum M Bieb. (Apiaceae): Kash.- Dhudhakh [W]. A hairless thorny herb of drylands and Karewa slopes. Stem bluish. Root used for skin irritations (Fig. E 7).

Eryngium campestre L (Apiaceae): Eng.- Field Eryngo [W]. A hairless thorny perennial inhabiting dry uplands of Kashmir.Leaves tough, stiff, whitish-green; basal leaves long stalked, pinnate and spiny.Infusion of the plant given for cough and urine infections.

Erysimum sisymbroides C A Mey. (Brassicaceae): Eng.- Blister Cress; Ladakhi – Chesulhang Inaku [W]. Perennial herb on rocky cliffs, common on alpine slopes above 2500m. Flowers yellow. Extract of the aerial portion of the plant is used as cooling drink and diuretic.

Eschscholzia californica Cham. (Papaveraceae): Eng.- California Poppy. [NI] A popular ornamental herb producing bright orange yellow poppy like flowers which tend to close in cool and dull weather. Leaves are blue greenish saucer shaped; orange yellow flower appear from June to September. Several varieties with shades of dull white, yellow, orange yellow are cultivated as garden ornamentals.

Euonymus fimbriatus Wall. (syn. Euonymus lacerus Buch –Ham.) (Celastraceae): Eng.- Spindle Tree; Hindi- Barphali; Kash.- Tran/ Chol [W]. A small sub-alpine deciduous tree with reddish-brown compressed branchlets. Leaves broadly ovate, finely double-toothed. Flowers cream-coloured, in lax flat-topped clusters, fruit enclosed in a red aril, with 4 conspicuous long tapering wings. Common on scrubby slopes near forest areas at 1700- 2400 m. Its wood has been used in making pen boxes and for tablets used in liu of slates in schools.

Euonymus fortunei (Turcz.) Hand. -Mazz (Celastraceae) [NI]. A climbing shrub with aerial roots along the ground, broad and tall growing, with oval, toothed, thinly leathery, dark green leaves. Flowers greenish-white, 5-12 in short cymes; followed by spherical white fruit, contain seeds with orange arils. Three varieties commonly used for hedging include *emerald gaiety (*compact and bushy, bearing bright green leaves with white margins, tinged pink in winter.), *emerald gold (*bushy and bears bright green leaves with broad, bright yellow margins tinged pink in winter.) and s*ilver queen (*bushy and upright, with white margined dark green leaves).

Euonymus hamiltonianus Wall. (syn. Euonymus thomsoniana) (Celastraceae): Kash.- Choel [W]. A deciduous shrub bearing thicker, oblong ovate to elliptic, short petioled mid green leaves, turning yellow, pink or red in November. Flowers larger, greenish-white in axillary branched clusters. Fruit pink, top shaped, contains blood-red seeds with orange arils. Grows near forest areas, in cultivated forms also. Roots were used as detergent for hair washing by ladies.

Euonymus japonicus Thunb. (Celastraceae): Eng.- Golden Pillar [NI]. A dense, bushy, evergreen shrub or small, erect tree popularly grown as a hedge plant in gardens and public parks. Leaves obovate to oblong, leathery tough and glossy dark green. Flowers greenish-white in long cymes appearing from early July to September. Commonly grown varieties include *albomarginatus* (leaves dull green margined white), *aureomarginatus* (leaves yellow bordered), *aureopictus* (leaves dark-green with central golden mark), *macrophyllus* (leaves elliptic) and *ovatus aureous* (leaves dark green with golden yellow margins).

Euphorbia helioscopia L (Euphorbiaceae): Eng- Sun Spurge / Mad Woman's Milk; Hindi- Hirruseeah; Kash.- Gure Sou–chal / Sarfa doude [W]. An annual herb growing is waste places across Kashmir Himalayas. Stem contains white latex. Flowers enclosed within a rosette of leaves at the top of the stem. Extract of the plant used to treat skin diseases. Root is anthelmintic (Fig. E 8).

Euphorbia prostrata Ait. (Euphorbiaceae): Eng- Sun Spurge; Hindi- Hirruseeah; Kash.- Gure Sou–chal [W]. Prostrate annual herb of sub-alpine grasslands; stems several from base, slender, purplish, sparsely hairy. Leaves opposite. Capsules small, green. Extracts of the rootstock used for skin problems.

Euphorbia thomsoniana Boiss. (Euphorbiaceae): Eng.- Sun Spurge; Ladakhi- Titri; Kash.- Hibri [W]. Erect glabrous herb, growing on sub-alpine dry slopes.Leaves alternate. Stem many branched, basally purplish, glabrous. Used in *Amchi* system of medicine in Ladakh. Roots used as purgative and also for hair washes.

Euphorbia tibetica Boiss. (Euphorbiaceae): Ladakhi- Gonbu [W]. Annual semi-prostrate perennial fleshy herb endemic to Ladakh. Stems many from the base, sub-erect. Inflorescence a pseudo-umbell or cyme. Cures constipation and acts as purgative (Fig. E 9).

Euphorbia wallichii Hook. f. (Euphorbiaceae): Eng.- Himalayan Sun Spurge; Kash.- Sarfgand [W]. Tall perennial shrubby herb of alpine meadowlands. Leaves sessile, obovate, obtuse. Involucres hemispheric, capsule smooth. Stem yields yellow poisonous latex. Used as folk medicine against skin problems. Root extract used as antiseptic.

Euphrasia oakesii Wettst. (syn. Euphrasia officinalis L.) (Orobanchaceae): Eng.- Eye Bright; Kash.- Sulai; Hindi– Barque Chashim; Ladakhi- Angste / Kangchun [W] Small branched annual herb on damp and shady sites in forests, waysides. Leaves deep cut. Flowers in terminal spikes like clusters of white and purple-tinged 2 –lipped flowers with yellow throats.The extract of fresh tops used against heart burning sensation. Dried herb is also used for eye conjunctivitis.

Euphrasia vulgaris Benth. (Orobanchaceae): Ladakhi- Kangchun [W]. A herb of alpine meadows. Leaves bluish-green, dentate. Flowers axillary, white. Plant used for heart ailments.

Euryale ferox Salisb. (syn. Euryale indica Planch.) (Nymphaeaceae): Eng.- Gorgan Nut/Fox Nut; Hindi- Makhana; Kash.-Juwar /kenae boube [W]. A prickly acquatic, rhizomatous herb, growing in shallow lakes of Kashmir. Fruits edible, eaten after roasting; also taken as laxative.

Fig. E1. Elaeagnus rhamnoides

Fig. E2. Elaeagnus umbellata

Fig. E3. Ephedra gerardiana

Fig. E4. Epilobium latifolium

Fig. E5. Equisetum arvense

Fig. E6. Erigeron umbrosus

Fig. E7. Eryngium caeruleum

Fig. E8. Euphorbia helioscopia

Fig. E9. Euphorbia tibetica

F

Fagopyrum acutatum (Lehm.) Mansf. ex Hammer (syn. Fagopyrum cymosum (Trev.) Meissn (Polygonaceae): Eng.- Medicinal Buckwheat / Perennial Buckwheat; Hindi – Kutu; Kash.- Tsok Hakh [RC]. A tall branched, puberulous annual growing wild in Drasss, Kargil, Gurez areas at 2800 to 3300 m amsl. Leaves tri-angular; perianth white. Grains acutely 3- cornered. The plant is an excellent fodder plant and is supposed to be an ancestor of the cultivated *Fagopyrum* species. The tender leaves of the plant are used as vegetable. The grains are locally used as a cereal and also in treating, collic, chloraic diarrhoea, fluxes of all kinds and abdominal obstruction (Fig. F 1).

Fagopyrum esculentum Moench (Polygonaceae): Eng.- Sweet Buckwheat; Kash.- Truamba [RC]. An annual herb, cultivated as a food cum fodder plant. Leaves are hastate and acute. Flowers in axillary or terminal cymes, pinkish white and fragrant. Fruit a 3- cornered achene and brown in colour. The plant was earlier grown in Kashmir as a substitute to rice. The grains being sweeter used to be ground into flour for bread making. Presently its cultivation is confined to Ladakh and harvested in October and is used mostly in the form of flour for making bread and porridge. Leaves are cooked as vegetable. The immature plants are also used as a pasture supplement for cattle (Figs. F 2,3)

Fagopyrum tataricum (L.) Gaertn. (Polygonaceae): Duckwheat / Bitter Tartary; Hindi- Kaspat; Kash.– Truamba [RC]. An annual plant cultivated as a cereal in the Ladakh at an altitude of 3000 - 4200 m amsl, especially in its coldest parts like Zanskar and western Tibet. Flowers are small, greenish or yellowish, self- fertile flowers; fruits ovoid and conical with more or less wavy out line, brownish grey to black in colour. The grains are however inferior and the flour has a darker colour and somewhat bitter. Grains are either ground in mills and made into bread or cooked and eaten as porridge. Leaves are used as vegetable (Fig. F 4).

Fagus sylvatica cv. purpurea Aiton. (Fagaceae): Eng.- Copper Beech [NI] A spreading ornamental tree with elliptic-ovate wavy margined leaves, silky and ciliate when young, become purple-brown in July, turning brown-green during September. Flowers unisexual, male catkins pendulous, hairy, bell- shaped perianth; female flowers single or in pairs, pedunculate; fruit cup 4-parted, soft, prickly exterior, three sided and grey.

Fatsia japonica (Thunb.) Decne. & Planch (Araliacea): Eng.- Paper plant [NI] A spreading, suckering, rounded evergreen ornamental shrub with thick stems bearing hairless, palmately lobed, dark-green and glossy leaves. Flowers 5-petalled creamy white long-stalked compound umbels followed by small spherical black, globose, fleshy fruits.

Ferula jaeschkeana Vatke. (Apiaceae): Eng.- Wild Asfoetida; Kash.- Hapat Kouphar [W]. Tall perennial stout tuberous herb common on dry forest slopes at 2000 - 2500m. Leaves pubescent, large, pinnate-decurrent, margins serrate. Flowers yellow in compound umbels. Latex of the plant used to treat wounds (Fig. F 5).

Ferula narthex Boiss. (Apiaceae): Eng.– Asafoetida; Hindi – Heenge; Kash.– Yanga [W]. A tall perennial herb growing in alpine grasslands of Kashmir upto 3000 m amsl. Stem succulent. Leaves tri-pinnate. Flowers yellow in large umbels. Roots are used in perfumery and for flavouring food products and also used in medicines in treatment of asthama, whooping cough, as an antispasmodic and nerve stimulant. As spice, its use is however restricted to local Hindus only (Fig. F 6).

Festuca kashmiriana Stapf. (Poaceae): Eng.- Meadow Fescue [W]. An endemic tufted, grass species found growing in alpine grasslands of Kashmir Himalaya. Culms slender, erect. Used for hay making.

Festuca levingei Stapf. (Poaceae): Eng.- Meadow Fescue [W]. Loosely tufted perennial endemic grass species growing in Kashmir Himalayas. Culms upto 50 cm tall, ascending from a spreading base. Lemmas awnless and hairless.

Ficus carica L. (Moraceae): Eng.- Common Fig; Hindi – Injir; Kash.- Anjur [NI] A small soft-wooded much-branched deciduous tree or large shrub with thick, long-petioled, strongly palmately ribbed leaves, 3-5 lobed. Fruit single axillary and subterminal, pear shaped variable in size, green when young, maturing to dark- green, purple or dark-brown.

Ficus pumila L. (Moraceae): Eng.- Creeping Fig [NI]. It is a root-clinging, evergreen climber. The leaves of the climbing stems are asymmetrically ovate, thinly leathery, dark green. Fruits oblong to elliptic, dark green, pear-shaped, densely hairy figs, green with white dots, ripening purple. Grown to cover walls of houses and ground.

Filago arvensis L. (Asteraceae): Eng.- Cud Weed; Ladakhi- Panksa [W]. A grey tomentose herb, commonly growing on dry rocky areas. Flowers whitish in cottony heads. Leaves alternate with entire margins, sharp pointed and woolly. A water extract of the plant is taken for gout and rheumatism.

Filipendula vestita (Wall. ex G Don) Maxim. (syn. Spiraea vestita Wall.) (Rosaceae) [W]. Tall shrubby herb on shady alpine / sub-alpine slopes. Leaves pinnatisect, lateral leaflets small, terminal very large, tomentose beneath. Medicinal.

Fimbristylis falcata (Vahl) Kunth. (Cyperaceae): [W]. Erect perennial grass like plant with a creeping rhizome. Stem usually solitary, also tufted. Rhizome used for intestinal troubles.

Foeniculum vulgare Mill. (Apiaceae): Eng.– Fennel; Hindi– Saunf; Kash.- Jungli Badyana [W]. An aromatic perennial herb mostly growing wild/ in semi-cultivated state, probably after getting introduced, in sub- alpine grasslands and orchards. Leaves green, finely divided into segments. Flowers yellow in large umbels. Fruits are used as spice and condiment. Leaves used for treating rheumatism and gastric disorders (Fig. F. 7).

Forsythia geraldiana Lingelsh (Oleaceae): Eng.- Golden Bell; Kash.- Tahri posh [NI]. A deciduous compact shrub, having ovate- oblong serrate leaves. Flowers bell shaped, deep yellow, long styled, cylax and corlla deeply 4 parted. Fruits a bivalved, beaked capsule. Very common, early spring flowering shrub with flowers appearing well before leaves. The Popularly grown varieties grown for heding/ Spring flowering include var. *densiflora* (sharp teethed leaves flowers solitary, very large, densely compact, light yellow) and var. *specetabilis* (entire leaves, flowers densely arranged, dark yellow, solitary).

Forsythia ovata Nakai (Oleaceae): Eng.-Early Forsythia [NI]. A compact deciduous shrub with ovate to broad ovate, dark green leaves. Flowers small, amber yellow. Branches grey-yellow, twisted.

Fragaria x ananassa (Duchesne ex Weston) Duchesne ex Rozur (syn. Fragaria hybrida Duchesne) (Rosaceae): Eng.- Chilean Strawberry [NI]. A spreading perennial. Many garden strawberry varieties producing good quality fruit with good aroma have been introduced in the valley very early and now thrive well and represent part of local horticultural biodiversity. Some of the popular for fruit are, *Senga, Sengana, Gorala* and *Cavalier.*

Fragaria nubicola (Lindl. ex Hook. f.) Lacaita (syn. Fragaria vesca var. nubicola Lindl. ex Hook. f.); Potentilla nubicola (Lindl. ex Hook.

f.) Mabb. [W]. A low growing, softly hairy perennial herb native to Himalayas. Leaves trifoliate, long stalked, leaflets ovate, coarsely toothed; runners long, rooting at nodes. Flowers white with five broad ovate petals. Fruits non- commercial. Rhizomes were being used as substitute to tea (Fig. F 8).

Fragaria vesca L. (Rosaceae): Eng. Alpine Strawberry; Kash.- Range raeche [W] Perennial stoliniferous herb, commonly growing in forests and on humus rich soils in sub-alpine pastures. Leaves tri-foliate, with toothed margins.Corollaregularandwhite.Fruitsred,succulentandsmall(Fig.F9).

Fraxinus americana L. (Oleaceae): Eng.- White Ash [NI]. A tall deciduous, broadly columnar tree bearing pinnate, dark green leaves with 5-9 oblong-lance shaped to ovate, tapered leaflets, turning yellow in mid-late October. Flowers apetalous, produced before the leaves in early April.

Fraxinus excelsior L. (syn. Fraxinus hookeri Wenzig.) (Oleaceae): Eng.- Common Ash; Kash.- Hgom / Kulem [W /UC]. A vigorous, spreading, deciduous handsome ornamental shrub growing on scrubby slopes and in forests at 1700-2300 m. Leaves pinnate, leaflets ovate oblong, sessile, serrate, dark green turning yellow in mid October. Flowers deep purple; flowering before the leaves in early April.

Fraxinus floribunda Wall. (Oleaceae): Eng. – Himalayan Ash; Hindi-Angan; Kash.- Hum [W]. A deciduous ornamental tree, 10-15 m tall, with broad ovate crown growing in dense alpine forests above 2500 m altitudes. Branches quadriangular. Leaves large, sessile, opposite, pinnate with 7-9 leaflets; leaflets serrate, dark green above, lighter beneath. Buds densely brown, tomentose. Flowers white and appear in long terminal panicles in early April. Fruits winged samaras (Fig. F. 10).

Fraxinus xanthoxyloides (Wall. ex G Don) DC (Oleaceae) [W]. A small deciduous tree isolated in forests. Bark grey; leaves opposite, midrib winged. Wood used for making tools. Flowers in dense heads. Aerial parts used medicinally.

Fritillaria cirrhosa D Don. (syn. Fritillaria roylei Hook.) (Liliaceae): Eng.- Himalayan Fritillary; Hindi-Pilyari; Kash.- Praneh Posh /Sheathkar [W].A perennial tall, bulbous herb of ornamental value, found in Kashmir Himalaya from 1800 - 3300 m amsl as forest undergrowth. Leaves linear, lance like. Flowers nodding, borne singly, campanulate, greenish yellow/brownish purple and pendulous. Dried bulbs are used as tonic and for the treatment of asthama and bronchitis (Fig. F. 11).

Fritillaria imperialis L. (Liliaceae): Eng.- Imperial Fritillary; Hindi- Bulatini; Kash.- Prenik /Pilyari / Bani hale [W/UC]. A hardy bulbous plant found wild in sub alpine grasslands on forest cover and also cultivated as border ornamental in gardens. Leaves narrowly lanceolate, glossy green carried in whorls along the erect stem. The tulip shaped flowers open in early April and are yellow to rich bronze red in colour and borne in a terminal pendent cluster topped by a crown of small leaves (Fig. F 12).

Fuchsia corymbiflora Ruiz & Pav. (Onagraceae): Eng.- Fuchsia [NI]. A deciduous tender shrub grown mostly as pot plants for its attractive pendulous bell shaped flowers. Stem with arching branches carrying oblong lanceolate leaves. Flowers crimson borne on pendulous corymbs from June to September.

Fumaria indica (Hausskn.) Pugsley. (syn. Fumaria parviflora Lamk. subsp. vailantii (Loisel) Hook.f. (Papaveraceae): Eng.- Fumitory; Hindi- Pitpapara /Shahtari; Kash.- Shahtar [W]. A spreading much branched annual herb, as a weed of orchards and waste lands. Stem glaucous and leafy, 5-30 cm long. Leaves divided into narrow pointed segments. Flowers pale-pink to whitish in apical clusters, spurred. Whole plant used as best liver tonic in treating chronic liver troubles. Also used as blood purifier and for skin ailments (Fig. F 13).

Fig. F1. Fagopyrum acutatum

Fig. F2. Fagopyrum esculentum

Fig. F3. Fagopyrum esculentum (grains)

Fig. F4. Fagopyrum tataricum

Fig. F5. Ferula jaeschkeana

Fig. F6. Ferula narthex

Fig. F7. Foeniculum vulgare

Fig. F8. Fragaria nubicola

Fig. F9. Fragaria vesca

Fig. F10. Fraxinus floribunda

Fig. F11. Fritillaria cirrhosa

Fig. F12. Fritillaria imperialis

Fig. F13. Fumaria indica

G

Gagea lutea (L) Ker Gawl (syn. Gagea elegans Wall ex G Don.): (Liliaceae): Kash.- Koker neeje [W]. A bulbous perennial herb inhabiting open sub-alpine slopes and forming groups. Leaves narrowly lanceolate. Basal leaf single, angular flat, exceeds the inflorescence, lanceolate linear, longitudinally furrowed at base. Flowers solitary on long pedicles, tepals yellow. Young leaves used as vegetable (Fig. G 1).

Gagea gageoides (Zucc.) Vved. (Liliaceae): Ladakhi- Gudgi [W]. Perennial bulbous herb growing in moist pastures. Bulbs obliquely drop shaped with thin light brown tunics. Flowers yellow. Groups of bulbils are in leaf axils of inflorescence. Decoction of powdered bulbs is used for curing whooping cough.

Gagea serotina (L.) Ker Gawl. (syn. Lloydia serotina (L.) Rchb. (Liliaceae): Eng.- Common Alp lily/ Snowdon Lily. A bulbous perennial growing above 2500 m altitudes inhabiting grassy slopes, alpine grasslands and forest thickets. Leaves filiform. Flowers white, with purple or reddish veins along the tepals. Capsules sub-ovoid with peresistant style.

Galium aparine L. (Rubiaceae): Eng.- Goosegrass, Catchgrass [W]. A rambling scabrid annual of alpine areas.Leaves narrowly oblong, tapering at both ends, apiculate, sessile, 6-8 in a whorl. Flowers white. Fruit clothed with spreading hooked bristles. Used to treat insect bites. Also used to treat skin ailments.

Galium mahadevensis G Sing (Rubiaceae): Eng.- Bedstraw [W]. Perennial herb with several stems, 5-15 cm tall arising from a slender creeping rootstock. Leaves in whorls of four, lanceolate, single nerved. Flowering peduncle axillary bearing three white flowers. Fruits dry and covered with long hairs.

Galium patzkeanum G.H. Loos (Rubiaceae): Eng.- White Bedstraw; Ladakhi- Rangchekarpo [W]. Perennial herb, on scruby slopes. Stem slender, erect, unbranched. Flowers minute, white. Gargles of extract of the plant used to treat throat infections. Fresh plants used as a supplement to alfalfa.

Galium spurium L. (syn. Galium pauciflorum Bunge.) (Rubiaceae): Eng.- Yellow Bedstraw; Ladakhi- Rangche [W]. An erect or reclining herb of Alpine slopes, upto 50 cm tall, endemic to Kashmir Himalaya. Leaves in whorls of 6-8, narrowly lanceolate. Flowers in multi-flowered cymes, white or yellow green. Whole plant used for throat infections.

Gardenia jasminoides J Ellis (syn. Gardenia jasminoides var. grandiflora (Lour.) Makino) (Rubiaceae) Eng.- Cape Jasmine; Hindi- Gandha Raj (NI). A medium to large sized evergreen shrub or small tree with ovate elliptic, or lance-shaped, glossy deep green leaves. Produces 5 to 12 lobed slaver form, strongly fragrant white to ivory flowers, borne singly or in few flowered cymes.

Gasteria carinata var. verrucosa (Mill.) Van Jaarsv (syn. Gasteria verrucossa (Mill.) Haw. (Xanthorrhoeaceae):Eng.-Keeled Gasteria [NI]. A green house succulent perennial, stemless with tough glossy marked leaves arranged in two ranks. Mostly grown as pot plants.

Gentiana capitata Buch. Ham ex D. Don (Gentianaceae): Eng.- Clustered Gentian [W]. An annual herb inhabiting forest scrubs and clearings at 4500 m altitude. Stem erect. Leaves broadly ovate, closely clustered under the flower head and stem usually leafless below. Flowers tubular, pale- blue, purple or white crowded at the top.

Gentiana carinata (D Don.) Griseb. (Gentianaceae): Eng.- Dark blue Gentian; Kash.- Gule Maudan [W]. Low compact annual herb of alpine grasslands at 2800-3500 m. Flowers blue, funnel shaped. Plant used for post natal bath of ladies (Fig. G 2).

Gentiana decumbens L. f (Gentianaceae).: Ladakhi – Sonpo/ Spangyan [W]. A perennial herb, 20- 30 cm tall growing near forest clearings, grassy slopes. Stems ascending, stout, glabrous, simple. Flowers dark blue. Stem extract used for stomach cramps.

Gentiana kurroo Royle (Gentianaceae): Eng.- Himalayan Gentian; Hindi- Kutki; Kash. –Neel kanth [W]. A perennial tufted herb distributed in grass-lands of N.W. Himalayas from 1500-3300 m amsl. Corolla narrow, funnel like, deep blue, paler in throat. Roots are used as blood tonic stomachic and for urinary affections. The plant is also useful in syphilis and leucoderma. Other *Gentian* spp. growing in Ladakh are very popular in *Amchi* medicine being used to treat bile troubles and throat ailments (Fig. G 3).

Gentiana pyhllocalyx Clarke (Gentianaceae): Eng.- Leaf-sepal Gentian; Kash.- Phungri [W]. A perennial herb of Kashmir Himalayas found in grasslands. Leaves crowded towards base. Flowers borne at branch ends, solitary, sessile and blue. Whole plant used for stomach disorders in cattle.

Gentiana prostrata Haenke (Gentianaceae): Eng.-Pygmy Gentian; Ladakhi-Wanglucktsetz [W]. Annual stoloniferous herb of high altitude meadows near moist places. Stems ascending to erect, branched from base, glabrous or pappilose. Flowers deep blue, calyx tubular. Extract of the plant taken as tonic.

Gentiana stipitata Edgew. (Gentianaceae): Ladakhi- Lucktsetz [W]. Perennial herb with spreading stems. Stem leaves elliptic, basal ones larger and pointed. Flowers tubular, pale –blue to light mauve coloured, terminal, solitary, sessile; calyx tube obconic, corolla lobes distinctly awned. Powdered roots, dissolved in luke warm water, are taken three to four times a day to cure diarrhoea.

Gentianella moorcroftiana (Wall. ex Griseb.) Airy Shaw (Gentianaceae): Ladakhi- Chumbi tikta [W]. Erect annual herb, upto 20 cm tall, endemic to Ladakh. Leaves sessile, basal leaves obovate, cauline linear –lanceolate with indistinct veins. Flowers in cymes on long stalks. Corolla blue, yellow in throat. Capsules ovoid-ellipsoid. Whole plant used for nausea, giddiness, as blood purifier and for cough and cold (Fig. G 4).

Gentianopsis detonsa (Rottb.) Ma (syn.Gentianella detonsa (Rottb.) G. Don. (Gentianaceae): Eng.- Windmill fringed Gentian; Ladakhi-Sheeti [W]. Biennial herb of alpine region. Flowers blue, margins of corolla lobes fringed. Flowers used for checking nausea, headaches and fevers.

Gentianopsis stracheyi (Clarke.) Kitam. (syn. Gentianella strachyei (Clarke) Kit. (Gentianaceae): Eng.- Swamp Gentian; Ladakhi-Wanglucktsetzenshe [W] Annual herb of alpine Himalayas above 2500 m. Flowers are blue/yellowish-white. Leaves spoon shaped. Powdered plant, dissolved in *Changue,* is taken to cure renal infections and to relieve painful micturition.

Gerbera jamesonii Bolus ex Hook. f. (Asteraceae): Eng.- Transvall Daizy [NI]. A tender half hardy perennial flowering plant. The lobed leaves are hairy throughout. The daisy like attractive, orange scarlet flowers are borne from May to August. Single and double- flowered varieties and hybrids, in attractive pastel colors, are grown in gardens as well as in green house conditions for cut flowers.

Gerbera maxima (D.Don.) Beauverd. (syn. Gerbera nepalensis (DC) Sch. Bip. (Asteraceae): [W]. Low perennial of alpine meadows. Leaves lyrate pinnatifid, glabrous above and woolly beneath.

Geranium gracile Ledeb ex Nordm (syn. Geranium grandiflorum Sweet) (Geraniaceae):Eng.- Garden Geranium; Hindi– Bhanda [NI]. A bushy herbaceous perennial with round mid-green long stalked leaves. Flowers blue-purple and appear from June to July. Grown as border perennial or a poted plant.

Geranium kishtvariensis R. Knuth (Geraniaceae):Eng.- Geranium; Hindi– Bhanda [W]. Perennial glandular pubescent herbn of shady places, endemic to Kashmir Himalaya. Leaves large maple like; flowers pink. Whole plant used as medicinal.

Geranium nepalense Sweet. **(Geraniaceae)**: **Eng.- Nepal Geranium; Hindi-Panch- patri** [W]. Perennial, creeping spreading herb commonly growing in moist and shaded places. Leaves are palmately cut in to 5-7 lobes. Flowers pale pink. Plant has been used as a source of dyes (Fig. G 5).

Geranium pratense L. (Geraniaceae): Eng.- Meadow Geranium [W]. Perennial glandular, pubescent herb, usually common in moister shaded places. Leaves 6-9 partite, pedicels densely hairy. Flowers bluish pink, petals spreading. Whole plant medicinal.

Geranium rivulare Vill. (syn. Geranium aconitifolium L. Herit.) (Geraniaceae): Ladakhi- Palto / Legkatin [W]. A glandular herb of alpine grasslands. Flowers violet veined purple. Flowers, root and leaves used for ulcers, insect bites and for wounds of cattle.

Geranium rotundifolium L. (Geraniaceae): Eng- Round leaved Geranium/Crane's Bill.; Hindi- Bhanda [W]. Annual herb, fairly common on open scrubby slopes, forests and in wastelands. Flowers pink. Plant used to treat burns and wounds (Fig. G 6).

Geranium wallichianum D Don ex Sweet. (Geraniaceae): Eng.- Cran's Bill; Hindi- Bhanda; Kash.- Ashude / Panch patri; Ladakhi- Perhi [W]. A perennial pubescent herb occurring all over temperate Himalaya at 1500-2700 m amsl. Leaves orbicular, palmately trilobed. Flowers purplish white. Roots used to colour medicinal oils. Root/seeds of the plant used as an astringent, for tooth aches, and for curing renal diseases. In Ladakh crushed leaves are rubbed against skin to prevent cutaneous eruptions. Its tea is used for rheumatism (Fig. G 7).

Geum elatum Wall. (syn. Acomastylis elata (Wall.) F. Bolle (Rosaceae): Eng.- Herb Bennet; Kash.- Gogal mool [W]. A perennial clump forming herb, common on alpine and sub-alpine pastures at 3000 – 3500 m. Flowers yellow, from June to September. Root extract used for leucoderma and intestinal disorders.

Geum urbanum L. (Syn. Geum roylei Wall ex Bolle) (Rosaceae): Eng.- Royle's Avens [W]. A hairy perennial herb of shrubberies in forests. Basal leaves unequally pinnately lobed. Flowers yellow. Fruit bristly hairy. Herbalists use it for treating heart ailments, halitosis and mouth ulcers.

Ginkgo biloba L. (Ginkgoaceae): Eng.- Maidenhair Tree; Hindi- Bal-kunwari [W/UC]. A coniferous deciduous upright tree, columnar with furrowed dull grey bark. Leaves light-green, flat, fan-shaped, with numerous parallel veins, turning golden-yellow in October - November. Dioecious, male catkin pendulous, cylindrical with yellow flowers, borne in clusters; female flowers produce plum like, yellow green fruit. Also grown as ornamental (Fig. G 8).

Gladiolus carneus F. Delaroche (syn. Gladiolus blandus Aiton.) (Iridaceae): Eng.- Gladiolus [NI]. Half hardy bulbous ornamental flowering plant grown as border plant as well as for cut flowers. Leaves are slender, dark-green, ribbed. Flowers arranged on long spikes white, red, and yellow rose, pink colour flowered varieties are grown mostly as cut flowers.

Gladiolus communis L. (syn. Gladiolus byzantinus Mill.) (Iridaceae): Eng.-Sword Lily [NI]. A fully hardy species grown for its wine red florets, appearing on long spikes. Large flower hybrids are grown for cut flowers by commercial floricultural units, mostly under green house conditions.

Glaux maritima L. (Primulaceae): Ladakhi- Spengcha [W]. Annual herb inhabiting alpine swamps of Ladakh. Root used as antiseptic and anthelmintic.

Gleditsia triacanthos L. (Fabaceae): Eng.- Honey Locust [NI]. A spreading deciduous tree with a long trunk and attractively arranged branches, densely covered with simple or compound flattened thorns. Leaves pinnate with 14-24 leaflets or bi-pinnate with 4-16 pairs; leaflets acute oval-oblong, bright green, turns golden yellow at fall. Flowers greenish, very short stalked, borne in racemes. Pods flat sickle shaped and twisted glossy dark brown.

Glycine max (L.) Merr. (syn. Phaseolus max L.) (Fabaceae): Eng. – Soybean; Hindi – Bhat; Kash. –Moughte [RC]. An annual with erect stem, cultivated up to 2200 m amsl, reaching a height of 45-150 cm, densely clothed with hairs; leaves trifoliate, ovate – lanceolate, long petioled; fls small, inconspicuous, borne on short axillary racemes, white or purple to redish purple; pods densely hairy, 2-4 seeded, seeds elliptical long, hilum compressed, chocolate or black/ dark yellow. Local types have smaller seeds and are cultivated chiefly for animal feed and not as a pulse (Fig. G 9).

Glycine max subsp. soja (syn. Glycine soja Sieb. & Zucc.) (Fabaceae): Eng.- Wild Soybean [W]. A selender twinner, pubescent herb growing among scrubs in forest openings in ravines. Flowers in axillary clusters of 1 to 3; pods smaller; seeds 2 to 3 per pod, oval, black with white hilum. Seeds medicinal.

Gnaphalium stewartii (Holub.) Clarke ex Hook. f. (Asteraceae): Ladakhi- Gnatzemetz [W]. Perennial, caespitose short stemmed herb, growing on scuby areas of Ladakh. Flowering stems 1-3 cm tall, densely floccose-tomentose. Capitula campanulate. Flowers pale yellow. Decoction of the whole plant taken as antipyretic.

Gnaphalium thomsonii Hook.f (Asteraceae): Eng.- Cotton Weed / Cat's Foot [W]. Erect branched annual alpine herb covered with white soft tomentum.Leaves linear,falcate with acute apex. Capitula terminal subtended by slender white woolly leaves. Whole plant used medicinally.

Gomphrena globosa L. (Amaranthaceae): Eng.- Bachelors Button; Kashi.- Mahraz Posh [NI] A bushy annual with erect stems, and hairy light green leaves that are oblong- ovate. Orange, yellow, purple, pink or white, avoid flower heads appear from July to September. Selected single colour varieties are usually cultivated as garden ornamentals.

Grielum grandiflorum (L.) Druce. (syn. Geranium grandiflorum L.) (Neuradaceae): Eng.- Cranes Bill; Hindi- Bhanda [NI]. A bushy herbaceous perennial with round, mid-green long stalked leaves. Flowers are blue- purple and appear from June to July. Grown as border perennial ornamental (Fig. G 10).

Fig. G1. Gagea lutea

Fig. G4. Gentianella moorcroftiana

Fig. G2. Gentiana carinata

Fig. G5. Geranium nepalense

Fig. G3. Gentiana kurroo

Fig. G6. Geranium rotundifolium

Fig. G7. Geranium wallichianum

Fig. G9. Glycine max

Fig. G8. Ginkgo biloba

Fig. G10. Grielum grandiflorum

H

Halerpestes tricuspis (Maxim.) Hand maz. (Ranunculaceae): Ladakhi-Chitaka [W] Annual herb of alpine zone of above 3500 m altitudes. Whole plant used as stomachic and for gastrointestinal problems.

Halogeton glomeratus (M. Bieb.) Ledeb. (Amaranthaceae): Eng.-Saltlover; Ladakhi- Ischermang [W]. Tall annual herb growing in alpine meadows and also as a weed in orchards. Stem curved at base and tinged reddish or purple. Leaves alternate, sessile, semi-succulent. Paste of leaves used to cure itching and athletes foot.

Hedera algeriensis Hibberd. (syn. Hedera algeriensis var. variegata Paul Hibberd.) (Araliaceae): Eng.- Algerian Ivy [NI]. A strong-growing climber. Leaves kidney-shaped, sometimes obscurely three lobed. Leaves deep green and pale green mottled creamy white. Flowers greenish yellow, small, in umbellate racemes. Fruit a 3 or 5 seeded, yellow berry.

Hedera canariensis Willd. var. gloire de marengo (Araliaceae): Eng.-Canara Island Ivy [NI]. A vigorous climber with 3-5 lobed, ovate-triangular leaves, deep green in the centre, merging into silvery-grey and margined white, borne on smooth, wine- red leaf-stalks. Flowers greenish yellow, small in umbellate racemes; fruit a 3 or 5 seeded, yellow berry.

Hedera helix L. (Araliaceae): Eng.- Common Ivy [W/UC]. A most adaptable and variable evergreen climber with hairy pubescence; with 3-5 lobed leaves on vegetative branches, dark green above, often with whitish venation, yellowish dark green beneath; leaves of the flowering branches oval-rhombic, entire. Flowers in globose umbels grouped into racemes, greenish yellow; fruit globose, black. Commonly grown cultivars include *atropurpurea* (leaves dark purplish- green), *eva* (leaves grey- green with creamy white margins), *glacier (*leaves silver grey with white margins), *gold child* (leaves pale green with golden yellow margins), *gold heart* (leaves with conspicuious central splash of yellow), *ivalace* (leaves bright green and curled at margins), *manda* *crested* (leaves wavy edged, bronze) and *sagitifolia variegata* (leaves grey with cream margins).

Hedera nepalensis K. Koch (syn. Hedera helix var. cinerea Hibberd; Hedera helix var. himalaica Hibberd.) (Araliaceae): Eng.- Himalayan Ivy; Kash.- Pal Walnu [W] A strong-growing, evergreen woody self-clinging climber, climbing on trees and rocky cliffs; with ovate-triangular leaves. Flowers tiny, many, yellowish- green, in stalked globular umbels arranged in domed clusters. Fruit globular shining yellow, turn into black. Plant has distinct scaly pubescence and yellow fruits. Poultice of leaves used for wounds and sores (Fig. H 1).

Helianthus tuberosus L. (Asteraceae): Eng.- Sun Root; Hindi – Kandmool; Kash.- Farhan gogije [W]. A perennial tuberous herb, about 2 m tall, growing on waste lands, orchards preferring dry locations. Leaves simple, fast green. Flowers golden yellow. Tubers edible and specially given to diabetic patients (Fig. H 2).

Heliotropium eichwaldii Steud. (syn. Heliotropium dunaense R.Degen) (Boraginaceae): Hindi- Nilkatti [W]. A wooly-tomentose herb, usually on Karewa slopes. Leaves used for number of ailments- from boils, ulcers to bites of snakes and mad dogs.

Hemerocallis fulva (L.) L (Liliaceae): Eng.- Orange Daylily [UC/W]. A herbaceous perennial ornamental growing from tuberous roots,mostly along drains; also grown as ornamental in gardens. Plant consists of a basal rosette of leaves and a flowering stalk of 20 – 35 cm tall. Flowers orange red. Fruit 3 valved capsule.

Heracleum candicans Wall ex DC (Apiaceae): Eng- Cow Parsnip/ White Leaf hogweed; Kash.-Phakij [W]. A sub-alpine tall villous herb,upto 2m tall, common on scruby and savana slopes at 1800- 2000 m. Flowers white in large umbels. Fruits flattened and abconic.The fruits of the herb used in treating leucoderma. Also used as aphrodiasic and nerve tonic (Fig. H. 3).

Heracleum pinnatum Clarke (Apiaceae): Ladakhi- Resho [W]. A pubescent herb, 30-40 cm tall, in grasslands of cold desert of Ladakh. Leaves basal, long petioled, pinnate with serrate margins. Powdered seeds are taken with water to avoid vomiting during travel.

Heracleum taylorii Norman (syn. Heracleum thomsonii Clarke) (Apiaceae): Ladakhi- Thukar [W]. A villous herb, 50-80 cm tall, on mountain slopes of cold desert areas of Ladakh. Dried and powdered plants are given to new borne cattle to give them strength and stamina.

Herminium monorchis (L.) R. Br. (Orchidaceae): Eng.- Marsh Orchid; Ladakhi – Patiksket [W]. A greenish-yellow orchid found in Western Himalayas. Leaves 2-4, oval, keeled, placed at the base of the stem. Flowers small, greenish–yellow, bell shaped and scented. Extract of the tubers used to treat wounds (Fig. H 4).

Herniaria hirsuta L. (syn. Herniaria hirsuta subsp. cinerea (DC) Cout.) (Caryophyllaceae): Eng.- Hairy Rupturewort [W]. Annual herb, forming semi-woody rootstock, usually on rocky slopes of Kashmir. Flowers green with hirsute calyx. Decoction of root is given to horses suffering from hot and cold. Also used in soar throat.

Hibiscus syriacus L. (Malvaceae): Eng.- Syrian Hibiscus / Rose of Sharon; Hindi- Swet jaba, Gulhar / Gule nasrin [NI]. A medium large-sized deciduous shrub with ovate to loom shaped shallowly to palmately 3-lobed, coarsely toothed, dark green leaves. Flowers solitary and axillary, broadly campanu!ate appear in succession between July to October. *Blue bird* (blue flowers), *caeruleus plenus* (rosy purple fls with maroon blotch at base), *diana* (white flowers) and *red heart* (white fls with red eye) are the commonly cultivars grown for ornamental value.

Hibiscus trionum L. (Malvaceae): Eng.- Trailing Holyhock/ Bladder Hibiscus [W]. Annual herb, usually in moist places, usually along rice fields. Flowers with yellowish white petals with a purple base. Leaves palmately compound. Leaves and flowers used medicinally.

Hieracium umbellatum L. (Asteraceae): Eng.- Canada Hawkweed [W]. An erect hirsute perennial herb on open alpine and sub-alpine slopes above 2500 m altitudes. Leaves pointed, ovate-lanceolate acute, entire with toothed margins. Flower heads yellow.

Hieracium virosum Pall (Asteraceae) [W]. A perennial herb, with thick rhizome, inhabiting alpine grasslands. Stem solitary or few fascicled, hairy. Flower heads yellow. Leaves cauline, ovate, remotely dentate and sessile. Synflorescence paniculate to nearly corymbose with many capitula.

Hieracium vulgatum Fries (Asteraceae): Eng.- Yellow Hawkweed [W]. Erect perennial on open alpine slopes, densely villous. Radical leaves broadly ovate. Heads yellow. Achenes black. Used in religious ceremonies.

Hierochloe laxa Hook. f. (Poaceae): Eng.- Lax Sweet Grass [W]. A perennial, rhizomatous, sweet scented grass of alpine meadows above 3000 m altitudes. Culms erect 40-75 cm tall. Leaf blades smooth and pointed. Inflorescence a panicle bearing grass green florets.

Hippochaete debilis (Roxb. ex Vaucher) Ching (syn. Equisetum debile Roxb. ex Vaucher) (Equisetaceae): Eng.- Field Horse tail; Ladakhi- Zetsmanpo [W] A perennial pteridophyte of moist and shaddy places of Kashmir Himalaya at 1800- 3500 m. Stem hollow and ribbed. Leaves greatly reduced and non-photosynthetic, arranged in whorls on nodes. Strobuli oblong, placed at the end of branches. Plant is used in the treatment of dropsy, kidney affections and gonorrhoea. In ladakhi system of medicine decoction of the plant is given to cure hyperacidity.

Hippolytia senecionis (Jacquem ex Besser) Poljakov ex Tzvelev (syn. Tanacetum senecionis (Jacquem ex Besser) J.Gay) (Asteraceae): Ladakhi- Purnak / Pumakh [W]. Perennial herb on alpine slopes. Heads yellowish. Plant used for colics, earaches and otorrhoea.

Hordeum aegiceras Nees ex Royle (Poaceae): Ladakhi- Grime [RC]. A six rowed barley type grown traditionally in Ladakh for grain and fodder. Unlike barley, it has peculiar lobbed appendaged tip of lemma (Fig. H 5).

Hordeum murinum subsp. glaucum Tzvelev (syn. Hordeum glaucum Steud. (Poaceae) [W]. A wild grass distributed in North-west Himalaya including Kashmir. Spikes very dense and axis readily disarticulating at maturity. An important fodder plant.

Hordeum vulgare L. (Poace*ae*): Eng.- Barley; Hindi – Jau; Kashmiri *Vishka*; **Ladakhi –Nus [RC].** A herbaceous annual grass grown mostly in Ladakh region both for grain and fodder purposes. Stem with few tillers each with five to eight nodes; leaves linear. Inflorscence a terminal cylindrical spike. Fruit a caryopsis, not adhered to lemma and palea as in other barley types. The grain is naked, smaller than that of wheat and pointed at both ends. The straw constitutes about 70% of the total biomass and serves as an excellent fodder for cattle.

In Ladakh the barley has been an important cereal crop having been used as a staple food and thus has received full attention during its cultivation since decades. The barley grains have religious and social significance for local Hindus as they are used during fire worshiping and offerings to God. Barley straw was also used for resting of dead bodies before their crematiom.

Some of the traditional and recently released varieties being largely cultivated in the Ladakh region are:

- **Tibetian Awnless:** A traditional naked barley variety grown in cold arid zone of Ladakh and higher hills of Kashmir up to altitudes of 3400 m. The plants are medium tall with naked grains. The grains of the variety have been a staple food for the people residing in Ladakh area since ancient times. Its flour is superior than other barley varieties, similar to wheat and suitable for bread preparation.

- **Ladakhi Vishka:**A traditionally grown barley variety of Ladakh. Plants are tall with weak stem and hence well prone to lodging. Grains selender, dull amber coloured suitable for *sattu* and bread preparation. Traditionally the extract of the grains was to given to patients suffering from renal problems. The grains are not esteemed as a food and are often mixed with wheat by millers (Fig. H 6).
- **Nurboo (SBL-4):** A recently released variety of naked barley being cultivated in barley growing areas of Ladakh under cold arid climatic conditions after being released by SKUAST-K during 1999. The plants are tall with ear neck curved. The grains are smaller, bold and shining, amber coloured suitable for *sattu* preparation. The variety gives average grain yield of 3.8 t/ha and straw yield of 6.5 t/ha. Plants are moderately resistant to yellow rust and pests.
- **Sindhu (NBL-11):** Naked barley variety eleased by SKUAST-K in 1999 for cultivation in barley growing areas of Ladakh. It is popularly grown in Leh and Kargil areas of Ladakh region during normal crop season. The plants are semi-tall, with large droopy leaves. Grains are bold, elliptical and slightly bluish in colour. Crop duration is from 100-110 days. Plants are tolerant to yellow rest but moderately resistant to loose smut. Average productivity potential ranges from 2.8 to 3.5 t/ha and straw yield of 5 - 7 t/ha. Grains suitable for *sattu* and *Changue* preparation.

Humulus lupulus L. (Cannabaceae): Eng.- Hop; Kash.- Hops [W/UC]. A rhizomatus, twining, perennial with roughly hairy shoots and deeply 3 to 5 lobed, coarsely toothed, light green leaves; bears broadly ovoid, fragrant green, straw-coloured spikes of female flowers. Flower cones used for medicinal purposes.

Hyacinthus orientalis L. (Asparagaceae): Eng.- Common Hyacinth; Kash.-Sumbul [W/UC]. A bulbous plant; leaves strap shaped, 15-35 cm long, with a soft, succulent texture and produced in a basal whorl. Flowering stem is a raceme, bearing fragrant bluish- pink, bell shaped six lobed perianthed flowers. Commonly grown as early spring ornamental. Varieties with dull white, light pink and sky blue flowers are grown in most of the gardens. Can also be seen growing in shaded,moist situations around some graveyards and shrines (Fig. H 7).

Hydrangea arborescens L. var. grandiflora (Hydrangeaceae): Eng.- Hydrangea [NI] A rounded, deciduous shrub grown for its foliage and flowers. Leaves long stalked ovate, elliptic or broad ovate acute to acuminate serrate, dark green above and paler, tomentose beneath. Flowers greenish white on branched inflorescences /corymbs.

Hydrangea aspera D. Don. (Hydrangeaceae): [NI]. An upright, deciduous shrub, up to 3 m tall, with large, lance-shaped to narrowly ovate darkgreen leaves. Flowers borne on flattened corymbs, blue to purple fertile flowers, surrounded by white, sometimes pink mauve-tinged sterile flowers.

Hydrangea macrophylla (Thunb.) Ser. (Hydrangeaceae): Eng.- Common Hydrangea. [NI]. A rounded, deciduous shrub with broadly ovate, coarsely serrate leaves. Flowers are borne on flattened corymbs with few pink sterile flowers and numerous pink or blue fertile flowers. Commonly grown varieties include G. *V.de Vibraya* (rose purple flowers) *maculate* (leaves with creamy white margins) and *M.E. Monillere* (serrated sepals with pink or blue eyes).

Hyoscyamus niger L. (Solanaceae): Eng.- Henbane; Hindi- Khurasani Ajvayan; Kash.-Bazar bang; Ladakhi- Phagun /Gyalamtang [W]. A pubescent herb occurring in waste lands in temprate Himalaya from Kashmir to Garwal at 2000-2500 m altitudes. Stem and and leaves coarsely hairy, sticky. Capsule enclosed in persistent globular base of enlarged calyx. Leaves/seeds are a source of drug used as sedative, narcotic and also in the treatment of bleeding gums, asthama, wooping cough and in reducing inflammations and pain of joints (Fig. H 8).

Hyoscyamus pusillus L. (Solanaceae): Eng.- Henbane; Hindi- Khurasani Ajvayan; Ladakhi- Phagun /Gyalamtang [W]. A pubescent herb of Ladakh; leaves petioled, ovate, sinuate toothed. Calyx teeth mucronate. Leaves used for joint pain and as sedative.

Hypecoum leptocarpum Hook. f. & Thomson. (Papaveraceae): Ladakhi– Parpapata [W]. Annual herb of mountain slopes of Ladakh. Flowers purple. Leaves ablanceolate, pinnate. Flowers white or pale- lavender, apically green. Flowering stems few to many, dichotomously branched. Decoction of the leaves is taken as stomachic.

Hypericum hookerianum Wight. & Arn. (Hypericaceae): Eng.- St. John's Wort. [NI] An evergreen or semi-evergreen ornamental shrub with erect branches. Leaves oval- oblong, rounded, acuminate, leathery, dark and bluish-green above, lighter beneath with reddish mid vein. Flowers bright yellow, deep cuped, borne in terminal cymes.

Hypericum patulum Thunb. (Hypericaceae): Eng.- Saint John's Wort; Kash.- Bassant. (Hypericaceae) [W/UC]. A bushy, evergreen or semi-evergreen shrub with spreading branches. Leaves often 2 ranked, ovate to oblong lanceolate bright green light bluish to whitish green beneath with short petiole. Flowers solitary or in few flowered terminal cymes, golden yellow. Leaves used for treating wounds.

Hypericum perforatum L. (Hypericaceae): Eng.-Perforate Saint John's wort; Hindi Bassant; Kash.- Chai-gause [W]. A herb found throughout temperate western Himalaya from 1800-2700 m amsl. Flowers yellow with small black dots on petals. Essential oil obtained from flowers is used externally for sores, wounds, ulcers, swelling and for rheumatism. Leaves have been used as substitute to tea leaves (Fig. H 9).

Fig. H1. Hedera nepalensis

Fig. H2. Helianthus tuberosus

Fig. H3. Heracleum candicans

Fig. H4. Herminium monorchis

Fig. H5. Hordeum aegiceras

Fig. H6. Hordeum vulgare

Fig. H7. Hyacinthus orientalis

Fig. H8. Hyoscymus niger

Fig. H9. Hypericum perforatum

I

Iberis amara L. (Brassicaceae): Eng.- Candy Tuft [NI]. Half hardy annual ornamental herb. Leaves light green, lanceolate. Flowers borne in corymbs of light fragrant flowers, ranging in colour from white through pink to red purple.

Ilex aquifolium L. (Aquifoliaceae): Eng.- Common Holly [NI]. A pyramidal or oblong, evergreen ornamental shrub or tree with grey bark. Leaves leathery tough, dark green and glossy above, light-green beneath, ovate to oblong margins undulate, coarse thorny toothed. Flowers dioecious, arranged in the leaf axils, usually in small cymes on male plants, solitary on the female; corolla white. Fruit an ovate drupe with coral-red pod.

Impatiens balsamina L. (Balsaminaceae): Eng.- Garden Balsam; Hindi- Gul Mehendi; Kash.- Maenze Poshe [NI]. An annual ornamental herb, 20-70 cm tall, with thick soft stem, grown in gardens. Leaves spirally arranged. Flowers pink, lilac or red.

Impatiens brachycentra Kar.& Kir (Balsaminaceae): Eng.- Spurless Balsam; Ladakhi- Patiksketse [W]. A tall glabrous annual herb common on humus rich soils along shady slopes at 1700-2700 m. Flowers spurless, smaller than other species and white. The extract of the seeds is used against infections of eyes and ear.

Impatiens glandulifera Royle (syn. Impatiens roylei (Walp.) Ser.) (Balsaminaceae): Eng.- Himalayan Balsam; Ladakhi- Ganglitz / Mava; Kash.- Trul [W] Robust, tall herb, found growing along moist places, native to W Himalayas. Leaves oblong, serrated. Stem redish, translucent and succulent. Flowers large, pink. Fresh plants are fed to ponies before and after long journies to give them strength and stamina (Fig. I 1).

Impatiens parviflora DC (Balsaminaceae) [W] Impatiens thomsonii Hk. F. (Balsaminaceae) [W]. Erect glabrous herb of Ladakh, inhabiting moist humus rich soils. Leaves narrowly ovate, acuminate, crenate-serrate. Flowers long, reddish pink. Used in *amchi* system of medicine.

Impatiens themsonii HK.F. (Balsaminaceae): Eng.- Thomson's Balsam [W]. Erect glabrous herb of Ladakh, inhabiting moist hamus rich soils. Leaves narrowly ovate, acuminate, crenate-serrate. Flower long, redish-pink. Used in amchi system of medicine.

Indigofera heterantha Brandis (syn. Indigofera gerardiana Baker) (Fabaceae): Eng.- Hiary Indigo; Kash.- Kotz /Kautche; Ladakhi- Keiche [W]. A silvery pubescent small shrub of alpine slopes, with branchlets *argento-canescent.* Leaves short stalked, leaflets opposite, ovate-oblong, pale green, tiny pubescent above and *argento-canescent* beneath. Flowers borne on racemes, calyx obliquely campanulate, petals pale-red or purple. Occurs wild as deciduous shrub on dry open slopes, also in forests. Twigs of the plant used for making baskets and *Kangris* (Fig. I 2).

Indigofera himalayensis Ali (Fabaceae): Eng.- Himalayan Indigo; Kash.- Kaetche [W]. A deciduous shrub, found on forest slopes. Leaves less canescent greenish, pinnate. Flowers purplish-pink borne in racemes. Ovary dark purple, corolla, calyx teeth exceeding the cup, the standard larger and broader. Branches and twigs used for making baskets and *Kangris* (Fig. I 3).

Inula obtusifolia Kerner (Asteraceae): Ladakhi – Minchen-nakpo [W]. Perennial herb, common on rocky cliffs. Flowes yellow. Powdered seeds and shoots are used to cure fevers.

Inula racemosa Hook. f. (Asteraceae): Eng.- Spikenard; Hindi – Poshkar; Kash.- Poshkar mool. [W/UC]. A tall annual herb with woody rootstock, of alpine W. Himalaya growing at 1500-3500 m amsl. Flower heads yellow.Dried roots are used to protect valuable garments from insect attack. Flowers and roots are used in medicines as diuretic, expectorant, to cure heart, liver and spleen ailments and in preventing hairfall (Fig. I 4).

Inula rhizocephala Schrenk var. rhizocephaloides (Asteraceae):Ladakhi- Minchienkarpo [W]. A stemless perennial herb of alpine meadowlands. Leaves radical, rosulate and sessile. Flower heads sessile, light yellow in central clusters. Powdered flowers used to stop dysentery.

Inula royleana DC. (Asteraceae): Eng.- Himalayan Elecampane; Kash.- Poshkar/ Zahelniilkohee [W]. Perennial stout, unbranched herb (upto 65 cm tall), usually on exposed dry slopes at 3000- 3500 m. Leaves large, elliptic- lance shaped. Heads large, golden yellow. A paste of the dried shoots used to cure dermatitis. Roots used to kill lice, fleas and ticks.

Ipomoea batatas (L.) Lamk. (syn. Convolvulus batatas L.; Batatas edulis (Thunb.) Choisy (Convolvulaceae): Eng.- Sweet Potato; Hindi – Shakar Kand, Mitha aloo [NC]. A diffusely spreading, tuberous perennial herb cultivated for its edible roots.

Ipomoea eriocarpa R. Br. (syn. Ipomoea hispidus Vahl.; Convolvulus hispidus Vahl.) (Convolvulaceae): Eng.- Morning Glory; Hindi- Boota [W/UC]. A robust, twinning, hairy annual,growing in grasslands, pathways and also in gardens for its flowers. Leaves covered with prostrate hairs. Flowers short stalked, pinkish, bell shaped.

Ipomoea nil (L.) Roth. (syn. Convolvulus nil L.) (Convolvulaceae): Hindi- Kala-dana [W]. A hairy twining and spreading herb inhabiting orchards and grasslands. Flowers purplish. Seeds medicinal and used as purgative.

Ipomoea tricolor Cav. (syn. Ipomoea hookeri G.Don.) (Convolvulaceae): Eng.- Morning Glory; Kash.- Ashqa paechan [NI]. An annual twining ornamental herb grown for its red purple or blue trumpet shaped flowers. The plant flowers freely producing blooms from July to September.

Iris albicans Lange (Iridaceae): Eng.- White Flag Iris /Cemetry Iris; Kash.- Mazar Muand [W]. A perennial rhizomatous herb. Grows occasionally in graveyards, stream banks and grasslands. Flowers from last week of May to middle of June. Rhizome light brown, leaves ensiform striated, pale green; scape solid, terminating in a head of 4-5 flowers; flowers pinkish lavender scented; spathes, green; falls obovate, cuneate, blade off-white, haft bearded with light hairs, standards blade pinkish purple.

Iris crocea Jacq. ex Foster (Iridaceae) [W/UC]. Rootstock a stout rounded rhizome which leaves upright stiff, dark green, ribbed, linear–ensiform; flowers 2-4, yellowish purple; fall blades with bluish-purple veins against a yellowish background; standards bluish yellow or golden-yellow. Grows in gardens, graveyards and hilly slopes.

Iris decora Wall. (syn. Iris nepalensis Wall ex Lindl. (Iridaceae): Eng.- Graceful Himalayan Iris; Hindi Chilochi; Kash.- Sonzal [W]. Perennial, growing in orchards, graveyards and sub-alpine slopes above 1800 m altitude. Flowers pale-pink lavender. Standards are bent outwards and downwards like falls; falls are spreading, obovate, marked with deep redish-purple veins and a wavy yellow central crest. Root powder used against rheumatic pain (Fig. I 5).

Iris ensata Thunb. (syn. Iris graminea Thunb.) (Iridaceae): Hindi– Sosun; Kash.- Krishame; Ladakhi- Krastz [W]. A widely distributed Iris species found growing in alpine, sub-alpine meadows along roadsides, river and canal banks, grasslands, orchards, shady places throughout the

Himalayan region. Rootstock a compact, closely tufted; leaves ensiform, linearly ribbed, with purplish basis, tips acute; scape small bearing a single terminal and occasionally lateral head of two flowers; perianth tube short, scape small; falls vary from pure white (with a few greenish veins) through dark blue to red purple (with white veins). Flowers from last week of April to end of May. Leaves of the plant are used as fodder, for making ropes, mats, snow shoes (*pulhoore*), baskets and for thatching roofs. In recent past the roof thatchings were made from its leaves, making them slippery on which the accumulated snow could slip down easily (Fig. I 6).

Iris germanica L. (Iridaceae): Eng. – Orris / Blue Flag Iris; Hindi– Sosan; Kash.- Mazar Munde [W]. A perennial herb found in Kashmir at 1600-3000 m amsl. A wide spread and completely naturalized sp. seen growing commonly in grasslands, stream banks and in all muslim/ christian cemeteries and graveyards. It is also planted on roof tops and mud walls for soil binding and beatification. Rhizomes stout, brown, trailing over long distances; leaves, ensiform, margins scarious, with purple sheathing bases; terminal head bears two flowers and each of the lateral branches bears one or two flowers; spathes scarious; flower bearded with purple or dark blue falls, standard obovate. Flowers in the month of April. Rhizomes are used in medicines for treating bronchitis, dropsy and liver complaints **(Figs. I 7, 8).**

Iris halophila Pall. (syn. Iris aurea Link.) (Iridaceae): Eng.- Golden Flag Iris [W]. A well adapted Iris species growing in gardens, graveyards and grasslands. Rootstock a slender, dull brown rhizome, branching at short intervals; leaves dark green, ensiform, flowers golden yellow, perianth tube yellow- green (Fig. I9).

Iris hookeriana Foster (Iridaceae): Eng.- Hooker's Iris [W]. Grows in alpine meadows above 3000 m amsl. Rootstock slender, light brown compact rhizomes; leaves yellowish green round tipped. Flowers two per scape, purple in colour, parianth tube green and spotted all over; falls blue purple, rounded, blades bearded with white hairs which are conspicuously orange or yellow tipped; standards of deeper colour than blades. Flowers in June. Rhizomes used for treating swollen joints and frost bites (Fig. I 10).

Iris kashmiriana Baker (Iridaceae): Eng.- Kashmir Iris [W]. Endemic to Kashmir and abundantly growing in all grave yards /cemeteries as well as in grasslands. Rootstock a stout rhizome with prominent leaf scars; leaves, pale green with pointed tip and scarious margins; flowers cream coloured, fragrant; spathes green perianth tube light green; falls obovate, cuneate, blades creamy white, bearded with white hairs; standards white with faintly yellowish veins. Bloom period lasts from second week of April to middle of May (Fig. I 11).

Iris pallida Lam. (Iridaceae): Eng.- Sweet Iris. [W]. Growing in orchards and rocky alpine slopes. Rootstock a stout thick light brown rhizome, leaves ensiform thick and leathery, greyish green; scape greyish green; flowers mauve puple, fragrant, perianth tube long, green; falls spreading, haft mauve purple blade, bearded with hairs having white basis and yellow orange tips; standards pale mauve purple, blade orbicular, keeled, deeply coloured along centre. The plant flowers during May. The dried and powdered rhizomes of the plant are being used by villagers, against rheumatic pain, acidity and diarrhoea (Fig. I 12).

Iris reticulata M. Bierbstein (Iridaceae): Eng.- Netted Iris; Kash.- Pyare [W]. Grows as a specific weed of saffron fields of Kashmir. Its association is believed to provide protection to saffron corms from rodent attacks. Rootstock a slender underground bulb with reticulate outer scales; leaves 2-4, quadrangular; scape long; flowers solitary, enclosed in three lanceolate, clasping spathes of variable size with transparent margins; perianth tube long, falls broad, rich violet with raised orange ridges, and with black blotches on upper surface; standards narrow, oblanceolate. Flowers in March (Fig. I 13).

Iris spuria L. (Iridaceae): Eng.- Spuria Iris [W/UC]. Rootstock a stout, prostrate, creeping rhizome; leaves ribbed, dark green, ensiform with acute tips; scape stout and terminating in a flower head consisting of 2-4 flowers; flowers blue purple or lilac purple; falls bluish purple; standard as long as falls. Growing in grave yards and in some sacred groves. Flowers in the last week of May to middle of June.

Iris tectorum Maxim (Iridaceae): Eng.- Roof Iris [W/UC]. Rootstock a slender, pale buff coloured, dichotomously branched and greatly spreading rhizome; leaves thin, pale green ribbed, flattened; flowers borne in one or two heads, dark purple, spathes green 2-3 flowered, parianth tube brown purple, falls with deep lilac veins, crested, standards obovate, blade lilac; rarely seen growing in grasslands and on *Karewa* / mountain slopes.

Iris variegata L. (syn. Iris flavescens Delile. (Iridaceae): Eng.- Lemon yellow Iris/ Hungarian Iris [W]. Rootstock buff coloured fleshy rhizomes, branching at short intervals; leaves ensiform glaucous; flowers pale yellow, spathes 2-3 flowered; parianth tube cylindrical, rounded, falls pale yellow, beard hairs have orange tips and yellow basis, standards obovate, pale- lemon-yellow with clarit brown veins. Flowers from last week of April to ending May. Grows in orchards and grasslands and also cultivated in gardens as an ornamental.

Iris xiphium L. (Iridaceae): Eng.- Spanish Iris / Small bulbous rooted Iris [NI] Cultivated as garden ornamental for its showy flowers. Rootstock an ovate bulb which remains covered by membranous, dark brown, smooth tunics; leaves channelled, linear, deeply furrowed, spathe green, margins scarious each enclosing, 1-2 flowers; parianth tube indistinct; falls broad with a sub orbicular blade, white to dark blue with yellow or orange streaks, standards oblanceolate. Flowers from Last week of May to middle of June.

Isodon rugosus (Wall. ex Benth.) Codd. (syn. Plectrunthus rugosus Wall. ex. Benth.; Rabdosia rugosa (Wallich ex Benth.) H Hara) (Lamiaceae): Kash.-Soule [W] A much branched, aromatic shrub growing in orchards and wastelands. Stem erect with slender quadric-angular branches, leafy with an indumentums of small stellate dendroid hairs. Leaves broad ovate coarsely toothed conspicuously white or grey –woolly beneath in contrast to green wrinkled upper surface. Flowers fragrant, tiny, white spotted pink. Plant used in the management of rheumatism and hypertension. Leaves also used against abdominal pain and round worm expulsion. The shrub is of great significance in sustaining honey bee flora.

Ixiolirion tataricum (Pall.) Schult. & Schult. f (Ixioliriaceae): Eng.- Lavender mountain Lily [W]. A perennial cormous herb of sub-alpine grasslands. Leaves linear. Inflorescence 3-6 flowered raceme. Flowers deep blue. Corms used in the management of rheumatic pain and hypertension.

Fig. I 1. Impatiens glandulifera

Fig. I 2. Indigofera heterantha

Fig. I 3. Indigofera himalayensis

Fig. I 4. Inula racemosa

Fig. I 5. Iris decora

Fig. I 6. Iris ensata

Fig. I 7. Iris germanica

Fig. I 8. Iris germanica (inhabiting graveyard)

Fig. I 9. Iris halophila

Fig. I 10. Iris hookeriana

Fig. I 11. Iris kashmeriana

Fig. I 12. Iris pallida

Fig. I 13. Iris reticulata

J

Jaeschkea oligosperma Knobl. (Gentianaceae): Ladakhi- Tikta [W]. Annual herb, along scrubby alpine slopes of Ladakh. Leaves lanceolate, blunt. Flowers bluish to reddish purple. The plant is consumed raw as blood purifier.

Jasminum humile L (syn. Jasminum revolutum Sims.) (Oleaceae): Eng.- Yellow Jasmine; Hindi- Chamele; Kash.- Yasmin [NI]. An evergreen or semi evergreen, erect or arching much branched ornamental shrub with green angular twigs. Leaves alternate, with 3-9 leaflets, oval to elliptic, deep green above, lighter beneath, glabrous. Flowers yellow, slightly fragrant, grouped 5-10.

Jasminum nudiflorum Lindl. (Oleaceae): Eng.- Winter flowering Jasmine [NI] A slender, deciduous shrub with pendulous green shoots. Leaves opposite, trifoliate, leaflets oval-oblong, ciliate, deep green; solitary, axillary, bright yellow flowers produced in leaf axils, before the leaves, from late March to late April (Fig. J 1).

Jasminum officinale L. (Oleaceae): Eng.- White Jasmine / Poet's Jasmine; Hindi- Motiya; Kash.- Yasmin [W/UC]. A twining, deciduous climber, with opposite, pinnate, mid green leaves; terminal leaflet long stalked, side leaflets ssessile. Flowers white, fragrant in terminal cymes. Grows wild in sub-alpine woodlands as a climber.The name of the plant has been symbolically used by Kashmiri sufi poets in their poetry (Fig. J 2).

Juglans cinerea L (Juglandaceae): Eng.- Butternut [NI]. A vigorous, spreading tree with grey, deeply furrowed bark. Pinnate aromatic leaves; leaflets 11-19, oblong-lanceolate, appressed, serrate, pubescent on both sides, glandular beneath. Fruit nut, oval oblong, black brown, with 8 rather distinct rough, sharp angled ridges.

Juglans regia L. (Juglandaceae): Eng.- Persian Walnut; Hindi- Akhrot; Kash.- Doon; Ladakhi- Starga [WC]. A large deciduous tree with aromatic leaves and bark cultivated throughout the valley, upto 2500m altitude, for its fruit. Leaves imparipinnate, leaflets 5-9, the terminal largest ovate-lanceolate. Male spikes lateral often superposed to one leaf-scar, bracts stipitate, stamens 10-20, apiculate, female flowers 1-3, sessile, in a short terminal spike. Drupe subglobose, green, pericarp leathery, aromatic; nuts 2 valued, 2-4- celled. Two sub species commonly cultivated for wood, fruit and as avenue trees are *kumaonica* (fruit smaller and hard) and *duelouxianica* (soft shelled and edible). Wood of the plant used for furniture and wood carving works. Male catkins of the plant are cooked as vegetable in villages and especially blended with fish. Bark peelings used for cleaning teeth. Extract of fruit epicarp used to colour wool. In Ladakhi system of medicine the roasted kernel are given with tea to cure constipation.Bark in the form of powder is used as tooth powder. Plant has also a religious significance as its nuts are used by Hindus in their religious ceremonies.Varieties under commercial cultivation include:

- **Kagzi Doon:** Trees large bearing very thin shelled nuts; kernel sweet and of finest quality.
- **Burzol Doon:** Trees large bearing medium thin shelled nuts; kernels sweet.
- **Wonth Doon:** Trees very large size and form large sized shady trees on foothills and mountain slopes. Fruits smaller and hard to break. Bark silvery grey.
- **Hamdan:**The variety has been developed by SKUAST-K through selection from local seedling origin germplasm. The tree is semi-dwarf with spreading nature, protandrous, intermediate female flower abundance; nut weight 14 g with 54% shelling.
- **Suliaman:**The variety developed by SKUAST Kashmir after selection from local germplasm of seedling origin. The tree is semi- dwarf and spreading in nature, protandrous with high female flower abundance. Nut weight more than 21gm with 52% shelling. Nuts are round with light coloured kernels. Average nut yield is 36kg/ tree (4,560 kg ha-1).

Juncus inflexus L. (Juncaceae): Eng.- Blue Rush /Blue Mohawk [W]. A perennial rhizomatous and tufted herb, inhabiting wetlands and ditches. Stem cylendrical, blue-green. Stalks sheathed.Tepals redish-brown, lance-shaped.

Juniperus communis var. sexatilis Pall. (Cupressaceae): Eng.–Common Juniper; Hindi – Hauber; Ladakhi.- Shukpa [W]. Evergreen coniferous shrub growing in forest areas above 3000m. Leaves needle like with a broad bluish white band above. Seed cones are berry like, purple- black with a blue woody coating at maturity. Powdered fruits are believed to act as carminative, diuretic and stimulant. In Ladakh it is used as incense during worship and has a religious significance.

Juniperus horizontalis Moench. (Cupressaceae): Eng.- Prostrate Juniper [NI] A dwarf or prostrate shrub with long, sometimes procumbent branches with green foliage, scale-like adult leaves with prominent gland on the back. Bears ovoid dark blue fruit. Grown as an ornamental.

Juniperus indica Bertol (syn. Juniperus wallichiana Hook.f. Thomson ex Parl.) (Cupressaceae): Eng.– Himalayan Black Juniper; Hindi – Hauber; Ladakhi.- Shukpa [W]. A spreading, gregarious shrub to small, ovoid or columnar tree distributed in N-W Himalayas above 2500 m. Leaves needle-like sharp-pointed with a broad bluish white bands on the inner faces, borne in whorls of three; male cones ovoid, female cone blue-black when ripe. In Ladakhi system of medicine the berries are believed to be tonic, carminative, anthelmintic and stimulant for treating diseases of spleen and abdomen. The powdered leaves and fruits are given with milk as blood purifier (Fig. J 3).

Juniperus macropoda Sm. (Cupressaceae): Eng.- Pashtun Juniper; Ladakhi- Shukpa [W]. A tree upto 18 m tall with light-brown bark, open foliate with spreading, sharp pointed leaves on lower branches. Ripe fruits are bluish-black resinous. Extract of the fresh seeds along with seed extract of *Polygonum hydropiper* is used as a diuretic. Grows wild in alpine woodlands above 2400 m altitudes. Wood used as fuel in Ladakh.

Juniperus recurva Buch- Ham. ex D. Don (Cupressaceae): Eng.- Drooping Himalayan Juniper; Hindi- Bettar [W /UC]. A low spreading alpine shrub or broadly columnar tree of Himalayas with smooth, orange-brown bark which peels in strips and is fissured in old trees; branches tail-like and curving separately in various directions. Leaves awl shaped green or greyish-green. Fruit purplish-brown to black, shining when ripe, ovoid containing a single seed. Grows in alpine woodlands above 3000 m. Also grown in gardens for its drooping foliage.

Juniperus squamata var. meyeri Rehder. (Cupressaceae): Eng.- Scaly leaved Nepal Juniper [W/UC]. A popular coniferous evergreen shrub and easily recognized juniper of semi-erect habit with flaky bark, stout ascending and angular branches and with arching and nodding shoot tips. Leaves awl shaped, incurved in whorls of 3; fruit ellipsoid, reddish–brown to black, single seeded. Plant native to Himalayan woodlands above 3000 m. Also grown as ornamental.

Jurinea ceratocarpa (Dcne.) Benth. & Hook. f. (syn. Jurinea ceratocarpa Benth. var. depressa Clarke ex Hook.f) (Asteraceae): Ladakhi- Turzit [W]. Small clump forming pernnial herb growing in high altitude rocky slopes of Ladakh at 3000 – 4500 m altitudes. Powdered roots are used against constipation, lumber pain and renal colic. Flower long-stalked, purplish pink.

Jurinea dolomiaea Boiss (syn. Jurinea macrocephala (Royle) C B Clarke (Asteraceae): Kash.- Dhupe [W]. Perennial clump forming, prostrate herb, with a rosette of leaves radiating from a tap root; growing on stony alpine slopes at 3000- 3800 m. Leaves white woolly beneath. Flower heads, lilac pink, enclosed in clustered leaves in centre. An extract from the roots is used as an incense. Bruised roots are applied to eruptions. Decoction of roots given in colics, also applied on affected parts in gout and rheumatism (Fig. J 4).

Fig. J 1. Jasminum nudiflorum

Fig. J 3. Juniperus indica

Fig. J 2. Jasminum officinale

Fig. J 4. Jurinea macrocephala

K

Kerria japonica var. pleniflora Witte. (Rosaceae): Eng.- Jew's Mallow / Japanese Rose [NI]. A deciduous ornamental shrub with vigorous and upright green shoots; alternate, ovate, pointed, sharply toothed, bright green leaves; produces very attractive, solitary, pompon like, double yellow flowers in April.

Kniphofia marungensis Lisowski & Wiland. (syn. Kniphofia modesta Baker(Xanthorrhoeaceae): Eng.- Torch Lily [NI]. A hardy perennial herbaceous ornamental used mostly as a border floweing plant. The smooth flower stems arise about the foliage and produce poker like spikes of closely set tabular flowers which are yellow tiped red. Flowering from July to September.

Koeleria crassipes Lange. (syn. Koeleria cristata (L.) Bertol.) (Poaceae): Eng.- Crested Hair Grass / Prairie June Grass [W]. A particularly common tufted bunch grass growing on open alpine slopes of North west Himalaya and provides fodder in these inhospitable areas. The seed heads are narrow, having contracted panicles with dense spikes (Fig. K 1).

Koeleria permollis Steud. (Poaceae): Eng.- Hair Grass [W]. An annual grass species of grasslands and waste places, of good fodder value.

Koelpinia linearis Pall. (Asteraceae): Eng.- Koelpinia; Ladakhi- Vodar [W]. Low, weak branched annual, upto 20-35 cm tall on open dry alpine/ sub-alpine slopes. Stem solitary, slender, branched from base, hairless. Flower heads with 5-9 pale-yellow florets. Plant used for treating rheumatism (Fig. K 2).

Koelreuteria paniculata Laxm. (Sapindaceae): Eng.- Pride of India / China Tree [NI] A spreading deciduous tree with pinnate leaves each with ovate-oblong, scalloped leaflets. Flowers are yellow produced in large pyramidal panicles; followed by bladder-like pink-or red-flushed capsules. Grown for its handsome foliage and star shaped flowers.

Krascheninnikovia ceratoides (L.) Gueldesnt. (Amaranthaceae): Eng.- Pamirian Winterfat; Ladakhi- Gabsan [W]. A shrub, 30- 100 cm tall, hairy with dendroid hairs growing along alpine pastures. Flowers minute in axillary clusters. Fruit obovoid, horned and covered with apressed hairs. Decoction of leaves is given against hyperacidity (Fig. K 3).

Fig. K1. Koeleria crassipes

Fig. K2. Koelpinia linearis

Fig. K3. Krascheninnikovia ceratoides

L

Laburnum anagyroides Med. (Fabaceae): Eng.- Golden Rain Tree [NI]. An upright, small sized ornamental tree with hairy, grey-green young shoots. Leaves dark- green above and grey-green beneath, having 3-elliptic-obovate leaflets. In late April to early May, produces bright yellow flowers in dense racemes.

Lactuca dissecta D Don. (Asteraceae): Eng.– Milk Thistle; Kash.- Dodije [W]. Annual herb, usually a weed of cultivation, also common on dry scrubby slopes. Stem solitary, erect, simple and glabrous. Flower heads yellowish. Young shoots and leaves used as vegetable.

Lactuca dolichophylla Kitam. (Asteraceae): Eng. –Milk Thistle [W]. Tall perennial, glabrous herb inhabiting mountain slopes of Ladakh; leaves long, lanceolate- linear, essile. Heads small, peduncled. Leaves used as vegetable.

Lactuca lessertiana (Wall. ex DC) Wall ex C B Clarke (Asteraceae):Eng.- Lessert,s lettuce; Ladakhi- Chhuasseo [W]. A herb of alpine grasslands, with stout rootstock. Leaves inverted lanceshaped, entire or toothed. Flower heads blue or purple. Leaves are sour and are used after boiling for renal colic.

Lactuca orientalis (Boiss.) Boiss (Asteraceae): Eng.- Splitleaf Lettuce; Kash.- Dodije; Ladakhi– Tchatis [W]. A biennial sub-shrub, common in alpine grasslands. Heads with 4-5 yellow ray florets only. In Ladakhi system of medicine paste of the powdered heads applied on foreheads to relieve headaches. Fried sliced roots are used as a substitute to potato chips (Fig. L 1).

Lactuca sativa L. (syn. Lactuca scariola L. var sativa Moris) (Asteraceae):

Eng. -Lettuce; Kash Panjaub Palakh; Ladakhi- Dums [WC]. An erect, glabrous herbaceous biennial grown for its crisp edible radical leaves. Among the varieties under cultivation include *Capitata* and *Crispa. Amchis* give its crushed leaves for promoting apetite.

Lactuca serriola L. (syn. Lactuca scariola L.) (Asteraceae): Eng.- Milk Thistle/ Prickly lettuce; Kash.- Kande Dodije [W]. A leafy biennial herb, usually on orchards as a weed. Leaves oblong or lanceolate, waxy gray-green with fine spines along veins and margins. Flower heads yellow. Dried seeds used as sedative. Leaves have milky sap and used for cough and abdominal pain (Fig. L 2).

Lactuca tatarica var. tibetica Hook. f. (Asteraceae): Eng.- Blue Lettuce; Ladakhi- Dumstsez / Tarnu [W]. Beinnial erect herb common in sub-alpine forest areas. Flower heads blue. The decoction of the fresh leaves is used against hyperacidity and as stomachic.

Lagenaria siceraria (Molina) Standley (syn. Cucurbita siceraria Molina) (Cucurbitaceae): Eng.– Bottle Gourd; Hindi– Loki; Kash.- Kashir Aale [WC] A common cucurbit grown for its bottle or horn shaped fruits. Fruits long, pale- greenish in colour. Large quantities of fruits are also sundried after slicing for subsequent use during winters. Commonly cultivated varieties under cultivation- include *Kashmiri local, Pusa Prolific Long* and *Arka Long.*

Lagerstroemia indica L. (Lythraceae): Eng.- Crape Myrtle; Hindi-Phurash [NI] An upright, deciduous ornamental tree or large shrub with peeling, grey and brown bark. Leaves short-petioled, obovate to oblong, dark green. Flowers pink, white or purple. Fruit a capsule.

Lagotis cashmeriana (Royle.) Rupr. (Scrophulariaceae): Eng.- Kashmir Lagotis [W] A perennial herb, 15 cm high, on moist alpine slopes. Leaves small, cuneate, toothed. Flowers dark-blue, tubular borne on dense cylindrical spikes. Roots used for fever, and dyspepsia (Fig. L 3).

Lamium album L. (Lamiaceae): Eng.- White Dead Nettle: Kash.- Baeri-gausse [W] A perennial glabrous herb of alpine/sub-alpine areas, usually among scrubs on humus rich soils, shady places. Leaves nettle like, triangular, softly hairy with serrated margins. Flowers white; used as mild astringent, haemostatic and hyptonic in bleeding piles (Fig. L 4).

Lamium amplexicaule L. (Lamiaceae): Eng.-Dead Nettle [W]. An annual herb, 10-30 cm tall, with soft hairy stems. Leaves opposite with lobed margins, gasping the stem. Plant used as stimulant and antirheumatic.

Lancea tibetica Hook. f. & Thomson. (Phrymaceae): Ladakhi– Depgul/ Tukche [W] Small rhizomatous mat forming herb of alpine meadow lands. Leaves rosulate, sub-leathery. Flowers dark-blue. In Ladakhi system of medicine roots, after roasting, are used as a narcotic to activate a person. Extract of the shoots is used orally against fevers.

Lapsana communis L. (Asteraceae) Eng.- Nipple Wort [W]. Annual herb with branching hairy stem,usually in forest shade. Heads yellow.Leaves alternate and spirally arranged. Latex of the plant used for treating ulcers and for nipple soothing.

Lathyrus aphaca L. (Fabaceae): Eng.- Wild Vetch / Yellow Vetchling; Hindi- Jungli matter; Kash.- Kaetche Karae [W]. A wild herb, occurring as a weed in cultivated fields and grasslands of Kashmir upto 2000 m altitude; flowers greenish white, pods and seeds smaller and brown. Wild pea types are also found growing in protected areas and sacred groves (Fig. L 5).

Lathyrus humilis (Ser.) Spreng. (syn. Lathyrus altaicus Ledeb.) (Fabaceae): Eng.- Vetchling; Kash.- Morkan [W]. Perennial glabrous herb on alpine forest slopes. Flowers purple. An excellent fodder plant.

Lathyrus odoratus L. (Fabaceae): Eng.- Sweet Pea. [NI]. A herbaceous annual grown as an ornamental for its pink and purple flowers. Leaves have two ovoate leaflets the stalk ending in a tendril. The flower appear from June to September.

Lathyrus pratensis L. (Fabaceae): Eng.- Black Pea /Meadow Vetchling [W]. Perennial herb of, meadowlands and grasslands; with runner like rhizomes,unwinged erect stem. Leaves alternate, short stalked, with large stipules; leaf blade pinnate with a single pair of broad lanceolate leaflets. Flowers yellow.A fodder plant.

Lathyrus sativus L. (Fabaceae): Eng.- Common Vetch /Grass Pea; Hindi- Khesari; Kash.- Krehne Karae [RC]. An annual or biennial herb 60-90 cm high grown as minor pulse crop. Also found in cultivated fields as an occasional weed. Leaves pinnate with a terminal tendril leaflets, 4-8 pairs, obong to oblanceolate, truncate, apiculate; fls sessile reddish blue/purplish, solitary or paired in the leaf axils; pods 8-10 seeded; seeds brownish and mottled. It is an excellent fodder plant for sheep. Seeds sometimes used as a pulse especially in villages.

Lathyrus sphaericus Retz. (Fabaceae): Eng.- Wild Vetch; Hindi– Jungli Mattar; Kash.- Kaetch Karae [W]. Annual climbing herb, growing among scrubs. Flowers purple. Seeds chocolate brown, small and edible.

Launaea tibetica Hook. f. & Thomson (Asteraceae): Ladakhi- Kara-roche / Riakse [W]. An annual herb inhabiting cold arid habitats. Seeds after boiling, used to increase apetite.

Laurus nobilis L. (Lauraceae): Eng.- Sweetbay; Hindi- Hab-el ghar [NI]. A conical tree or large evergreen shrub with aromatic, narrowly ovate glossy dark green leaves. Bears axillary clusters of greenish-yellow flowers producing broadly ovoid black berries. Two varieties namely *Angustifolia* and *Crispa* are recognized on the basis of foliage, size and shape.

Lavandula angustifolia Mill. (syn. Lavandula officinalis Chaix) (Lamiaceae): Eng.- Lavander [NI]. A compact, bushy shrub with linear to narrowly lanceolate, obtuse sessile, entire, silver- green leaves.

In late June to late July, long, unbranched stalks produce fragrant, pale to deep purple flowers in dense spikes. Cultivated for its fragrant flowers and aromatic foliage.

Lavatera kashmiriana Mast. (Malvaceae): Eng.- Wild Holyhock; Kash.- Van sotchal /Jungli saze-posh [W]. A perennial puberulous herb,growing on humus rich and open slopes. Flowers bluish pink. Leaves round heart shaped, used as vegetable. Root extract used for urinary irritation and throat problems. Flowers used as anti-phlogistic (Fig. L 6).

Lemna gibba L. (Lemna polyrhiza L.) (Araceae): Eng.- Duck weed; Kash.-Boude mangole [W]. Floating thalloid herb of water bodies. Flowers greenish. Fed to horses and cows.

Lemna minor L. (Araceae): Eng.- Duck weed; Hindi- Bur; Kash.- Loukote Mangoule [W]. Floating, freshwater thalloid herb, usually in rice swamps and shallow ponds with 2-3 leaves, each with a single root hanging in the water. Flowers greenish. Thullus fed to ducks and horses.

Lens culinaris Medik. (syn. Lens esculenta Moench.) (Fabaceae): Eng.- Lentils; Hindi – Masoor; Kash.- Musor [RC]. A partially twinning herb grown for its seeds. Common cultivated varieties include:

- **Kashmiri Masoor:**A herbaceaous annual bushy with light green foliage. Leaves pinnately compound. Flowers pinkish on axillary peduncles; pods short, two seeded. Seeds smaller, light brown in colour and lens shaped.
- **Shalimar Masoor-1:** The variety developed and released by SKUAST-K through single plant selection from exotic genetic resource (EC-2216). It is suitable for cultivation under rainfed marginal *Karewa* lands of Kashmir upto an altitude of 1650 m amsl.

The plant type is erect with lush green foliage, white flower and amber pink seed colour. The variety exhitibs medium maturity (200 -205 days) and resistance to root rot, leaf spot and pod blight. The seeds are bold (*macrosperma* type) with light pink, stain and pink cotyledons. The protein content ranges from 20-21% and average seed yield potential is about 8 q ha -1.

Leontopodium himalayanum DC (Asteraceae): Eng.- Himalayan Edelweiss [W]. A clustered perennial growing in alpine meadows above 3000m altitude with densely haired stems. Leaves linear, white, woolly. Flower heads white borne in globular dense flat topped clusters (Fig. L 7).

Leontopodium leontopodium (D C) Hand. -Mazz. (Asteraceae): Eng.- Edelweiss; Ladakhi- Tscha [W]. Tufted perennial on open alpine slopes. Stem and leaf lower surface woolly. Heads are in clusters covered by leafy bracts. Powdered aerial parts, as poultice, are used on cuts and wounds as antiseptic.

Leontopodium nanum (Hook. f. & Thomson. ex Hook. f. Thomson) Hand. -Mazz. (Asteraceae): Eng.- Edelweiss; Ladakhi- Tzigma [W]. Perennial stemless, densely tufted, woolly perennial herb growing in high altitude alpine pastures above 3500 m. Capitula solitary. Powdered roots used to cure dysentery (Fig. L 8).

Leontopodium nivale subsp. alpinum (Cass.) Greuter. (syn. Lentopodium alpinum Colm. ex Cass.) (Asteraceae): Ladakhi- Apanktsa [W]. A perennial hairy herb inhabiting alpine slopes, preferring rocky limestone places at about 1800 -3000 m altitudes. Leaves and flowers tomentose. Each flowering bloom consists of 5 to 6 small yellow clustered spikelet florets surrounded by fuzzy white bracts in a double star formation. Root extract used for abdominal problems, coughs, fevers and general weakness.

Leonurus cardiaca L. (syn. Cardiaca vulgaris Moench.) (Lamiaceae): Eng.-Common Motherwort; Hindi.-Halkusha; Kash.- Guma [W]. Erect stout perennial herb, upto 100 cm tall,, alpine sub- alpine, usually on hums rich and moist soils. Leaves opposite, with serrated margins, palmately lobed. Flowers densely villous, whitish. Dried leaves and flower tops used as diaphoretic, stamachic, tonic and antispasmodic.

Lepidium capitatum Hook. f. & Thomson. (Brassicaceae): Kash.– Crutch; Ladakhi- Chewlapu [W]. An annual, sub-erect branched herb endemic to alpine areas of N-W Himalayas. Racemes of 20-30 small, white flowers, densely congested above. Decoction of seeds and leaves used as an antidote for poisoning.

Lepidium didymum L. (syn. Coronopus didymus (L.) Smith) (Brassicaceae) [W]. Eng.- Lesser Swine Cress [W]. Annual herb of wastelands. Stem repent, decumbent or ascending, multiple from the base, radiating from the central part. Leaves pinnate and alternate. Flowers in small compact racemes, white inconspicous. Seeds medicinal.

Lepidium latifolium L. (Brassicaceae): Eng.- Broadleaved Pepper Weed; Kash. –Crutch; Ladakhi- Seoji [W]. A herbaceous perennial, about 100 cm tall, growing in abandoned areas. Stems many, woody with alternating waxy leaves and cluster of white flowers. Pasted leaves applied to relieve rheumatism. Seeds used for joint pain and wounds. Leaves used as vegetable.

Lepidium virginicum var. kashmericum Naqshi & Javed (Brassicaceae): Eng.- Verginea Pepperweed [W]. A herbaceous perennial. Leaves on the stem are sessile. Flowers in racemes coming from plants high branched stem. Flowers white. Fruits greenish. All plant parts have peppery taste.

Lespedeza juncea var. sericea (Thunb.) Lace & Hauech (syn. Lespedeza cuneata (Dum. Cours.) G. Don. (Fabaceae): Eng.- Perennial Lespedeza; Kash.- Ringa jac Gasse [W]. Deciduous, upright, semi-woody undershrub, common in moist shaded places and grasslands. Leaves thin, alternate, abundant and three parted. Flowers in clusters of 3-4 small, creamy white with purple throats. A fodder plant.

Leucas cephalotes (Roth.) Spreng. (Lamiaceae): Eng.- Head Leucas [W]. An annual weed of cultivated fields. Stem erect, hairy with spreading and adpressed retrose hairs; leaves narrow, oblong-elliptic, cuneate entire or slightly toothed. Flowers white. Leaves edible and used as vegetable.

Leucas lanata Benth. (Lamiaceae): Eng.- Woolly Leucas [W]. A densely woolly-haired perennial herb found on alpine stony slopes at 1800 – 3000m altitudes. It has dense white lanate tomentum on all parts and spreading stem hairs. Flowers white, small, appearing in leaf axils. Sepal cup tube like. Used in colds and coughs.

Ligularia fischeri (Ladeb.) Turcz (Asteraceae): Eng.- Leopard Plant [W]. A herbaceous perennial plant inhabiting damp localities in alpine grasslands. Stem erect, 80- 160 cm tall. Capitula numerous. Heads are orange-yellow with brown or yellow central disc. Plant used medicinally for its anti-inflamatory properties.

Ligustrum japonicum Thunb. (Oleaceae): Eng.- Japanese Privet; Kash.- Privet [NI] An upright, dense, evergreen ornamental shrub with broadly ovate to oval-oblong, glossy, dark green leaves. White flowers in pyramidal panicles appear in late June to early July, followed by ovoid - oblong black fruit.

Ligustrum lucidum A T Aiton. (Oleaceae): Eng.- Chinese Privet [NI]. A conical, evergreen small tree or large ornamental shrub. Leaves with ovate glossy dark green. Flowers cream white tubular, borne in panicles during August to September.

Ligustrum ovalifolium Hassk. (Oleaceae): Eng.- California Privet [NI]. An upright, deciduous to semi evergreen ornamental shrub with elliptic-oblong, glossy-ovate rich green leaves, with broad, bright yellow margins. Flowers yellowish-white; fruits spherical, shiny black.

Ligustrum vulgare L. (Oleaceae): Eng.- Wild Privet. [NI]. A bushy semi-evergreen or deciduous ornamental shrub, with glabrous, obovate-oblong to lanceolate, dark green leaves. Flowers in pubescent panicles corolla tube white and fragrant.

Lilium henryi Baker (syn. Lilium x holandicum Bergmans) (Liliaceae): Eng.-Lily [NI]. A bulbous perennial herb grown in gardens and parks for its showy flowers which are crimson -red and bell shaped and pendulous. Many hybrid varieties have been introduced and produce flowers of different colors ranging from bronze yellow to golden yellow to light pink.

Lilium polyphyllum D Don. (Liliaceae): Eng.- White Himalayan Lily [W] A white rare lily species growing in Himalayan meadowlands above 2000 m. Flowers large, pendant, fragrant, waxy,bell shaped with lower half of the tepals strongly recurved.Tepals white with many purple spots. Anthers large, yellow to orange. Bulbs long, narrow, white and used as tonic and as a source of energy (Fig. L 9).

Limnanthemum nymphoides (L.) Hoffmans & Link. (syn. Nymphoides peltatum Kuntze. (Menyanthaceae): Eng.- Fairy Water Lilly; Kash.- Bume [W]. Perennial, fibrous rooted, proliferous aquatic herb, growing in Dal Lake. Leaves are floating, ovate, deep green and heart shaped at the base. Flowers yellow borne in the axils from the petioles which are fleshy and brittle and easily snap off. The leaves along with petioles are very much palatable to cattle and serve as good fodder for milking cows. Decoction of leaves and that of *Adiantum venustum* have been used for headaches (Fig. L 10).

Limonium macrorhabdon O. Kuntze (Plumbaginaceae): Ladakhi-Stspakchek [W] An alpine, densely prickly shrub. Flowers rosy. In Ladakh the dried leaves are used in a local narcotic preparations called *Staspakchek.* Leaves also used as vegetable during winters.

Linaria reticulata (Sm.) Desf. Willd (Plantaginaceae): Eng.- Toad Flax [NI]. A hardy annual herbaceous ornamental. leaves pale green. Flowers purple with a orange or yellow blotch on the lower lip, borne in clusters, from May to July.

Lindelofia longifolia (Benth.) Baill. (Boraginaceae): [W]. Erect perennial herb of open slopes. Stem simple. Leaves ovate-lanceolate, lower petioled. Flowers deep blue. Leaves used medicinally.

Lindernia anagallis (Burm. f.) Pennel. (Scrophulariaceae) [W]. Annual herb of wet places, especially in rice fields. Flowers axillary, light purple. Leaves used medicinally for nephrites and for gonorrhoea.

Linum usitatissumum L. (Linaceae): Eng.- Linseed; Hindi- Alsi; Kashmiri- Aliesh [RC]. A biennial, bushy, highly, branched oilseed plant; stem greyish green and flowers in terminal racemes of bluish violet colour. Fruit an indehiscent capsule, globouse in shape with persistent calyx, brown in colour and with pointed acumen; seeds compressed, shining light brown to yellowish with oil content of about 34%. Seed cake is fed to animals especially to milking cows. Oil is exclusively used for edible purposes. The variety is low yielder and yields about 5 to 7 q ha-1. The seeds, in the form of poultice,have also been used for treating ulcers.

Lithospermum arvense L. (Boraginaceae): Eng.- Field Gromwell/ Corn Gromwell [W]. Annual herb of orchards, wastelands. Leaves alternate, lower leaves stalked, upper ones stalkless, ovate- lanceolate, margins hairy. Flowers white. Leaves used medicinally.

Lolium perenne L (Lolium brasilianum Nees.) (Poaceae): Eng.- Perennial rye- grass; Ladakhi- Jamak [W]. A wide spread tufted, hairless grass of North Western Himalayas. Leaves dark-green, smooth and glossy on lower side. This species is very valuable grazing and hay grass.

Lolium temulentum L. (poaceae): Eng.- Darnel [W]. Annual grass of wastelands, orchards and as a weed of cultivation. Inflrescence in ears; flowers green; grains purple. Whole plant medicinal.

Lomatogonium caeruleum Hary Sm. ex Burtt. (Gentianaceae): Ladakhi– Tikta [W] A perennial upto 30 cm tall gowing on open alpine slopes at 3300 – 4000 m altitudes. Flowers violet to mauve purple. Whole plant used as antipyretic against coughs.

Lonicera asperifolia Hook.f. Thomson (Caprifoliaceae): Eng.- Honey Suckle [W]. An erect shrub, up to 1.5 m tall with branches and leaves hispid all over. Leaves ovate or elliptic, sub-acute, base more or less rounded, margins bristly ciliate, with scattered bristly hairs on both surfaces. Flowers tubular, style and stamens exerted. Berries in pairs with peresistant calyx. Plant inhabits alpine forest slopes above 3500 m altitudes.

Lonicera japonica Thunb. (Caprifoliaceae): Eng.- Japanese Honey Suckle [NI]. A vigorous, woody, evergreen or semi-evergreen, twining climber garden hedge with paired ovate dark green leaves. Flowers white long tabular, 2 lipped, very fragrant ageing to yellow. Fruit blue –black berries. Three varieties are common is Kashmir gardens *aurea reticulata* (leaves yellow veined) *halliana* (leaves with golden yellow reticulate venation, pure white fls) and *repens* (purple tinged foliage, fls white and red flushed).

Lonicera ligustrina var yunnanensis Franch (syn. Lonicera nitida E H Wilson) (Caprifoliaceae): Eng.- Box Honey Suckle [NI]. A dense-habited, small-leaved evergreen shrub with paired, ovate-oblong, glossy dark green leaves. Flowers creamy white and mildly fragrant. Fruits glossy, blue purple berries. Grown as a hedge plant in gardens.

Lonicera obovata Royle ex Hook. f. Thomson (Caprifoliaceae): Eng.- Small leaved Honey Suckle [W]. A small leaved, dwarf shrub with prostrate or nearly erect branches, inhabiting sub-alpine Himlayas. Leaves inverted egg shaped or elliptic with a roundish tip and a wedge shaped base. Flowers in pairs, pale yellow, bell shaped. Berries nearly spherical, dark bluish-purple.

Lonicera periclymenum L. (Caprifoliaceae): Eng.- Woodbine Honey Suckle; Kash.- Hey Thaere [NI]. A vigorous species climbing, scrambling or trailing in habit. Leaves oval oblong, narrower, dark-green. Flowers appear in spikes, deep purple outside, turning paler in time, abundantly fragrant. Fruit bright red berries.

Lonicera quinquelocularis Hardw (Caprifoliaceae): Eng.- Himalayan Honey Suckle; Kash.- Pakhur / He Thaere [W]. An upright, large deciduous shrub with spreading branches and reddish young shoots. Leaves broad ovate to elliptic grey-green. Flowers short stalked in axillary pairs, creamy white, later turning yellow. Fruits translucent whitish with violet seeds. Occuring throughout forest slopes of Kashmir Himalaya above 2000m altitude (Fig. L 11).

Lonicera spinosa (Dcne.) Jacq ex Walp. (Caprifoliaceae): Ladakhi-Tukche [W]. Deciduous, rigid shrub growing on alpine highlands of Ladakh. Corolla purplish red, tubular. Bark splitting. Its twigs are used used as fuel in Ladakh. Fresh crushed leaves are used as poultice on burns and are believed to have antiseptic and healing effects.

Lonicera webbiana Wall ex DC. (syn. Lonicera heterophylla Decne.) (Caprifoliaceae): Eng.- Webb's Honey Suckle [W]. A large erect shrub, up to 3 m tall inhabiting coniferous forest slopes above 1800 m. Leaves elliptic to ovate, glabrous above and pale-green and reticulately veined beneath. Flowers are carried on stout stalks which are glandular velvety. Flowers bilobed, glandular- hairy both inside and outside, strongly gibbous at the base, greenish yellow, diffused deep red at base. Berries ovoid and red.

Lotus corniculatus L. (Fabaceae): Eng.- Bird's Foot / Trefoil; Kash.- Gur-masur [W] A small perennial ascending herb growing on open alpine/ sub-alpine slopes; leaflets 5; flowers yellow.Used as a forage plant (Fig. L 12).

Luffa acutangula (L) Roxb. (syn. Cucumis acutangulus L.) (Cucurbitaceae): Eng.- Ridge Gourd; Hindi – Kali Toria [RC]. A large climber with palmately 5 to 7 angled or lobed leaves, cultivated for its fruit. Fruits cylindrical with 10 prominent longitudinal ridges. Ridge gourd is a popular vegetable in the valley and is eaten when tender.

Luffa cylindrica (L.) M. Roem. (syn. Luffa aegyptica Mill.) (Cucurbitaceae): Eng.- Dish Cloth Gourd; Hindi –Ghia Toria [RC]. A weak climber cultivated for its fruit. Fruit fibrous, smooth and cylindric.

Lupinus polyphyllus Lindl. (Fabaceae): Eng.- Large Lupin / Meadow Lupin; Hindi- Tarmiss [W/UC]. Herbaceous perennial grown for its long spire like flowering racemes of light pink and blue. Leaves digitate. Plants are cultivated as well as growing in wild in alpine grasslands of Gulmarg (Fig. L 13).

Lycopersicon esculentum Mill. (Solanaceae): Eng.- Tomato; Hindi- Tamatar; Kash.- Tamatar [WC] Tomatoes, after having been introduced, are one of the most favourite vegetables in the region. Its fruits are cooked, in fresh form as well as after sun-drying. Common tomato types under cultivation include:

- **Shalimar-1:** A tomato variety bred by State Deptt. Agri. and released for cultivation in 1984 under the name Shalimar-I. Plants are semi determinate and fruit profusely. Fruits are large, round and red, appealingincolourandmaturein102-110days, yielding250-275/qha-1.
- **Shalimar-**2: The variety evolved by State Department of Agriculture, during 1984 and released for cultivation in the State. An early maturing semi dwarf and erect growing, potato leaved plant with high yielding potential at higher elevation. It yields about 200-210q / ha-1.
- **Shalimar Tomato Hybrid -1:** A single cross hybrid between a male sterile and pollen parent (SH-FMS-1 and SH-T-11). The hybrid released by SKUAST-K during 2010 has very high yielding potential of 704 q/ ha, possessing 83% yield increase over Shalimar-1. Plants indeterminate and matures within 70-75 days. Tolerant to early blight, leaf spot and buck eye rot. Long fruiting period.
- **Shalimar Tomato Hybrid-2:** A single cross hybrid released by SKUAST-K during 2010 and developed between a male sterile and pollen parent (SH-FMS-1 and SH-T-IS-1). Its average yield is 649q/ ha with yield superiority 69% over Shalimar-1. Plants indeterminate; mature within 67-75 days and are tolerant to early blight, leaf spot and fruit rot. Long fruiting period.

Other lately introduced tomato varieties include Marglobe, Oxheart, Pusa Ruby, Punjab Chaura, S-2 and several hybrids.

Lycoprsicon esculentum. var. cerasiforme A. Gray (Dun) (Solanaceae): Eng.- Cherry Tomato; Kash.- Kaushre tamatar [WC]. A spreading, much branched, pubescent herb with indeterminate growth and a strong characteristic odour. Leaves greyish green and pinnate. Fruits villose when young, glabrous and shining when mature, small cherry sized to walnut, smooth, spherical, uniformity deep red at maturity. Fruits being thin skinned and less pulpy are liked by the local people owing their taste, flavour and being easily sundried. Leaves of the plant are also cooked as vegetable, mostly by villagers (Fig. L 14).

Lycopodium clavatum L. (Lycopodiaceae): Eng.- Stag's horn Clubmoss [W]. A perennial evergreen spore bearing plant inhabiting woodlands, pastures, moors under moist situations. Stem 1 mm long, much branched and covered densely with small spirally arranged microphyll leaves. Branches bearing strobili turn erect and produce yellow –green spore cones on sporophylls. Spores and whole plant used against fevers, and kidney problems.

Lycopus europaeus L. (Lamiaceae): Eng.- Gypsywort; Hindi- Kasni vahshi; Kash.- Handhi Gausse [W]. Perennial herb, usually along watery situations along sub-alpine areas. Leaves opposite, sessile, hairy with serrated margins. Flowers pale-pink. Fruits have been used as mordant for dying of wool (Fig. L 15).

Lysimachia maritima (L.) Galasso, Banfi & Soldano (syn. Glaux maritima L.) (Primulaceae): Eng.- Sea Milkwort; Ladakhi– Spengcha [W]. A perennial herb, with creeping rootstock, of Ladakh ranges. Leaves opposite, sessile, fleshy, glabrous and bluish green. Flowers campanulate, light red and dark spotted. The plant is fed to goats to increase the quality and quantity of pashmina wool. Its powdered roots are taken as antiseptic and as anthelmintic.

Lythrum salicaria L. (syn. Lythrum cashmerianum Royle.) (Lythraceae): Eng.- Purple Loose strife [W]. Perennial herb of damp situations, often along mountain drains. Flowers bluish purple. Stem redish-purple. Leaves lanceolate, arranged in whorls of three. Whole plant used medicinally, as astringent.

Fig. L 1. Lactuca orientalis

Fig. L 2. Lactuca serriola

Fig. L 3. Lagotis cashmeriana

Fig. L 4. Lamium album

Fig. L 5. Lathyrus aphaca

Fig. L 6. Lavatera kashmeriana

Fig. L 7. Leontopodium himalayanum

Fig. L 8. Leontopodium nanum

Fig. L 9. Lilium polyphyllum

Fig. L 10. Limnanthemum nymphoides

Fig. L 11. Lonicera quinquelocularis

Fig. L 12. Lotus corniculatus

Fig. L 13. Lupinus polyphyllus

Fig. L 14. Lycopersicon esculenturn var cerasiforme

Fig. L 15. Lycopus europaeus

M

Magnolia aquilfolium (Pursh.) Nutt. cv. Orange Flame (Magnoliaceae) [NI]. An open suckering erect shrub. Leaves ovate to elliptic, spiny-toothed, leaflets dark-green changing in colors during seasons. Flowers are yellow borne in densely clustered racemes in late March to mid April.

Magnolia grandiflora L. (Magnoliaceae): Eng.- Southern Magnolia/ Bull bay; Hindi- Anda champa [NI]. The tree is prized for its showy gragrant flowers, fruits and attractive ever-green foliage and called the aristocrat of gardens of temperate regions. The leaves are oval-oblong to obovate glossy, leathery and dark green. The flowers are delightfully fragrant, creamy white and bowl-shaped. Planted in many gardens and lawns of government institutions.

Magnolia kobus DC. (Magnoliaceae) [NI]. A broadly conical small tree or large shrub with broad obovate, aromatic mid-green leaves. Bears saucer-shaped white fragrant flowers, from mid March to early April before the leaves.

Magnolia liliiflora Desr. (syn. Magnolia liliiflora var. nigra (G. Nicholson) Rehdr.) (Magnoliaceae) [NI]. A wide spreading medium sized shrub with obovate or obovate-oblong to nearly oval leaves. Flowers erect, campanulate, dark purple outside but light purple within. Flowers from early April to May.

Magnolia x soulangeana Soul. – Bod. (Magnoliaceae) [NI]. It is one of the most liked deciduous flowering shrub or small tree. The leaves are obovate to elliptic, dark green, tapering to the apex. Flowers are erect, campanulate, interior usually white, exterior deep crimson to rosy-purple, 10 cm across, borne before the young leaves during late March to early April.

Magnolia stellata (Sieb. & Zucc.) Maxim. (Magnoliaceae) [NI]. A charming, slow growing deciduous small tree, forming a compact rounded specimen usually wider than high. Leaves obovate-oblong to inversely lance shaped, dull green. Silky buds open to a star-shaped pure white fragrant flowers before the leaves from mid March to Early April.

Mahonia aquifolium Pursh. (Berberidaceae): Eng.- Oregon Grape [W]. Evergreen shrub of open sub-alpine slopes growing 1 m tall. Leaves pinnate with spiny leaflets. Flowers yellow, in dense clusters, produced in early spring. Berries dark bluish-black.Plant used for fevers and skin problems.

Mahonia japonica (Thunb.) DC. (Berberidaceae): Eng.- Japanese Grape [NI]. An erect ornamental shrub, with upright branches and magnificent dark green pinnate leaves, ovate-oblong to lance shaped leaflets. Produce pendulous racemes of fragrant, lemon-yellow flowers from early to mid April, followed by ovoid blue-purple berries.

Malaxis muscifera (Lindl.) Kuntze. (Orchidaceae): Eng.- Fly bearing Malaxis [W] A terrestrial orchid of Himalayas found growing at 2500-4000 m altitudes. Stem 20-50 cm tall; pseudobulbs ovoid. Leaves are broad, stalkless, paired unequal, elliptic arising from base of the stem. Flowers, borne on a spike, are yellowish-green and small. Used for stomach and intestinal problems and rheumatism.

Malva alcea L. (Malvaceae): Eng.- Cutleaved Mallow; Kash.- Saze Posh. [W/UC] A border perennial growing along paths, waste places on drier soils. Flowers funnel shaped, mauve-pink coloured. Leaves are light green, deeply lobed and toothed, covered with stellate hairs. Flowers used as skin- scalp conditioner.

Malva neglecta Wallr. (Malvaceae): Eng.- Dwarf Mallows; Kash.-Sotchal; Ladakhi- Sheomou [W/UC]. A perennial herb abundantly branched and growing in wastelands. Flowers white; leaves alternate, long stalked stipulate and with toothed margins. Large quantities of leaves are consumed as vegetable with medicinal properties (Fig. M 1).

Malva parviflora L. (Malvaceae): Eng.-Marsh Mallows; Hindi- Panirak; Ladakhi- Khubazi; Kash.- Sotchal [W]. A biennial herb of alpine areas – orchards, vegetable fields. Stem tough and woody upto 100 cm tall. Leaves alternate, palm shaped with shallow lobes. Leaves used as vegetable. Seeds used in cold, coughs and ulcers in bladder.

Malva pusilla Sm. (syn. Malva rotundifolia L.) (Malvaceae): Eng.-Round Leaved Mallow; Hindi- Khubazi; Kash.- Sotchal [W]. A wild biennial herb, usually along roadsides and in waste places. Leaves orbicular, with serrate margins. Leaves are used as vegetable and to treat piles. Seeds are demulcent, used in bronchitis, coughs, ulceration of bladder and haemorrhoids. Flowers white, used as blood purifier.

Malva sylvestris L. (Malvaceae): Eng.- High Mallows; Hindi- Gulkhair; Kash.- Sotchal [W]. An erect biennial herb of fallow fields. Leaves alternate, palmately lobed. Flowers mauve-pink with dark veins. Leaves are rich in mucilage and used as laxative. Plant posseses demulcent, cooling and emollient properties.

Malus baccata (L.) Borkh (syn. Pyrus baccata L.) (Rosaceae): Eng.- Siberian Crab Apple; Hindi- Jangli Seb [W]. A small to medium-sized tree of rounded habit. Leaves ovate, acuminate, fine and scabrous, serrate, glossy above, light green down. Flowers white to creamy-white in small corymbs, calyx glabrous, calyx teeth long acuminate; fruit more or less globose, yellow with a red cheek, calyx abscising.

Malus domestica Borkh. (Pyrus malus L.) (Rosaceae): Eng.- Apple; Hindi – Saebe; Kash.- Choonght [UC]. A short trunked, round crowned tree, more open, twigs usually thornless, young shoots tomentose. Leaves more elliptic to ovate, acute to obtuse. Flowers white, turning pink; calyx lobes acuminate, longer than the corolla cup. Fruit globose, green, indented on both sides.

The commonly grown cultivars of apples in Kashmir are given below and have been grouped on the basis of maturity and ripening time. It is pertinent to state here that at present due to commercial preferances, introduction of new varieties and change in land use pattern only some of them are in cultivated state while majority of them have been marginalized and replaced by high yielding commercial types.

A: Early Season Apple Varieties

- **Hazratbal Saharanpuri (Benoni)** [UC]: Trees medium tall with curved upright branches. Bark dark brown. Lenticels prominent and brownish red. Leaves ovate, serrate, dark- green and upcupped. Flowering occurs from 2nd to 4th week of April and bearing is regular and annual. Fruits medium sized roundish and symmetrical, skin pale greenish, shaded with carmine, red stripped. Flesh yellowish white, juicy, sub- acid and aromatic.
- **Kashir Kirmiz Trel (Sops of Wine) [NC]:** Tree medium to large, branches spreading, upright. Bark dark-brown. Lenticels oblong. Leaves oval crenate. Flowering period commences from Ist to 3rd week of April. Bears annually with light and heavy crops alternatively. Fruits mature by July-August; fruits, round, slightly conic, skin shining, greenish-yellow covered with purplish red and whitish bloom; flesh pinkish white, juicy, aromatic and sub-acid.
- **Laere- Choongth / Safarkand (Early Victoria) [RC]:** Tree vigorous with extensive branching. Bark greyish with scurfy skin.Leaves pale green, coarsely crenate; flowers during Ist to 3rd week of April. Bearing is annual and fruits maturing by July ending. Fruits conical, irregularly conical, yellowish green; flesh greyish white, sweet and juicy.

- **Safed Saharanpuri (White Transparent) [RC]:** Trees medium, spreading and upright. Bark greyish brown, olive green on young branches. Lenticels oval and greyish brown. Leaves ovate, crenate, obovate, green pubescent. Flowering commence from 2nd to 4th week of April. Bearing is annual but light and heavy crop alternately; fruits mature by ending July to August. Fruit conical, skin smooth, pale milky yellow in colour. Flesh yellowish white, juicy and acidic.
- **Shanda Gounde (American Mother)** [RC]: Tree medium tall and spreading with redish- brown bark. Lenticels free raised and roundish. Flowers in last fortnight of April but bearing is biennial. Leaves ovate, light green and serrate. Fruit matures in July/August. Fruit roundish oblong, conic oblate, irregularly elliptical; cavity deep, narrow lipped and rusted; eye closed; basin narrow; skin smooth bright, lemon yellow with brown red flush and streaks. Flesh yellowish -white, juicy, aromatic and mild acid.
- **Sur Saharanpuri (Irish Peach) [UC]:** Medium tall trees, branches spreading and slightly drooping. Bark dull brown with shades of olive green. Lenticels long, raised and greyish brown. Flowers during Ist fortnight of April and bears annually. Leaves ovate, serrate with acute tips dull green. Early maturity. Fruits small sized, roundish, pale yellowish, skipped with carmine with a milky red wash all over. Eyes closed, stem short, flesh yellowish white, juicy, sub-acid.
- **Tontal Saharanpuri (Mr Gladstone) (RC):** Trees medium tall and spreading. Bark dark brown, scaly. Lenticels conspicuous and roundish. Leaves thin, ovate and yellowish green. Flowering during middle of April;bears annually and matures from July to August. Fruits are medium to large, conical, cavity deep and acute, skin greenish yellow covered with dark red stripes flesh greenish white, juicy and sub- acid.

B. Mid Season Apple Varieties

- **Balpuri Trel (Early Julyan) [UC]:** Trees dwarf, stout trunk and spreading. Bark dark grey on old wood; and olive-green on new wood; lenticels round/oblong, appressed, reddish-brown. Flowering is from 2nd to 4th week of April. Bearings is annual with heavy and light crops alternate. Fruits mature by August. Fruit smaller roundish, conical, skin waxy, pale golden yellow, flesh crisp, yellowish, juicy, sub-acid.
- **Bihi Saharanpuri (Dutches of Oldenberg) [RC]:** Trees medium to large, branching dense and slightly drooping. Bark brown with olive green patches. Lenticels long and dark brown. Leaves ovate, crenate, dull green. Flowering period is mid April. Bears annually and fruits maturity by end of August to September. Fruits medium to large, roundish oblate,

pale greenish yellow streaked with red/ crimson shade. Flesh yellowish white, juicy, sub acid.

- **Bombchoongtch Trel (Blushed Caliville) [RC]:** Trees medium, spreading and open branching bark brownish with olive-green patches. Lenticels small, round are raised. Leaves ovate, green, and serrate. Flowering from 2nd to 4th week of April; maturity period-August. Fruits medium to large, conical/roundish; skin light green/ yellowish-green, blushed on sunny sides; flesh crisp, juicy, sub-acid.
- **Bomburi (Summer Golden Pipin) [UC]:** Trees medium, upright and vigorous. Bark greenish brown on old wood and olive-green on new wood. Lenticels scattered, raised and oval. Leaves obovate, serrate, yellowish green, up-cupped with thick pubescent petiole. Flowers from Ist to 3rd week of April. Periodicity is annual but light and heavy bearing alternate. Fruits mature by August. Fruits are oblong/roundish, greenish yellow with occasional orange flush and red stripes; flesh yellowish-white, crisp, juicy, aromatic and highly acidic.
- **Bombori Trel (Paraquet) [RC]:** Trees medium, spreading branches with very short internodes. Bark dark- brown. Lenticels oblong, dark-brown and few. Leaves small, ovate, serrate, light- green with long petioles. Flowers from 3rd week of April and fruits mature by August. Bearing is annual with heavy and light crops alternating. Fruits conical, asymmetrical in size and shape, greenish yellow, with rich crimson red stripes and streaks; flesh yellowish white, juicy, mild sub-acid.
- **Hendwund Trel (Blenhim Pippin) [RC]:** Trees medium, with compact branching. Bark olive greenish- brown, lenticels few, whitish and raised. Leaves obovate, dark-green, serrate with highly pubesent petioles. Flowering occurs in 2nd to 3rd week of April and fruits mature by August- September. Bearing is annual, fruits oblate, greenish yellow-lemon; russet dots conspicuous. Flesh white, tender, sweet and juicy.
- **Hengal Lal (Cramoise- De- Gascogne) [RC]:** Trees medium with spreading and drooping branches. Bark redish brown on old wood while on branches it is olive- greenish. Lenticels round, greyish brown. Leaves ovate, serrate, deep- green. Flowers from 2nd to 4th week of April. Bearings annual with light and heavy crops in the alternate years. Fruits mature by September- October; fruits are medium sized, conic, yellowish- green covered with bright- carmine white. Flesh yellowish white, crisp and sub-acid.
- **Kagzi Khoure (Pine Apple Russet) [RC]:** Trees medium, spreading with open top. Bark redish brown on old wood and obovate– green on young wood. Lenticels oblong/round, redish and greyish. Leaves deep-

green, crenate and upheld. Flowers from 2nd to 4th week of April. Bearing is alternate. Fruits mature during August. Fruits are oblong conic, greenish- brown covered with fine russet; flesh yellowish- white, sub-acid and with good flavour.

- **Kesari Terel (Cox's Orange Pippin) [UC]:** Trees show vigorous growth and spreading. Bark dark brown with olive green tinge. Lenticels prominent and dark green. Leaves ovate, serrate and green. Flowering period is from second to third week of April. Bears annually and matures by September. Fruits medium sized, round to conical, golden yellow, flushed with brownish red stripes. Flesh aromatic, juicy and sub-acid.
- **Malmu (Golden Spice) [RC]:** Trees are large, upright and spreading. Bark redish- brown on old wood and redish green on new wood. Lenticels oval and redish brown. Leaves big, obovate, crenate, yellowish green and slightly hanging. Blooms from Ist week of April and fruits mature by September/October. Periodicity is annual but light and heavy crops alternate. Fruits are round medium sized, yellowish- green, flesh whitish, sub –acid but sweet at full maturity. Used to be cooked and sundried for longer keeping.
- **Noupora (Oline Muri) [RC]:** Trees upright, spreading. Bark brownish mingled with dark-green patches. Lenticels round, dark-brown. Leaves ovate, serrate, dark-green. Flowers from Ist to 3rd week of April. Bearing is annual but light and heavy crops alternate. Fruits mature by September- October. Fruits medium sized, conical/ elliptical, bright yellowish; mottled with pinkish red, marked with many stripes of bright carmine; flesh yellowish- white, juicy, sub-acidic and aromatic.
- **Phokhla (Kerry Pippin) [RC]:** Trees medium, tall with strong and upright trunk. Bark chocolate coloured, white young branches look olive-green. Lenticels few round, greyish brown. Leaves ovate, up-folded, serrate and light green. Flowers from 2nd to 4th week of April. Bearing is biennial and fruits mature during October. Fruits smaller, oblong, creamy- yellow with conspicuous splashs and stripes; flesh aromatic, juicy and mild sub-acid.
- **Rekhal Lal (Calulle Rouge De – Hiver)** [UC]: Trees medium with dense and compact branching. Bark olive green on young wood and brownish red on old wood. Leaves pale- green and serrate. Flowers from Ist to 3rd week of April. Bearing is annual. Maturing period falls from November-December. Fruits medium, conical, greenish yellow covered with dark crimson flush and stripes; flesh crisp white with red shade, juicy and sub-acid (Fig. M2).

- **Saebi- Kalan (French Large Redish) [UC]:** Trees medium sized, and spreading. Bark redish brown with olive green patches. Lenticels round and greyish brown. Leaves ovate, crenate, green slightly upcupped. Flowers during mid April and fruits mature by ending August. Bears annually. Fruits are large sized, round slightly conic, skin faintly bronze blushed with redish brown strips. Flesh aromatic tender, yellowish-white, juicy and sub acid.
- **Safar Kand (Millux Seedling) [RC]:** Trees upright with spreading habit. Bark dark-brown. Lenticels few, elongated. Leaves ovate, serrate and light green. Flowers from 2nd to end of April. Fruiting periodicity is biennial, bearing heavy crops. Fruits mature by the end of August. Fruits, roundish conical, creamy yellow, striped and flushed with bright crimson on exposed side; flesh white, juicy, slightly aromatic and pleasantly sweet.
- **Shahanshah (Emperor) [UC]:** Trees medium with spreading and drooping branches. Bark greyish brown on old wood. Leaves oval, yellowish- green, bi serrate up cupped. Blooms in 2nd to 4th week of April. Bears annually but heavy and light crops alternate. Fruits mature by August. Fruits are large, conic, ribbed, greenish yellow covered with bright strips with deep carmine rays. Flesh greenish-yellow, crisp, juicy, sweet/ sub-acid.
- **Sheesh Khoure (Blenhim Orange) [UC]:** Trees medium to large, spreading. Bark dark redish brown. Lenticels round and brownish. Leaves lanceolate, dark green, sharply serrate with acute tips. Blooms in mid April and bears annually. Fruits mature during August, are medium to large, roundish, skin yellowish with dull pinkish streaks; flesh aromatic yellowish white, sub-acid.
- **Sil Trel (Hislop) [UC].**Trees medium in the stout, extensive branching. Bark redish brown on trunk with olive-green hue on branches. Lenticels oblong and conspicuous. Leaves smaller, greenish yellow, serrate/entire. Flowers from Ist week to 3rd week of April. Bearing is annual with light to heavy crops alternate. Fruits mature from September to October. Fruits are small medium, obovate, skin pale-yellow, covered with deep-redish bloom. Flesh yellowish white, with red tinge near skin, juicy and sub-acid in taste and flavoured.
- **Sultan Choongth (Lady Sudeley) [RC]:** Trees of medium size and spreading habit. Bark greenish –brown. Lenticels round, raised, greyish brown. Leaves obovate, serrate to biserrate, yellowish green. Flowering period is from 2nd to 3rd week of April; fruits mature during August.

Bearing is biennial. Fruits conical oblong, golden yellow with scarlet flushed stripes, flesh yellow with sweet delicious flavour.

- **Zarrad Trel (Precoce David) [UC]:** Trees medium and with spreading dense branching. Bark rough, scaly and dark-grey on the main trunk white branches look olive- green. Lenticels oval raised and greyish. Flowers from Ist to 3rd week of April; fruits mature by September to October. Bearing is biennial. Fruits smaller, oval, pale yellow with pinkish red flushes on exposed sides. Flesh white juicy, sweet and astringent.

C: Late Season Apple Varieties

- **American Trel (American Apirouge) [UC]:** Trees medium tall, spreading, vigorous. Bark redish brown, scarfy on old wood. Fruits bone on spurs, round, medium with red blush on exposed sides; flesh white, soft, crisp and sweet.
- **Batakh Tonthe (Bee-De-Olse) [RC]:** Trees medium tall, with short and dense branching. Bark yellowish brown on main trunk with redish green tinge on new shoots. Lenticels greenish brown, prominent and raised. Bearing is annual with light and heavy crops alternate. Fruits mature by October. Fruits are small to medium, conical, pale-yellowish; flesh whitish yellow, sub-acid and aromatic.
- **Bihi Saebe (Queen's Apple) [RC]:** Tree medium tall, spreading with curved strong branches. Bark olive-green mixed with redish brown scarf. Lenticels small, elongated/round. Leaves oblong, serrate, pale-green with small petioles. Flowering from by end of April. Bearing is biennial and fruits mature by November to December. Fruit large, angular, conic, ribbed, bright yellow when ripe; flesh firm crisp, juicy, pleasant acid, highly aromatic.
- **Blood Red (Gascoynes Scarlet) [UC]:** Trees medium, vigorous, spreading, knotty. Bark redish-brown/ olive-green. Lenticels few, brown, raised and oval. Leaves large yellowish pale, upcupped, margins irregular. Flowering time is Ist to 3rd week of April. Fruits mature in October but there is distinct alternate bearing. Fruits are medium, oval, slightly ribbed, entirely deep scarlet red coloured with crimson streaks. Flesh pale white, juicy, aromatic and pleasantly sweet (Fig. M3).
- **Bomburi Turush (Stones) [RC]:** Vigorous growing; trunk stout with reddish brown bark, and oblong, raised, brown lenticels. Leaves lanceolate, serrate, upheld, yellowish-green and larger in size. Flowering period extends between 2nd to 4th week of April. Fruits mature by October. Bearing is alternate. Fruits medium to larger size, oblong, pale-yellowish green with brown red flesh and strips. Flesh white, crisp, juicy and acidic.

- **Chakjal Choongth (Olivers Red) [RC]:** Trees medium tall, vigorous and spreading. Bark redish-brown mingled with olive-green lenticels oblong and raised. Leaves ovate, serrate, dark-greenish-upholded. Flowering time is from 2nd to 4th week of April. Periodicity of bearing annual. Fruits mature by November, are large, roundish, greenish yellow, striped and splashed with red and pinkish carmine; flesh whitish-yellow, sweet/ sub-acid.
- **Chamora (Chamure) [WC]:** Trees medium tall, spreading and upright. Bark dark redish-brown. Lenticels few, round and oblong, raised. Flowering period is from Ist to 3rd week of April. Bearing is annual but light and heavy crops appear alternatively. Leaves ovate, yellowish-green, serrate and some what twisted. Fruits mature by November-December and are medium to large sized, oblate, conic, greenish yellow, almost wholly covered with rosy-red flush.
- **Chithal Lal (Reinette De Cuzy) [UC]:** Trees big, strong and with spreading and upright branching. Bark is rough, scaly and dark-brown. Lenticels many, roundish/oblong raised. Leaves ovate, serrate, dark-green, up-held and undulating. Flowering commences from last week of March to middle of April. Bears light and heavy crop alternatively. Fruits mature by November- December, are oblong, tapering towards eye; skin greenish yellow, red stripped and streaked; flesh yellowish white, sub-acid, aromatic.
- **French Green Ambri (Court of Wick) [RC]:** Trees medium tall, spreading. Bark redish brown on main trunk, olive-green on branches. Lenticels round, redish brown, raised. Leaves oblong, sharply serrate, pale-green. Flowering period falls from 2nd to 3rd week of April. Bears annually and fruits mature by November. Fruits medium, conical, greenish yellow with white scarf, dots russet. Flesh yellowish white, richly flavoured, acidic and sweet (Fig. M4).
- **Golekharoo (Royal Russet) [RC]:** Trees medium tall, straggly branched. Bark redish-brown on old wood and olive-green on new-wood. Lenticels round, raised, greyish. Leaves oval, thick, dark-green, curved up. Flowering time is from 2nd to 4th week of April. Fruits mature by November to December. Bearing is annual with light and heavy crops alternatively. Fruits large, flattened and roundish, flesh greenish yellow, mild- sub acid and nearly sweet.
- **Gole Rekhal Lal (Chelmsford) [RC]:** Trees medium, wonder with spreading drooping branches. Bark is dark-brown on main trunk. Lenticels scattered, oval brownish and raised. Leaves ovate, serrate,

yellowish-green, down folded and pubescent. Flowering occurs during Ist to 4th week of April. Bearing is alternate. Fruits mature by October to November. Fruits medium to large, round and regular, pale-yellowish, flesh whitish or light yellowish, sub-acid and aromatic.

- **Golden Delicious (Starking delicious) [WC]:** Trees medium, spreading and strong. Bark redish brown. Fruits medium to above average, uniform in shape and regular, broad oblate and greenish yellow at maturity; flesh whitish, crisp and sweet. Long shelf life (Fig. M 5).
- **Grechhal Lal (Dutch Mignoune) [RC]:** Trees with vigorous growth, wide spreading, branching dense and slightly drooping. Barks redish brown on main trunk and olive –green on new wood. Lenticels raised small and oval. Leaves lanceolate, serrate, undulating, pale-green. Flowering is from Ist to 4th week of April. Bearing is annual with light and heavy crops alternating. Fruits mature by November. Fruit are medium sized, elliptical, straw-yellow with orange flesh, which deepens to orange red with age of fruit; flesh yellowish white, firm sub-acid. The fruits have long shelf life.
- **Hardu Turush (Nancy Jackson) [UC]:** Trees with the upright trunks. Bark is grey mingled with olive green patches. Lenticels are round of greyish brown colour. Leaves large, yellowish—green, coarsely crenate, folded upwards. Plants flower during second fortnight of April. Bearing is alternate. Fruit maturity by September/October. Fruits are conical to round, dull yellowish with bright red flush and crimson stripes; flesh white, acidic and with good shelf life.
- **Kadusari (Adams Pearmain) [RC]:** The variety is said to have been introduced for Kabul. Trees medium, upright, spreading. Bark redish brown. Lenticels round, few, scattered. Leaves oblong, sharply serrate, grey green, up-cupped. Flowers from Ist to 3rd week of April. Bearing is annual but bearing heavy and light crops alternative. Fruits mature by late October. Fruits conical greenish yellow, red striped and russeted and white dots scattered; flesh yellowish, sweet, sub-acid and aromatic.
- **Kadusari Balpuri (Bellow Flower) [RC]:** Trees medium, vigorous with spreading, drooping branching. Lenticels inconspicuous, oblong, greyish yellow. Bark redish brown, scarfy. Flowers from 2nd week to end of April. Bearing is alternate. Fruits mature in October. Leaves, ovate, boldly serrate, green, halves unequal, petioles pinkish and hairy. Fruits are large, oblong roundish, ribbed, conical in outline, russet dots all over; skin lemon yellowish with pinkish red blush; flesh pale- yellow/ whitish yellow, juicy, aromatic and subacid.

- **Kadusari Turush (Popes Scarlet Costard) [UC]:** Trees stout and upright. Bark is greyish on old wood and redish brown on new wood. Flowering time is second fortnight of April. Fruits mature by October /November. Bearing is alternate. Leaves are greenish yellow, oblong, shallow serrate, down curved with thick, pubescent petioles. Fruits are conical, uniform in shape and size, pale-yellow, splashed with broken dark-red crimson stripes; flesh creamy white, sub-acid, sweet. Long keeping quality.
- **Kagzi Khoure (Stalaphel) [RC]:** Trees medium tall, having redish-brown bark on trunk. Lenticels raised, round/ oblong. Leaves ovate, dark-green, serrate, up-held and stipulate. Flowering time commences from 2nd to 4th week of April and bearing is annual. Fruits large, round/ oblong, light yellow-greenish with brownish red checks. Flesh yellowish white and sub-acid.
- **Khoure (Yellow Newton) [UC]:** Trees large sized, dense, drooping and spreading branching. Bark redish brown mixed with olive-green. Lenticels roundish brown and many. Leaves oval, deep green, biserrate, broad, and upheld. Flowering time is from 2nd to 3rd week of April. Bearing is biennial. Fruits mature by November-December. Fruits medium to large, roundish, oblate, straw yellow, covered with russet dots; flesh yellowish- green, rich flavoured, sub-acid; has a long shelf life.
- **Kharroo Lal (Reinette De Britagne) [RC]:** Trees medium tall, spreading and upright. Bark greenish brown; lenticels oval, raised and greenish white. Flowering time lasts fortnight of April. Periodicity of bearing is annual, but light and heavy crops alternate. Fruits mature in November-December. Fruits yellowish-green, red streaked and splashed; flesh yellowish white, aromatic and sub-acid. Leaves ovate, dentate, yellowish- green, up-held with thick, pubescent petioles.
- **Maharaji (Calville Grand Due –De- Bade) [UC]:** Trees medium tall. Branching spreading and curved inwards. Bark redish brown. Lenticels round/oblong, raised and greyish- brown. Leaves lanceolate, dark-green, biserrate, up-held with long petioles. Flowering commences in first fortnight of April. Fruits mature by November-December. Bearing is alternate; fruits big, roundish, yellowish-green, red streaked, mixed with carmine; flesh greenish white, mild sub-acid (Fig. M 6).
- **Marchewangan Trel (Api- Rouge) [UC]:** Trees medium tall, dense and erect. Bark dark- brown. Lenticels round, conspicuous, yellowish brown. Flowering occurs during last fortnight of April. Bearing is annual but light and heavy crops appear alternatively. Fruits mature during

October. Fruits are small, flat, smooth and shining, skin bright and glory with pinkish red blush bloom, juicy and sub acid sweet. Leaf oblong, appressed towards the mother axis crenate and light green.

- **Methalal (Annie Elizebeth) [RC]:** Trees medium tall, upright and compact. Bark olive- green mixed with redish brown. Flowering commences from 2nd week of April to end of April. Periodicity is biennial. Leaf oval, serrate and dark green. Fruits mature by October. Fruits are large, oblong irregularly ribbed; skin brilliant red with streaks and splashes and dots of bright carmine; pale-yellow on shaded sides; flesh white, sub- acid sweet.
- **Michikharu (Golden Pippin) [RC]:** Trees are medium tall, vigorous in habit and spreading. Bark yellowish olive green to dark brown. Lenticels roundish, raised medium sized. Leaves obovate, serrate, dark-green, petiole with dark red stipules. Flowers appear by the last fortnight of April but fruits mature by October. Light and heavy bearer. Fruits roundish, entirely covered with golden russet/ yellowish russet; flesh yellowish, mild acid and aromatic.
- **Nawabi Turush (King of Pippins) [UC]:** Trees medium tall, upright with spreading all branches. Bark redish brown with olive green tinge on young branches. Lenticels raised, elongated and redish-brown. Flowers during Ist to 3rd week of April. Periodicity is annual, fruits mature during October. Fruits are round/oblong, pinkish white with yellow orange patches, flesh creamy yellow, sweet-bitter taste but with pleasant aroma.
- **Panze-trel (Wise Apple) [RC]:** Tree small to medium with spreading but drooping branches. Bark dark-brown and scarfy. Lenticels conspicuous, elongated and raised from stem surface. Bloom time is from end of April to Ist week of May. Periodicity is annual and fruits mature by October. Leaves are lanceolate, upheld, green, serrate to hairy stipules. Fruits are small, round, irregular, skin rough with russet, greenish yellow, flushed with red stripes; flesh yellowish, aromatic, sub-acid.
- **Pindh Choongth (Minkler) [RC]:** Trees of large size and with vigorous upright growth habit. Bark brownish red. Lenticels raised roundish and redish-brown. Leave oblong, biserrate, dark-green, with long petioles. Flowering time extends between Ist fortnight of April and fruits mature by October-November. Bearing is annual. Fruits medium, oblate-conic, green yellow/pale-yellow splashed with pinkish red strips and flushed with dark dull carmine. Flesh yellowish-white, tasty sweet. Fruits, because of their long keeping quality can be stored for many months.

- **Sabaz Choongth (Coleoptera) [RC]:** Trees medium, with spreading habit. Bark redish- brown with olive- green tinge; lenticels round, few, raised. Leaves long, acute, deep green, serrate. Flowering period commences during first fortnight of April. Bearing is alternate.

  Fruits mature by October. Fruits medium to large, roundish oblate and conical, skin greenish yellow with brownish red flushes on exposed side; flesh yellowish, crisp, juicy, sub-acid.
- **Tontal Rekhal (Champion) [RC]:** Trees upright and with dense branching. Bark dark redish brown. Lenticels numerous, round and oblong. Leaves large, lanecolate, yellowish green. Flowering time is 2nd to 3rd week of April. Fruits mature by October. Bearing is alternate. Fruits are medium to large, oblong conical, green yellowish, entirely covered with red flush and bright red, stripes, mingled with carmine. Flesh nearly white, very firm, crisp, juicy, sub-acid and aromatic.
- **Trelezard (Golden Harvey) [UC]:** Trees medium tall with compact branching. Trunk stout with brown and olive-green bark. Lenticels whitish, raised, round and scattered. Leaves small, serrate, acute tiped. Fruits light greenish yellow with dull red flush covered with russet; flesh firm, yellowish, sweet.
- **Zard Choongth (Edward VII) [RC]:** Trees upright, spreading, trunk knotty. Bark redish-green. Lenticels conspicuous, raised, brown, round and oblong. Leave dark green, ovate, crenate. Flowering period is from end of March to 2nd week of April. Bearing is annual and fruits mature by November. Fruits are large, oblong, pale yellow, faint brownish, red flushed; flesh creamy, sub- acid and juicy.

D. Indigenous Early Season Apple Varieties

- **Gilgati [RC]:** Trees medium sized, with long branches. Bark grey on old wood and olive-green on new wood. Lenticels round, raised and grey. Leaves ovate, up –folded, yellowish- grey, irregularly serrate with long thick pinkish petioles. Flowering time extends from ending April till middle of May. Fruits mature in August and periodicity of bearing is alternate. Fruits are oblong, deep yellow at maturity; flesh creamy white, juicy, sweet with bitterness.
- **Pindhwal [RC]:** Tree stout with vigorous growth. Bark is scaly, greyish to redish brown. Lenticels greyish raised and round to oblong. Plants flower by ending April and mature by August. Periodicity of bearing is alternate. Leaves are ovate, green, finely serrate with thick petioles. Fruits are roundish, medium pale- yellow splashed with pinkish red stripes; flesh white, very juicy and sweet, aromatic and pleasant.

- **Ramchoo [RC]:** Trees are medium sized. Bark olive-green on new wood and brownish on old wood. Lenticels are round to oval and conspicuous. Leaves lanceolate, fine serrate, yellowish green with long pinkish petioles. Plants flower in first fortnight of April and mature in June. Bearing is annual with light and heavy crops alternating. Fruits are small, flat round, regular, pale-yellow with white dots all over; flesh yellowish creamy, less juicy and less sweet.

E. Indigenous Late Season Apple Varieties

- **Ambri Kashmiri [UC]:** Trees medium and spreading. Regular bearing with fruits of long shelf like. Fruits medium to large sized, oblong elliptical in shape, tapering; skin yellowish green with pinkish streaks and shining; flesh white crisp, juicy and aromatic. Fruits ripen by September. Good cropper and regular bearer (Figs. M 7, 8).
- **Douda Ambri [RC]:** Finest of all local ambri types, ripens early but does not stand exportation.
- **Mauchi Ambre [RC]:** Trees large sized, vigorous and with dense growth habit. Bark scaly, dark- grey on old wood and olive green on new wood. Lenticels oblong, raised and brownish. Leaves ovate, yellow green, serrate and up held. Flowering period in 3rd week of April to Ist week of May. Fruits mature by October. Periodicity of fruiting is annual. Fruits are medium, oblong slightly conical, yellow- green splashed with bright deep red crimson; flesh white juicy, very sweet and aromatic, long keeping quality.
- **Mohi Ambre [RC].**Trees large sized, bearing redish, oblong, acid fruits liked much for their flavour and taste.
- **Kadur [RC]:** Trees big, vigorous, strong trunked having greyish bark with olive- green shade. Lenticels elongated and grey colored, numerous. Leaves ovate, dark- green with yellowish shade, sharply serrate, upheld with long petioles. Flowering time extends from 2nd to 4th week of April. Periodicity is biennial and fruits mature by October. Fruits are medium to large, conical yellowish- green, often splashed with light pink on exposed cheeks; flesh whitish, juicy and slightly aromatic, with good keeping quality.

F: Mid Summer Cider AppleVarieties

- **Baeel Cider [RC]:** Trees medium tall, spreading and upright habit. Bark reddish brown; lenticels many, raised and round. Produces flower in the first three weeks of April. Fruits mature in September. Periodicity of bearing is annual. Leaves smaller, oval, pale- yellowish, crenate with thick, pubescent petiole. Fruits are small, oval/ oblong, pale- yellow; flesh whitish juicy, sweet.

- **Chini Cider [RC]:** Trees medium, spreading and up-right branching with brownish bark. Lenticels elongated, greenish-brown. Leaves obovate, pale-green finely serrate, in up-right posture. Flowering time is from Ist to 3rd week of April. Fruits mature by August; bearing alternately light and heavy crops. Fruits are roundish, greenish pale yellow, often with pinkish cheeks; flesh white, juicy, sub-acid
- **Lal Cider [RC]:** Trees large sized, spreading habit with dark-red bark. Lenticels are round, small and numerous Leaves are ovate, sharply serrate, green with pinkish stipules. Flowering time extends from 2nd fortnight of April. Periodicity of fruting is alternate. Fruits mature by August. Fruits are medium, oblate/roundish, bright pale-yellow or greenish shaded, with pinkish red stripes. Flesh whitish, crisp, juicy, sub-acid and aromatic
- **Rekhal Cider [RC]:** Medium tall tree with scaly, rough and reddish brown/greyish bark on trunk and branches. Lenticels are oblong, raised and brownish. Flowering period falls in the first fortnight of April. Fruits mature by September; periodicity of bearing is biennial. Fruits are medium, oblong, tapering to eye, greenish yellow, splashed with long red streaks. Flesh white, juicy, astringent and sweet.
- **Safaid Cider [RC]:** Trees medium tall, compact habit, Bark redish brown to olive-green. Lenticels brownish-red, round and raised. Leaves ovate, green and serrate. Flowers in last week of April and periodicity is annual. Fruits mature by October. Fruits are small to medium, round and flattened, pale-yellow with red flush/stripes; flesh yellowish white, sweet and pleasant.
- **Tetha Cider [RC]:** Medium tall trees, with reddish brown bark on old wood, greenish red on new. Lenticels oblong scattered and somewhat sunken. Leaves dark-green, undulating crenate margins. Flowering time is from 2nd to 4th week of April. Fruits mature by September. Periodicity of bearing is annual. Fruits are medium, oval, tapering towards eye, greenish- yellow splashed with crimson red stripes all over the fruit; flesh greenish white, juicy, very bitter, usually watery cored.
- **Tetha Madur Trel [RC]:** Trees medium, strong, spreading and upright. Bark reddish brown on old wood, brownish on new wood. Lenticels few, round and raised. Produce flowers from Ist to 3rd week of April and fruits mature by October. Bearing is biennial. Fruits are small, roundish/ conic, greenish yellow with red stripes and splashes all over; flesh soft juicy, crisp and moderately sweet.

- **Tetha Shakur [RC]:** Trees medium to large with spreading and strong branches. Bark redish brown; lenticels many, oval/oblong, greenish white. Flowering time is Ist to 3rd week of April. Leaves narrow, serrate, green with long petioles. Fruits mature in September. Bearing is annual with light and heavy crops alternatively. Fruits are small to medium, greenish yellow; flesh white, soft, sweet, slightly bitter but refreshing.

G. Crab Apples

Malus prunifolia (Willd.) Borkh. (Rosaceae): Eng.- Himalayan Crab Apple The Crab Apple varieties include:

- **Batpuri Trel [RC]:** A wild carb apple cultivar with small light yellow sweet fruits.
- **Chouk Choonght [RC]:** Trees medium tall with spreading and drooping branches. Bark is reddish brown scaly and scarfy on old wood while new wood has a olive-green, shade. Lenticels few, round, olive-green raised. Flowering occurs from 2nd to 4th week of April. Periodicity of bearing is biennial. Fruits mature by September and are medium sized, tapering towards eye, greenish yellow, covered with pinkish red stripes/ entirely flushed, flesh greenish white, soft, juicy and sub-acid.
- **Chouk Gulab Trel [RC]:** Trees medium tall, branching spreading and slightly drooping. Bark reddish brown and scaly. Lenticels oblong and whitish- green. Leaves crenate, glabrous, down curved with thick, pubescent petioles. Flowering time is from 2nd to 4th week of April. Bearing is annual but light and heavy crops alternate. Fruits mature by August and are very small, cherry sized, round and flattish, skin whitish yellow covered with pinkish red flush and stripes on exposed sides; flesh soft, whitish, juicy, sub- acid.
- **Hapat Trel [W]:** Trees medium tall, strong trunk, with spreading curved branches. Bark reddish brown. Lenticels round, whitish-brown, raised. Flowering is from Ist to 3rd week of April. Bearing is biennial, leaves ovate, crenate, yellowish green, slightly down curved. Fruits are small, flattish, round, greenish yellow with russet patches; flesh whitish, soft, juicy sour to sub-acid.
- **Harda Choongth [RC]:** Trees medium tall with upright posture/ branching. Bark reddish brown and scaly on old wood. Trunk and branches bear many, small and oblong, slightly raised lenticels.

  During 2nd to 3rd week of April trees flower. Periodicity of bearing is annual.Fruits mature during August. Fruits are medium, irregular, roundish flat on both ends, greenish yellow; flesh whitish, crisp, juicy and acidic.

- **Janbazi Trel [UC]:** Trees medium, branching prolific, branches slightly drooping. Bark reddish brown on old wood with olive green patches on new wood. Lenticels few, big, greyish, raised and round in shape. Periodicity is biennial. Flowering time commences from Ist to 3rd week of April. Fruits mature by September. Fruits small, round to angular, covered with bright red flesh, flesh whitish, crisp, juicy, melting, sub-acid.
- **Khatoon Trel [RC]:**Tree medium tall, strong trunked. Flowering in middle of April. Bearing is biennial. Fruits medium, flavoured and light yellowish with pinkish blushes at maturity.
- **Khiram Trel [RC]:** Trees medium tall with slender branches. Bark scaly, reddish brown on old wood and olive-green on new; bears conspicuous grey-brown, oval lenticels. Leaves lanceolate, serrate, yellow- green with thick, pubescent petioles. Flowers from Ist to 3rd week of April. Bearing is annual. Fruits mature during August. Fruits are small to medium, greenish yellow; flesh soft, greenish white, sub-acid but turn bitter sweet on full maturity.
- **Lalbugi Trel** [RC]: A wild crab apple cultivar having pale yellow and sweet fruits.
- **Madur Shrawan Trel [RC]:** Trees large, spreading, vigorous, dense and with drooping branches. Bark reddish brown, scaly. Lenticels few, roundish, raised and greyish. Flowering time falls in 2nd to 4th week of April. Bearing is biennial. Maturity of fruits is August. Leaves are obovate, yellowish-green, serrate, with long petioles. Fruits are medium, round, flat on both sides, greenish yellow, covered with red stripes on exposed sides; flesh greenish white, juicy and sweet.
- **Nabed Trel [RC]:** Trees medium tall, spreading and branches drooping. Bark reddish brown. Lenticels round, raised. Leaves larger, green, ovate, finely serrate, with thick petiole. Flowering time is Ist to 3rd week of April, comes to bearing every year and fruits mature by September. Fruits are small, round and greenish yellow; flesh whitish, soft, sub-acid.
- **Nimioh Chouk Trel [RC]:** Trees medium with stout trunk. Branching spreading, upright. Bark reddish brown with olive-green tinge. Lenticels few, round and reddish brown. Flowering time is from Ist to 3rd week of April and fruits mature by late October. Bearing is annual. Fruits walnut sized, roundish/oval, flat on both ends, greenish yellow whitish dots all over. Flesh whitish, soft, juicy and acidic. Leaves long, lanceolate, twisted, green, bisserrate; petiolate with thick and pubescent petioles.

- **Sut Trel [RC]:** Trees medium tall and with reddish brown and scaly bark. Lenticels scattered, round, raised and yellowish brown. Leaves ovate, tapering on either side, crenate, yellowish green. Flowering period falls in the last fortnight of April. Bearing is biennial. Fruits mature by August. Fruits small, round, shining greenish- yellow with pinkish red cheeks.; flesh white, soft, juicy, sub-acid and flavoured.

H: Apples Varieties released by SKUAST-K

- **Akbar [RC]:** The variety developed by crossing Ambri x Cox's Orange Pippin.Tree moderately vigorous, wide spreading. Flowering regular; fruits medium to large in size, oblong, conical in shape with deep cavity, red coloured dominating on skin; average fruit yield is 245 kg per tree (Fig. M 9).
- **Firdous** [RC]: Trees with upright and spreading habit. Leaves ovate, serrate and light green. Fruit oblong conical, skin with bright red flush, yellow ground colour, whitish flesh and acidic in taste. Maturing by end of August; yields 55-60 kg /tree at 20 years age.
- **Shalimar Apple-1** [RC]: A mid season scab resistant apple variety developed by SKUAST-K for cultivation in the apple growing areas of the state. Trees vigorous and spreading. Leaves simple and deep green. Fruit roundish to oval yellowish with red blush, and with juicy and sweet flesh; yields 95 kg of fruit/tree of 20 years age (Fig. M 10).
- **Shalimar Apple -2** [RC]: A high yielding improved apple varieity with an average yield of 106 kg/tree of 25 years age. Fruits asymmetrical, conical, yellowish and red mottled. Flesh whitish green juicy and sweet. The variety has very quality fruits and has moderate resistance to major diseases and pests (Fig. M 11).
- **Shirien** [RC]:Medium tall tree with open tops. Leaves dull green and broadly ovate. Fruit oval-conical in shape, dark red blushed and with greenish yellow ground colour. Resistant to apple scab. Yields 65-70 kg/ tree of 20 years age (Fig. M 12).

I: Apples varieties of Ladakh Region

Apple is second important fruit crop of Ladakh region, after apricot. It especially grows well in its Kargill blocks, which have extreme cold-arid climates, like Baltal, Shagole and Samba. Most of the varieties have irregular bearing habit and are poor yielders. Common apple varieties include:

- **Bong Kushu** [RC]: Trees with spreading habit and irregular bearing. Fruits smaller, globose, and green in colour, sweet in taste and mature by mid of September (Fig. M 13).

- **Karkithoo** [RC]: Famous apple variety of Ladakh named after a village Karkichu were it grows abundantly. Tree habit is exceptionally upright. Fruits large, dark red skinned, sweet, juicy aromatic, medium sized and globose in shape. Bearing is biennial and fruits mature by end of October (Fig. M 14).
- **Mangol-kushu** [UC]: An apple variety probably introduced from Mangolia, resembling with that of Ambri of Kashmir in shape and size of its fruits. Fruits are dark-green colour and are smaller and mature in September but have low keeping quality. Trees have spreading habit and with irregular bearing.
- **Mar-kushu** [RC]: An early ripening apple variety of Ladakh maturing by September. Fruits light red when ripe, small to medium sized, and globose in shape. The variety has spreading habit, has irregular bearing and is poor yielder (Fig. M 15).
- **Menzay-kushu [RC]:** Average tall tree bearing small, oblong, and dark green, yellow spotted fruits.
- **Nas-kushu** [RC]: An early maturing medium tall apple variety maturing by end of August. Bearing is irregular; fruits small, oblong and light – greenish yellow at maturity and sweet in taste. Variety has red and green fruited variants (Fig. M 16).
- **Samarkand-kushu** [RC]: An introduced and well adapted apple variety maturing by September. Fruits small sized globose, sweet and cream based with red patches on maturity. Bearing is irregular. Plants have spreading and vigorous habit.
- **Sheur-kushu** [RC]: Trees are medium tall with spreading habit but have irregular bearing. Fruits nature by mid of September, are small ellipsoid green and acidic in taste. Variety with red and green types fruits are growing in different areas of Kargil (Fig. M 17).
- **Shing-kushu [UC]:** Woody apple variety of Ladakh with erect plant habit. Fruits small sized, oblong and dark-green with yellow spots. Maturity and bearing is irregular and fruits matures by end of September.
- **Tha-kushu** [RC]: The apple variety is grown around Indus valley. Its fruit are very sweet, juicy but with low shelf life.

Other indigenous apple varieties of the region include: **Femurkushu, Neeoto-Kushu, Papapali, Maspala, Yaqmapa and Shakark-kushu.**

Malus x purpurea (Barbie.) Rehder. (Rosaceae): Eng.- Flowering Crab Apple [RC] A beautiful small ornamental tree producing a wealthy of rosy crimson flowers and dark purplish green shoots and leaves. Fruit crimson purple. cv *Eleyi* (purple green leaves) and *lemoinei* (dark red purple leaves) are the two popular varieties planted in gardens.

Malus sylvestris (L.) Mill. (Rosaceae): Eng.- Forest Apple / European Crab Apple; Hindi- Jungli Saeb; Kashmiri – Jungli Choongth [W]. A thorny tree or shrub usually in forest thickets.Flowers large fragrant, light pink outsise and white inside. Inflorescence a cluter of 2 – 6 flowers. Leaves long petioled,ovate with a tapering tip,toothed. Fruits oval, sour, green pome.

Malus toringoides (Rehder.) Hughes. (Rosaceae): Eng.- Cutleaf Crab Apple [W] A spreading wild tree. Flowers creamy white slightly fragrant. Fruit globose, yellow red patches abundant and persistent.

Malus zumi (Matsum) Rehder. (Rosaceae) [W]. An upright, pyramidal to rounded tree growing in hardwood forests. Leaves ovate acuminate, long and crenate. Flowers white. Fruit fragrant, small, globose, red.

Marlea begonifolia L. (Alangiaceae) [W] A tree of Himalayan forests. Wood used for furniture works.

Marrubium vulgare L. (Lamiaceae): Eng.– White Hoarhound; Hindi- Pahari gandana; Kash.- Troupare; Ladakhi – Marphet [W]. A robust perennial herb, usually on waste places and orchards. Stem and leaves densely woolly haired. Flowers white. In Ladakhi system of medicine the extract of the plant used to cure eye infections. Also used as expectorant, diuretic, carminative and for colds and coughs (Fig. M 18).

Marsilea quadrifolia L. (Marsileaceae) [W]. A perennial fern of swamps and ponds. Stalks and leaves used as pot herb. Leaves resembling four leaved clover. Plant with anti-inflamatory, diuretic, febrifuge and refrigerant properties.

Matthiola revoluta Bunge ex. Boiss. (Brassicaceae): Ladakhi- Khil-chenagroo [W]. Perennial herb, hoary, branched. Flowers purple. Powdered roots are used orally every morning with milk or *Changue* as diuretic.

Matricaria chamomilla L. (Asteraceae): Eng- Chamomile / May Weed; Hindi- Gawe- chashme [W]. Non- aromatic herb growing in sub- alpine areas in orchards and grasslands of Kashmir. Leaves narrow, bi-pinnate or tri-pinnate. Flowers are borne in paniculate flower heads, which have white ray florets and dull yellow disc florets. Extracts of the plant is used to treat liver problems and gastric stimulant and for fevers (Fig. M 19).

Meconopsis aculeata Royle. (Papaveraceae): Eng.- Blue Poppy; Hindi- Kant Swanyus; Ladakhi- Achat/Surmum [W]. Prickly perennial herb growing in rocky shade in alpine grasslands at 3000 - 4500m altitudes; leaves irregularly pinnatifid, sparsely bristly haired; flowers bluish purple, 4 petalled with golden yellow stamens, borne in long racemes. Root considered narcotic and poisonous. Leaves used for renal colic and backache (Fig. M 20).

Meconopsis betonicifolia Franch. (Papaveraceae): Eng.-Himalayan Blue Poppy. [W] A semi-hardy perennial herb with yellow juice.Leaves oblong and cordate at base. Flowers blue to purple borne in groups of 3 to 4 at the top of stems. Plant is cultivated and also growing wild in sub-alpine forests lands of W Himalayas.

Medicago falcata L. (Fabaceae): Eng.–Yellow Lucerne; Hindi- Jungli Lucerne; Ladakhi- Oal, Bukshuk [W]. A deep rooted perennial forage herb growing wild in grasslands of Ladakh and Kashmir upto 4000 m amsl. Stem semi -erect and much branched. Flowers are in short clusters and light yellow (Fig. M 21).

Medicago lupulina L. (Fabaceae): Eng.- Black Medik.; Kash.- Poshe Gasse; Ladakhi- Oalbug [W]. Perennial spreading herb common in alpine grasslands and pastures. Leaves compound with three, oval hairy leaflets. Flowers yellow, small often grouped in tight bunches. Fresh plants are given to milking cows to increase their milk output.

Medicago sativa L. (Fabaceae): Eng. – Lucerne; Hindi- Viliati Gawuth; Kashmiri—Tripatre [UC]. A deep rooted perennial herb cultivated as well as growing across grasslands of Kashmir Himalaya from 1550 to 3000 m amsl. Stems are semi erect 15-35 cm high, arising from a crown which has woody base. Leaves are alternate, narrowly toothed and pinnately trifoliate. Flowers are in short clusters coloured from violet, to blue to light yellow. Fruit is a spirally twisted pod containing 1-8 small kidney shaped seeds (Fig. M 22).

Megacarpaea polyandra Benth. (Brassicaceae) [W]. An alpine herb endemic to Kashmir. Leaves edible.

Melia azedarach L. (Meliaceae): Eng.- Persian Lilac; Kash.- Drek [NI]. A fast growing much branched deciduous ornamental hedge. Leaves pinnate. Produce a profusion of star shaped fragrant lilac flowers in arching panicles.

Melica persica Kunth. (syn. Melica jacquimontii Decne.) (Poaceae): Ladakhi- Aswa [W]. A rhizomatous perennial herb, forming clumps. Culms erect, upto 60 cm tall. Leaves with glabrous surfaces, sheaths smooth, tubular. Panicles linear, spiciform. Spikelets cuneate and solitary. Leaf sap used for eye disorders, gout and rheumatic pains.

Melilotus alba Medic. (Lathyraceae): Eng.- White sweet Clover [W]. Biennial herb of grasslands and orchards. Flowers white. Used as a green fodder plant.

Melilotus officinalis (L.) Pall. (Fabaceae): Eng.- Yellow sweet Clover; Ladakhi- Owal [W]. An alpine /sub-alpine, biennial herb growing in moist situations. Leaf with three leaflets. Flowers yellow. Dried plants are fed to milking cows to increase their quantity and quality of milk. Flowers have been used for extraction of yellow dye.

Mentha arvensis L. (Lamiaceae): Eng.- Field Mint/ Wild Mint; Hindi- Jungli Pudina; Kash.- Jungli Pudne [WC]. A perennial wild herb of moist situations, known for its aromatic leaves. Used as carminative, refrigerant and stimulant. Leaves opposite. Flowers pale- purple in whorls at the base of leaves (Fig. M 23).

Mentha longifolia (L.) L. (Lamiaceae): Eng.- Horse Mint; Kash.- Vaena [W]. An aromatic, rhizomatous perennial herb, inhabitinhg damp places. Stem erect, whitish, straite, retrose short tomentose- villous. Leaf blade ovate to oblong-lanceolate, tomentose-villous, margins serrate-dentate,apex acute. Verticillasters in cylendric terminal spikes. Flowers purplish.Leaves used as carminative, antispasmodic and for improving digestion.

Mentha royleana Wall. ex Benth. (syn. Mentha longifolia subsp. royleana (Wall. ex Benth.) Briq.; Mentha longifolia subsp. kashmeriana Briq. (Lamiaceae): Eng.- Jungli Pudina; Kash.- Vena; Ladakhi- Phloling [W. A perennial strongly scented herb, usually along water channels. Leaves densely hoary, pale beneath. Flowers pale pink. Dried leaves used as carminative and stimulant. In Ladakhi system of medicine decoction of leaves used as an antiacid.Traditionally the bouquet of flowering shoots of the plant (along with roses and lotus flowers), are used as floral offering to God by local hindus (Fig. M 24).

Mentha spicata L. (Lamiaceae): Eng.- Garden Mint; Hindi- Pudina; Kash.-Pudna. [W]. An aromatic, rhizomatous, perennial grown in home gardens; used widely for treating various digestive troubles. Flowers bluish-purple. Leaves dark-green with coarse margins.

Menyanthes trifoliata L. (Menyanthaceae): Eng.- Bog Bean/Buck Bean; Hindi- Kachhashimbi; Kash.- Treveter [W]. A perennial acquatic herb with creeping rootstock and characteristic long petioled purplish trifoliate leaves. Flowers in terminal compact spikes are pinkish white. Plant used in rheumatism, gout, dropsy, scurvy and skin diseases.

Micromeria biflora (Ham.) Benth. (Lamiaceae): Eng.- Lemon Savory [W]. Perennial shrubby herb on dry slopes, usually along pathways. Flowers rosywhite. Leaves broadly ovate, entire pointed, gland dotted below. Leaves were being used as substitute to tea. Also used for treating wounds of cattle.

Microula tibetica Benth. (Boraginaceae): Ladakhi- Charokneoma [W] A small annual herb of alpine slopes of Ladakh. Whole herb is used for cough and various pulmonary diseases. Leaves prostrate, spatulate. Inflorescence terminal, crowded with flowers like a head. Flowers white.

Minuartia kashmerica (Edgew. & Hook.f) Mattf. (Caryophyllaceae): Ladakhi- Pangyan karpo [W]. A branched, tufted perennial herb of rocky alpine slopes of Kashmir above 3000 m altitudes. Flowers white. Whole herb used for headaches.

Momordica charantia L (Cucurbitaceae): Eng.- Bitter Gourd: Hindi- Karela; Kashmiri- *Karael [RC].* A week climber grown for its ridged, bitter fruits used as vegetable in immature stage. It is more used by people suffering from diabeties and billious affections. The fruits are bitter in taste and turn redish orange on maturity. Tender leaves are also used as vegetable.

Monochoria vaginalis (Burm.f.) Persl. Ex Kunth.(Pontederiaceae) [W]. Perennial herb, usually in rice swamps. Flowers blue. Leaves medicinal.

Morchella esculenta L. ex Gr. (Morchellaceae): Eng.- Common Morel; Kash.- Kanne guchsi [W]. An edible fungus growing in disturbed areas, hard woods and orchards,usually during spring. Fruiting body or caps are sub-globose, yellow brown/greyish, spongy with lighter ridges. Yellow brown caps are cooked as a vegetable (Fig. M 25).

Morina coulteriana Royle. (Caprifoliaceae):Eng.- Yellow whorl Flower; Hindi – Dhoop; Kash. –Dhup; Ladakhi- Agzaima [W]. Perennial herb of sub-alpine / alpine forests, in dry situations. Stem upto 1m long with a basal cluster of spiny-lobed thistle like leaves. Leaves stalkless with spinous margins. Flowers yellow in long terminal, interrupted spikes. Roots used as incense (Fig. M 26).

Morina longifolia Wall ex DC (Caprifoliaceae): Hindi – Dhoop; Kash. –Dhup; Ladakhi- Agzaima [W]. An aromatic herb found in Kashmir Himalaya above 2000 m amsl. Flowers pink or purplish. Rootstock used as incense and in the preparation of *dhup agarbhatis* generally used in religious ceremonies. The oil of the plant is useful in leprous ulcers. In Ladakh the crushed seeds are given to children as nuitritive.

Morus alba L. (Moraceae): Eng.- White Mulberry; Kash.- Choute Tull [W/ UC] A medium sized deciduous tree. Leaves variable, ovate, serrate or crenate-serrate. Flowers dioecious or monoecious, in dense spikes. Fruit ovoid, white or nearly black. Leaves used for feeding silkworm.

Morus indica L. (Moraceae): Eng.- Common Mulberry; Kash.- Wazul Tull / Chaeri Tul [W/UC]. A medium sized deciduous tree. Leaves truncate or rounded at base or slightly cordate, mostly variously lobed, apex acuminate; female spikes shortly ovoid. Fruit 1 cm long, black when ripe.Traditionally long twigs of the plant were used by local hindus in marriage functions – for lighting a fire and to comb hair of a bride a day prior to her wedding (Fig. M 27). Leaves used for feeding silkworm.

Morus *nigra* **L. (Moraceae): Eng.- Black Mulberry; Hindi – Krishna Toot; Kash.- Shah Tul [W/UC].** A large deciduous tree. Leaves ovate to oblong ovate, acuminate, truncate to subcordate at base, sharply serrate. Fruit 2-4 cm long, ovoid, becoming dark purple at maturity.

Popular mulberry varieties growing in Ladakh include Ba- Osay, **Bedana Osay, Khara Osay, Karpo Osay, Naqpo Osay and Photo Osay.**

Myricaria elegans Royle. (Tamaricaceae): Eng.- False Tamarisk; Ladakhi-Umboo / Umpre [W]. A glabrous under shrub with upright wand like branches. Flowers white/ pink. Grows wild in Ladakh and is an important fodder plant and being relished by cattle. Decoction of bark and twigs is used as blood purifier in Jaundice patients in Ladakh. Paste of the plant also used to relieve swollen joints (Fig. M 28).

Myricaria germanica (L.) Desv. (syn. Myricaria bracteata Royle.) (Tamaricaceae): Ladakhi - Umboo [W]. A narrowly upright deciduous shrub or sub-shrub of Ladakh, growing near river valleys, on sandy beds. Leaves scale like blue green, linear-lanceolate, dense clustered leaves. Flowers spike terminal and lateral and pinkish red; spikes elongating in to fruit. Paste of twigs used as an anti-inflamatory agent and a blood purifier. Also serves as a fodder plant.

Myrsine africana L. (Primulaceae): Eng.- Cape Myrtle; Hindi-Chapra: Kash.- Gudil / Gugil [W]. A small alpine shrub. Leaves leathery, dark-green. Flowers cream coloured. Berries bright purple. Fruits used medicinally as anthelmintic,especially for tape-worm infestation.

Myrtus communis L. (Myrtaceae): Eng. Common Myrtle; Hindi- Vilaiti Mehndi [NI]. An upright, evergreen ornamental bushy shrub bears entire smooth, glossy, glabrous, leathery leaves very aromatic when crushed. Produces solitary, 5-petalled flowers followed by oblong – ellipsoid purple black berries.

Fig. M1. Malva neglecta

Fig. M2. var. Rekhal Lal

Fig. M3. var. Blood Red

Fig. M4. var. Green Ambri

Fig. M5. var. Golden Delicious

Fig. M6. var. Maharaji

Fig. M 7. var. Ambri

Fig. M8. var. Ambri

Fig. M 9. var. Akbar

Fig. M 10. var. Shalimar Apple 1

Fig. M 11. var. Shalimar Apple 2

Fig. M 12. var. Sheerin

Fig. M 13. var. Bangkushu

Fig. M 14. var. Karkithoo

Fig. M 15. var. Markushu

Fig. M 16. var. Naskushu

Fig. M 17. var. Sheurkushu

Fig. M 18. Marrubium vulgare

Fig. M 19. Matricaria chamomilla

Fig. M 20. Mecanopsis aculeata

Fig. M 21. Medicago falcata

Fig. M 22. Medicago sativa

Fig. M 23. Mentha arvensis

Fig. M 24. Mentha royleana

Fig. M 25. Morchella esculenta

Fig. M 26. Morina coulteriana

Fig. M 27. Morus indica

Fig. M 28. Myricaria elegans

N

Nandina domestica Thunb. (Berberidaceae): Eng.- Chinese Sacred Bamboo [NI]. An evergreen or semi-evergreen ornamental shrub with upright shoots. Leaves mostly bi-to tripinnate, rich green purple on fall. Flowers star shaped, white with large yellow anthers, in erect panicles followed by long lasting, spherical bright red fruit.

Narcissus psuedonarcissus L. (Amaryllidaceae): Eng.- Wild Daffodil; Hindi-Nargis; Kash.- Yimberzoule [W/UC]. A hardy bulbous herbaceous perennial with narrowly strap-shaped leaves. Found growing wild in grasslands with variable flower colours. The species has pale-yellow flowers with deep yellow/ lemon yellow central trumpet (Fig. N 1).

Narcissus tazetta L. (syn. Narcissus canaliculatus Guss.) (Amaryllidaceae): Eng.- Bunch Flowered Narcissus; Hindi- Nargis; Kash.- Yemberzoule [W/UC]. A perennial bulbous herb growing in gardens, parks, orchards and graveyards. Leaves strap shaped. A variable and tall species, distinguished by the flowers appearing in branches on each stem. Flowers have white petals and short, cup- shaped lemon yellow coronas are strongly scented and flowers from February to March (Fig. N 2).

Nardostachys jatamansi (D Don.) DC. (syn. Nardostachys grandiflora DC.) (Valerianaceae): Eng.- Spikenard; Hindi- Jatamansi [W]. A rare perennial herb of alpine areas. Oil used for hair growth. Extract of the rhizome used for treating convulsions and epilepsy.

Nasturtium officinale R. Br. (syn. Rorippa nasturtium acquaticum (L.) Hayek. (Brassicaceae): Eng.- Water Cress; Hindi- Brami sag; Kash.-Nagi Baber/ Koule hakh [W]. Perennial acquatic herb, usually in standing and slow moving streams. Leaves pinnately compound. Flowers white. Leaves edible and used as vegetable (Fig. N 3).

Nelumbo nucifera Gaertn. (syn. Nymphaea nelumbo L.; Nelumbium speciosum Willd.) (Nelumbonaceae): Eng.- Indian Lotus; Hindi – Kanwal/ Nilofer; Kash.- Nadure / Khila Wather/ Pum posh [W]. A large aquatic herb growing in Dal and Anchar Lakes of Kashmir. Plant has thick underground stoloniferous rhizomes. Flowers white with pinky show. Roots are milky white to off-white and cooked as a vegetable in Kashmir especially during winters. The plant is sacred to Hindus (Figs. N 4, 5).

Neottia ovata (L.) Bluff & Fingerh (syn. Listera ovata (L.) Br. (Orchidaceae): Eng.- Common Twayblade/Eggleaf Twayblade [W]. A medicinal terrestrial orchid of Kashmir Himalaya inhabiting alpine meadows. Plant 20-60 cm tall with two large opposite basal leaves. Flowers are borne on the stems, small and yellowish-green.The sepals and the side petals form a fairly open hood (Fig. N 6).

Nepeta annua Pall. (syn. Nepeta botryoides Aiton.) (Lamiaceae): Ladakhi- Khamyu [W]. An aromatic herb of high altitude areas of Ladakh. Stem erect, branched from base,leafy with a dense vinous, spreading, eglandular indumentums. Leaves broad ovate, bipinnatisect. Flowers white. Extract of the plant used to treat arthritis.

Nepeta cataria L. (Lamiaceae): Eng.- Catmint; Kash.- Brari Gausse / Gandhe Soi [W] An erect, selender, softly tomentose herbaceous perennial, found in wastelands. Leaves are toothed, triangular to ovate. Flowers small bilabiate, white and finely spotted with pink. Leaves and flower heads are used as carminative, refrigerant and stimulant (Fig. N 7).

Nepeta discolour Royle ex Benth. (Lamiaceae): Ladakhi- Jimthegale [W]. A low growing perennial catmint species inhabiting alpine Himalayas. Leaves short petioled, ovate, densely puberulent, adaxially green, abaxially gray. Flowers sessile, purple. Plant used for treating various eye infections.

Nepeta eriostachys Benth. (Lamiaceae): Ladakhi- Zimthikle [W]. A perennial herb of Kashmir Himalaya inhabiting cold arid lands of Ladakh. Leaves sub-sessile. Flowers bluish. Plant used to cure redness of eyes and for weak vision.

Nepeta flocossa Benth. (Lamiaceae): Eng.- Woolly Catmint; Ladakhi – Shamalolo [W]. An aromatic, clump forming perennial herb, quite common along scrubs in forests in Kashmir and also on stony slopes of Ladakh above 3000 m altitudes. Leaves lemon scented, covered with woolly white hairs. Flowers bluish purple borne in densely rounded, widely spaced woolly whorls. Decoction of the leaves is used against malaria.Young shoots and leaves are used for making vegetable soups (Fig. N 8).

Nepeta glutinosa Benth. (Lamiaceae): Ladakhi- Jatukpa [W]. Aromatic clump forming perennial of Himalayan meadows with an all over dense glandular indumentum of short and long capitates glandular hairs. Leaves serrate-pectinate. Flowers mauve or purple borne on shortly pedunculate verticillasters, borne in the axils of upper leaves. Plant used for all fevers.

Nepeta longibracteata Benth. (Lamiaceae): Eng.– Longbract Catmint; Ladakhi-. Tiyanku/Gibshiang [W]. An aromatic, clumped perennial scented herb growing across N-W Himalayas, mostly around Ladakh where *Amchis* use it against variety of ailments such as abdominal and liver troubles, fever and colds. Corolla blue-voilet, appearing with lance shaped often purple bracts. Leaves gray, woolly.

Nepeta racemosa Lam. (syn. Nepeta salviaefolia Pers.) (Lamiaceae): Ladakhi- Zatukpa [W]. Perennial herb on rocky dry areas of alpine zone. Flowers pale lilac.Whole plant used against madness and as a cerebral tonic (Fig. N 9).

Nepeta raphanorhiza Benth. (Lamiaceae): Eng.- Catmint; Kash.-Van Gogije [W] Low growing perennial, tuberous herb common on open dry places. Stem several, prostrate, ascending. Leaves, concolorus, villous. Flowers purplish borne in a terminal compact cylindrical head. Leaves and seeds used in coughs and fevers. Tubers edible.

Nepeta thibetica Jacq. ex Benth. (Lamiaceae): Ladakhi- Zatukpa [W]. A perennial erect and pubescent herb endemic to Ladakh Himalaya. Leaves ovate, cordate. Flowers light blue.Plant used as cerebral tonic in Amchi system of medicine.

Nerine undulata (L.) Herb. (syn. Nerine flexuosa (Jacq.) Herb. (Amaryllidaceae) [NI]. A half hardy bulbous perennial ornamental. Leaves linearly strap shaped appearing after flowers appear. Stems thin leafless carrying about a dozen of pink flowers in the form of loose umbel. Flowers pink with large strap shaped petals that often are twisted.

Nerium oleander L. (syn. Nerium indicum Mill.) (Apocynaceae): Eng.- Oleander; Hindi– Kaner [NI]. An upright evergreen ornamental shrub with silver-grey bark. Leaves in whorls of 3, linear- lanceolate to oblong, shining dark green. Flowers borne on terminal branched clusters of fragrant, red pink or white funnel shaped flowers. Hairy *varigata* is the common cv. with leaves having creamy margins.

Notholirion thomsonianum (Royle.) Stapf. (Liliaceae): Eng.- Rosy Himalayan Lily [W]. A bulbous perennial of Himalayan forests above 2000m altitudes. Bulbs nearly ovoid, tunic thick, brown. Leaves flaccid.

Flowers are borne in dense spikes at the top of stem and are pale-purple or rose-purple, funnel shaped and fragrant. Petals are narrowly spoon shaped, curved outwards at the top.

Nymphaea alba L. (Nymphaeaceae): Eng.-European White Water Lily; Hindi- Pandharen kamal; Kash.- Khoure [W]. A perennial acquatic herb, abundantly growing in the shallow lakes of Kashmir. Flowers yellowish on long stalks.Tender stems are arranged in bundles, sundried and later cooked as vegetable especially with fish/eggs during winter. An infusion of flowers, fruits and rhizomes is diaphoretic and used for diarrhea. Leaves are palatable to cattle and fed fresh to milking cows (Fig. N 10).

Nymphaea candida C Persl. (Nymphaeaceae): Eng. Water Lily; Hindi- Nil kamal / Nilofur; Kash.- Keni boub [W]. A perennial rhizomatous, acquatic herb, growing in the shallow lakes of Kashmir. Leaf blade sub-orbicular, papery, abaxially glabrous. Flowers white, floating. Fruits semi-globose. Fruits and seeds edible.

Nymphaea stellata Willd. (Nymphaeaceae): Eng.- Water lily; Kash.- Bumposh [W] Perennial herb of shallow waters in lakes, ponds and channels. Leaves floating, smooth with diverging lobes. Flowers yellowish, floating. Fruit spongy berry.Seeds edible. Rootstock given in dyspepsia, diarrhoea and piles. The infusion of rhizomes is used for diseases of urinary tract.

Fig. N 1. Nepeta cataria

Fig. N 2. Narcissus tazetta

Fig. N 3. Nasturtium officinale

Fig. N 4. Nelumbo nucifera

Fig. N 5. A vegetable seller on the banks of Dal Lake

Fig. N 6. Neottia ovata

Fig. N 7. Nepeta cataria

Fig. N 9. Nepeta racemosa

Fig. N 8. Nepeta flocossa

Fig. N 10. Nymphaea alba

O

Ocimum basilicum L. (Lamiaceae): Eng.- Sweet Basil; Hindi- Ban tulsi; Kash.-Baber [RC]. A cultivated aromatic herb grown for its seed. Flowers showy, light blue, fragrant. Leaves used for flavouring purposes. Seeds used for preparing refreshing drinks/ sharbats.

Oenothera glazioviana Micheli (Onagraceae): Eng.-Large flowered Evening Primrose; Kash.- Shamme sundre [NI]. A hardy biennial, hairy herb growing in waste places and fruit orchards; the hairs with reddish glandular bases. Leaves lanceolate. The flowers are pale–yellow and produced in dense spikes from June to September (Fig. O 1).

Olea europaea L. (Oleaceae): Eng.- Olive; Hindi- Zaitun; Kash.- Zaitoon [RC] A slow growing, evergreen tree, developing a rounded head, with thrornless nearly terete branches,with opposite, leathery, elliptic to lance-shaped irregularly toothed leaves. Tiny fragrant creamy white flowers are borne in axillary panicles.

Olimarabidopsis pumila (Celak.) Al Shahbaz, O Kane & Price (syn. Sisymbrium pumilum Stephan.; Arabidopsis griffithiana (Boiss.) N.Busch.) (Brassicaceae): Eng.- Dwarf Rocket [W]. Annual herb, usually on mud wall tops and waste places. Stems erect, simple or few branched at base, pubescent with short stalked, stellate trichomes. Flowers yellow. Fruit linear terete, curved or straight. Seeds brown, used in curing eye sores.

Onosma hispida var. kashmirica (I M Johnst.) I M Johnst. (syn. Onosma kashmirica I M Johnst. (Boraginaceae); Kash.- Ratan Joyot; Ladakhi-Deemok [W]. An erect perennial herb with a thick rootstock distributed across Kashmir Himalayas at 2000-3500 m, on dry rocky slopes. Aerial parts used in medicine, as cosmetic and also in worship in Ladakh. In *Amchi* system of medicine it is used for checking blood womiting and lung troubles (Fig. O 2).

Onosma thomsonii Clarke. (Boraginaceae): Eng.- Thomson's Onosma [W]. A rare perennial woody and hairy herb of Kashmir Himalayas.

Flowers tubular, whitish to bluish, in nodding clusters.Used for profuse hair growth.

Ophioglossum vulgatum L.(Ophioglossaceae): Eng.–Adder's Tongue; Kash.- Chonchur. [W]. A small rhizomatous herbaceous fern growing in and around Dal and Anchar Lakes and forest covers,whose fronds are used as vegetable. Fronds have two parts- the rounded diamond shaped sheath enclosing a narrow spore bearing spike (Fig. O 3).

Opuntia microdasys (Lehm.) Pfeiff. (Cactaceae): Eng.- Prickly Pear; Hindi- Nagphani [NI]. Green house perennial mostly grown as pot plant. Fruits large globular. Stem is modified into pads, which have spines and also bear tufts of glochids on each areole. Leaves blue-green and covered with network of golden spines and hairs. The glochids are yellow brown.

Opuntia scheeri Weber. (Cactaceae): Eng.- Slipper Thorn; Hindi-Hathhathoria [NI] A slow growing cactus grown as pot plant and has long oblong joints up to 20 cm olong.

Origanum vulgare L. (syn. Origanum normale Don.) (Lamiaceae): Eng.-Wild. Potmarjoram; Hindi– Bantulsi; Kash. - Wan Baber / Marzan josh [W]. An aromatic branched perennial herb growing across Kashmir Himalayas in grasslands or open scrubs. Flowers blue purple crowded into a dome shaped inflorescence. Calyx glangular and hairy. Leaves and seeds are used as stomachic, diaphoretic against digestive problems and as cardio-tonic. Leaves are also used for post delivery bath of ladies (Fig. O 4).

Orchis latifolia L. (Orchidaceae): Kash.- Salem panja [W]. A tuberous perennial (60-70 cm tall) of Himalayas found at altitudes of 2500 – 5000m in damp places.Tubers are fleshy and have palm like appearance. Flowers dull purple. Tubers used as aphrodiasic and tonic (Fig. O 5).

Orobanche alba Steph. ex Willd. (Orobanchaceae): Eng.- Broom Rape [W]. Perennial glandular pubescent herb, usually parasitic on roots of *Origanum normale*. Stem erect thick. Flowers brown purple. Used against renal problems.

Orobanche solmsii C B Clarke (Orobanchaceae): Eng.- Broom Rape [W]. Perennial densely whit villous, glandulr pubescent herb, parasitic on the roots of *Ferula jaschkeana*. Flowers pale yellow-brown. Whole plant used medicinally.

Orobanche cernua Loeffl. (Orobanchaceae): Eng.- Broom Rape; Ladakhi-Orboutch [W]. A parasitic herbaceous plant, growing under forest covers. Stem straw coloured bearing bluish snapdragon like flowers. Leaves are present as scales. In Ladakh its shoots are used for making soups and pickles. Powdered seeds dissolved in curds are used to relieve renal pain and as diuretic.

Orthosiphon rubicundus (D. Don) Benth. (Lamiaceae): Eng.- Red Jave Tea [W] A perennial herb with woody rootstock. Stem erect slender, quadriangular. Leaves obovate, serrulate, sessile with acute apices. Flowers borne in verticellasters; corolla pinkish or white, calyx tubular. Tubers edible (Fig. O 6).

Oryza sativa L. (syn. Oryza sativa subsp. japonica S. Kato; Oryza sativa subsp. indica) (Poaceae): Eng.- Paddy; Hindi - Dhaan; Kashmir- - Dauni [UC]. A herb grown as food cum forage crop. After harvest straw is used as fodder for cattle and for thatching of houses and cow sheds. In recent past rice straw was used for making shoes and ropes and mats of different sizes and shapes for furnishing houses. (Fig. O 7). Among the three regions of the state Kashmir valley has prominently been a rice bowl feeding majority of its inhabitants. Diversity of rice in Kashmir is largely represented by introduced/locally bred old and new varieties, few land races and farmers varieties (belonging both to *indica* and *japonica* types) which are mentioned below.

A. Land races / Farmer's varieties

- **Aziz Beoule** [NC]. Plants medium tall. Leaf blade and basal leaf sheath green; ligule clefted and white; collar green; auricle pale green; flag leaf angle horizontal. Panicles short and fully awned; grains bold and slightly aromatic.
- **Baber** [NC]. Plants medium tall. Leaf blade green with purple leaf sheaths; horizontal leaf angle, ligule white and clifted, collar and auricle purple; flag leaf horizontal. Panicles compact with straw coloured awns. Grains bold and slightly scented. The variety has early maturity and intermediate threshability (Fig. O 8).
- **Bala-anzul** [NC]. Plants of average height. leaf angle horizontal; leaf blade and basal leaf sheath colour green. Ligule white and clefted; collar green; auricle purple. Flag leaf angle horizontal. Panicles compact and awned; lemma and palea glabrous and of straw colour. Grains bold and non-aromatic.
- **Bala Koune** [NC]. Plants short statured.Leaf blade and leaf sheath purple; leaf angle horizontal; flag leaf angle intermediate; ligules white and clefted; colar green; auricle purple. Panicles compact with purple awns. Threshability loose. Grains short and bold and non-aromatic.

- **Barat** [NC]. Plants tall. Leaf blade and leaf sheath colour green. Leaf angle horizontal. Ligule white and clefted, collar green, auricle pale green. Flag leaf angle horizontal. Panicles compact with black awns. Grains bold and non-aromatic. Lemma and palea glabrous and brown.
- **Batu Baber [NC].** Plants medium tall, leaf blade and leaf sheath green. Leaf angle erect. Ligule white and clefted; collar green, auricle pale green. Panicles compact awnless; lemma and palea purple. Grains bold, slightly scented and non –shattering (Fig. O 9).
- **Begum** [NC]. Plants medium tall; drooping leaf angle; leaf blade and basal leaf sheath green; ligule clefted and white in colour; collar green, auricle pale green; flag leaf angle descending. Panicles fully awned, with purple awns. Grains bold, non-aromatic and non-shattering.
- **Black Rice** [NC]. Plants tall. Leaf angle erect. Leaf blade dark green and pubescent; basal flag Leaf sheath green. Leaf angle horizontal. Collar and auricle colour pale green; ligule white with acute tip. Panciles compact, awnless. Lemma and palea brown to black in colour. Grains bold but selender with high length/ breadth ratio and sceneted.
- **Brez** [NC]. Plants of medium height. Leaf blade and leaf sheath colour green; ligule white and clefted; collar green; auricle pale green; flag leaf angle horizontal; panicles compact and partly awned. The variety matures early and has characteristic bold and slightly scented grains.
- **Chini Bara** [NC]. Plants have purple coloured foliage. Panicles are compact and fully awned. Grains are bold and non-aromatic. Flag leaf angle horizontal; ligule purple, colar and auricle with purple spots.
- **Gulla Bara [NC].** Plants tall statured with early maturity and purple coloured awns. Leaf angle horizontal; leaf blade and leaf sheath colour green; ligule white and clefted; auricle purple; colar green; panicles compact with long purple awns. Grains are bold, slightly scented and non- shattering.
- **Gulla Zag [NC].** The variety has red coloured glumes and purple awned panicles. Flag leaf angle is horizontal; colar green and auricle purple. The variety is average yielding and shattering susceptible. Grains are bold with good cooking qualities.
- **Guru-Kaune** [NC]. An early maturing variety characterised by compact panicles with black awns. Leaf angle horizontal; leaf blade and leaf sheath green; ligule white and clefted; auricle and collar green. Grains bold and slightly scented.
- **Kala Brear** [NC]. Leaf blade and leaf sheath colour green; ligules white with purple lines; green collar, pale green auricle. Panicles are compact

and awnless; flag leaf erect. Grains non-shattering and non-scented. Plants have profuse tillering.

- **Kamad** [NC]. Plants of medium height and good vegetative growth. Medium tillering; leaf angle erect; leaf blade green and pubescent; flag leaf angle inclined; ligule white and clefted; collar green; auricle pale green. Panicles well exerted, compact and awnless; lemna and palea purple and glabrous; Grains redish, non- shattering and scented. Early maturity, high mill rice recovery (Fig. O 10).
- **Kathwari** [NC]. Tall plants with vigorous growth; medium tillering, culms bend; leaves erect; leaf blade green and pubescent; green basal leaf sheath; flag leaf horizontal. Ligule white and clefted; collar green; auricle- pale green. Panicles well exerted compact and fully awned, lemma and palea with brown spots on straw and hair on lemma bed. Grains bold, non- shattering and non-scented. The land race is good yielder (3-3.5 t/ ha).
- **Kawa Krear [NC].** Plants of medium height and vigorous growth. Medium tillering; culms erect; leaf blade green and pubescent; basal leaf sheath green; flag leaf inclined; auricle- pale green; ligule white and clefted; collar green. Panicles partly exerted beyond flag leaf sheath, compact and fully awned; awns purple. Lemma and palea glabrous and with brown spots on straw. Grains selender, non- shattering and slightly scented.
- **Khouch** [NC]. Tall plants with erect angled leaves; flag leaf inclined. Panicles are semi-compact and awnless. Grains are bold, shattering type and slightly scented. Lemma and palea glabrous with golden furrows on straw.
- **Larbeoule [NC].** Tall plants, Leaf angle erect; leaf blade green and pubescent; green basal leaf sheath; flag leaf inclined; ligule white and clefted; collar green, auricle pale green. Panicles well exerted, compact and awnless; lemma and palea white and glabrous; Grains selender, non-scented and non- shattering. Quality rice with good cooking qualities.
- **Mazette** [NC]. Plants of average height and medium tillering. Leaf angle erect; leaf blade dark- green and pubescent; green basal leaf sheath; ligule white and clefted; collar green; auricle pale- green. Flag leaf erect; panicles well exerted, compact and partially awned; awns purple and short; lemma and palea redish to light purple and with short hairs. Grains bold, non–shattering and slightly scented. Early maturity period.

- **Meer Zag [NC].** Plants of medium height and vigorous growth. Leaf angle erect; leaf blade green and pubescent; green basal leaf sheath; ligule white and truncated; collar green; auricle purple; flag leaf – erect; panicles well exerted, awned and semi compact. Lemma and palea with purple spots on straw, glabrous. Grains bold, non- shattering and slightly scented. Early maturity period. High yielding.
- **Mehwan** [NC]. Plants of medium height and profuse tillering, culms erect; leaf angle erect; blade dark- green and pubescent; green basal leaf sheath; flag leaf inclined; ligule with purple lines and clefted, collar and auricle-pale-green. Panicles long, semi compact, awnless and well exerted; lemma and palea with brown spots and glabrous. Grains bold, shattering type and slightly scented.
- **Moughal** [NC]. Plants tall with medium tillering, culms erect; leaf blade green and leaf angle erect; flag leaf erect; ligule purple and clefted; collar green; auricle pale- green. Panicles partly exerted, compact and awnless. Lemma and palea purple and glabrous. Grains bold, non-shattering and slightly scented. Leaf senescence late. Early maturing.
- **Mushkandi** [NC]. Plants tall and with vigorous growth; poor tillering; leaf blade green and pubescent; leaf angle horizontal; green basal leaf sheath; ligule white and clefted; collar green; auricle – pale- green. Flag leaf horizontal. Panicles just exerted, fully awned and semi-compact. Lemma and palea white and glabrous. Leaf senescence late and slow. Grains bold, non- shattering and slightly scented. Cold tolerant and cultivated in hilly areas of the valley. Short maturity period.
- **Mushka–Buduji** [RC]. Plants of medium height and medium tillering. Flag leaf angle horizontal; leaf blade green and pubescent, green basal leaf sheath; ligule white and clefted; collar green; auricle- pale green. Panicles well exerted, awnless/partly awned and compact. Lemma and palea white and glabrous. Leaf senescence late and slow; Grains bold, non- shattering and scented. High mill rice recovery.
- **Nick-Cheena** [RC]. Medium tall and with normal growth; Medium tillering; culms erect; leaf angle erect. Leaf blade green; basal leaf sheath green; ligule white and clefted; collar – green; auricle–pale green. Flag leaf erect. Panicles compact, exerted and partially awned. Lemma and palea brown with hairs. Grains bold, non – shattering and non- scented.
- **Niver-Zag** [NC]. Short plant stature, with erect culms and medium tillering; leaf angle drooping; leaf blade green; purple basal leaf sheath; ligule white and clefted; collar green, auricle pale green. Flag leaf angle descending. Panicles well exerted, compact and awnless. Grains bold,

non-scented and non- shattering. Early maturing; late leaf senescence.

- **Niver** [NC]. Plants of short stature with thick culms. Grains bold, redish, sweet and nourshing (Fig. O 11).
- **Noor- Meri** [NC]. Plants of medium height; culms erect; leaf angle erect; green basal leaf sheath; leaf blade green and pubscent; ligule white and clefted; collar green, auricle- pale green. Flag leaf inclined. panicles well exerted, compact and awned. Lemma and palea with brown spots and glabrous. Grains shattering type and non scented; late leaf senescence.
- **Nunbeoule** [NC].Tall plants type and vigorous growth; Leaf angle erect; leaf blade green and pubescent; green basal leaf sheath; ligule white and truncate. Collar green; auricle purple; flag leaf erect; panicles compact, awned and fully exerted. Lemma ad palea glabrous and with purple spots on straw. Panicles compact, awned and non-shattering. Grains bold and slightly scented (Fig. O 12).
- **Prenie Baber** [NC]. Medium tall, profuse growth, medium tillering; culms erect, leaves horizontal; leaf blade glabrous with purple basal leaf sheath; erect flag leaf angle; ligule white and clefted; collar purple; auricle purple. Panicles compact, well exerted and fully awned, awns of golden colour. Grains bold, slightly scented and non-shattering. Early maturing.
- **Poot Brear** [NC]. Medium tall variety with profuse tillering; culm stout and erect; leaf angle horizontal, leaf blade green and glabrous; basal leaf sheath green; ligule white and clefted; collar green, auricle pale – green. Flag leaf descending. Panicles semi- compact, with awns of purple colour. Lemma and palea with purple spots and hairs on lemma bed. Grains selender, non- shattering and non-scented (Fig. O 13).
- **Prenie Zager** [NC]. Plants medium tall. Leaf blade pubescence purple; basal leaf sheath purple. Flag laf angle erect; ligule white and clefted, collar and auricle purple; panicles well exerted compact and awnless; lemma and palea glabrous with golden furrows on straw. Leaf senescence late and slow. Grains bold, slightly scented and non- shattering.
- **Punche- Wall** [NC]. Plants of medium height and vigorous growth; leaf blade pubescent and green; basal leaf sheath green; erect flag leaf angle; ligule white and clefted, auricle purple. Panicles compact fully awned with long purple awns. Medium tillering. Panicles threshability loose. Grains bold and non-scented.
- **Qadir Ganai** [RC]. A farmers variety with high yield, normal growth and profuse tillering. Culms erect; leaf blade pubescent and grey; basal leaf sheath green; leaves horizontal; ligule white and clefted; collar green,

auricle pale–green. Flag leaf horizontal. Panicles compact, awnless and well exerted. Lemma and palea with short hairs. Grains bold, scented and shattering type (Fig. O 14).

- **Qadir Baig** [RC]. Medium tall. Leaf angle horizontal; leaf blade pale-green, basal leaf sheath green; ligule white and clefted; collar green; auricle purple. Flag leaf horizontal. Panicles well exerted, compact, partially awned; awns short and purple; lemma and palea glabrous, purple. Grains bold, shattering prone and slightly scented.
- **Rama Hall** [NC}. Tall plant type, with medium tillering. Culm and Flag leaf angle erect; lemma and palea with golden furrous on straw and with long hairs; panicles compact well exerted and partially awned. Grains bold, non- shattering and non- scented.
- **Rehman Bhatti** [RC]. Tall plants, with culm angle erect; leaf angle horizontal. Flag leaf angle descending. Panicle partially exerted, loose and fully awned; awns long and purple. Lemma and palea with purple spots on straw and with hairs on lemma bed. Grains bold, shattering prone and non- scented.
- **Resham** [NC]. Medium tall plants with extensive tillering. Leaf blade green with purple margins. Flag leaf descending. Panicles long compact, awnless and well exerted. Lemma and palea of straw colour and with long hairs. Grains selender, non- shattering and non- scented.
- **Safed Buduji** [NC]. Plants tall, medium tillering and vigorous growth. Flag leaf angle erect. Panicles compact, partly exerted and with awns of straw colour. Lemma and palea with golden furrows on straw and with short hairs. Grains bold, scented and non-shattering.
- **Safed Khuch** [NC]. Plant of medium height and with vigorous growth; leaf blade green; basal leaf sheath dark- green. Flag leaf erect. Panicles compact with short awns of golden colour and partly exerted. Grains bold, non-shattering and non-scented. High head rice recovery. Late maturing.
- **Shahie [NC]**. Medium tall, culm angle erect, leaf angle horizontal. Flag leaf angle semi-erect; panicles compact and awnless and partially exerted. Lemma and palea white and glabrous. Grains bold, shattering prone and non- scented. Leaf senescence late and slow. Quality rice.
- **Shalla Kew** [NC]. Medium tall plants with dark- green leaves. Flag leaf angle intermediate between erect and horizontal. Panicles compact, fully awned long and of straw colour. Lemma and palea with brown spots and glabrous. Leaf senescence slow. Grains bold, shattering prone and non-scented.

- **Siga [NC]**. Tall plant type with medium tillering. Culm angle erect. Flag leaf angle erect. Panicles compact, well exerted and fully awned; awns of straw colour. Lemma and palea brown and glabrous. Grains bold, non- shattering and slightly scented. High yielding but late maturing.
- **Tilla Zag** [NC]. Plants of medium height with extra vigour and profuse tillering. Leaf blade purple, basal leaf sheath purple; ligule white and clefted; colar purple, auricle purple. Flag leaf erect. Panicles compact, fully awned and partially exerted from flag leaf sheath. Grains bold, glume colour dark red, non- shattering type and scented.
- **Trele zagir** [NC]. Tall plants with profuse tillering. Leaf blade dark–green; basal leaf sheath–purple. Ligule white and clefted; colar and auricle pale green. Flag leaf intermediate between erect and horizontal. Panicles compact, partially awned and well exerted. Lemma and palea with purple spots and glaucous. Grains bold, shattering prone and non-scented (Fig. O 15).
- **Tumlahall [NC]**. Medium tall plants, profuse tillering. Culms erect. Leaf angle erect; leaf blade purple and pubescent, basal leaf sheath purple. Flag leaf intermediate between descending and erect. Panicles long, semi compact, partially awned; awns short and purple. Leaf senescence early. Grains cylindrical, non- shattering and non –scented.
- **Tumla zag** [NC]. Medium tall plants with profuse tillering. Leaf blade purple with purple basal leaf sheath. Ligule purple and truncate; collar and auricle purple. Flag leaf angle erect. Panicles compact, fully awned with awns of purple colour. Lemma and palea purple and glabrous. Leaf senescence early. Grains bold with redish kernel, non- shattering and slightly scented.
- **Wata zag [NC].** Plants of medium height and profuse tillering. Leaf angle drooping, leaf blade purple blotched with light purple basal leaf sheath colour. Ligule with purple lines and clefted, colar and aricle- pale green; flag leaf intermediate between erect and horizontal. Panicles well exerted, awnless and open. Grains bold with red glumes, non- scented and non- shattering.
- **Yimberzuol** [NC]. Plants tall, with medium tillering. Leaf blade green with light purple basal leaf sheath. Ligule white and clefted; colar green, and auricle pale green. Flag leaf angle erect. Panicles semi-loose, awnless and well exerted. Lemma and palea reddish to light purple and pubescent. Grains selender, shattering prone, and non-scented (Fig. O 16).

A: State Released Varieties

- **China -972 [NC].** The variety was bred at Rice Research Station, Khudwani and released by State Varietal Release Committee (SVRC) in 1956. Non- shattering but with easy threshability. Grains short and bold, non-scented, and non-glutinous. Suitable for cultivation in the valley upto 1650 m altitude; matures in 145-150 days; yields 5 to 5.5 t ha-1. Non- lodging and cold tolerant at early stages of growth.
- **China-988 [NC].** The variety was bred at Rice Research Station, Khudwani, Anantnag and released by SVRC in 1956. Non-shattering with short and bold, non- scented and non-glutinous grains. Suitable for lower belts of the valley upto 1650 m; matures in 147-150 days; yields 5 to 6 t ha-1. Non-lodging, cold tolerant at early stages of growth (Fig. O 17).
- **China-1007** [RC]. Bred at Rice Research Station, Khudwani, Anantnag and released by SVRC in 1956. Variety is non- shattering; grains are short and bold, non- scented and non-glutinous. Suitable for the lower belts of the valley upto 1700 m; matures in 147-152 days; yields 5 to 6 t ha-1. Non- lodging and cold tolerant at early stages of growth.
- **China-1039 [WC].** A selection from introduced material; bred at Rice Research Station, Khudwani, Anantnag and released by SVRC in 1955. A medium tall variety with intermediate flag leaf angle. Husk-straw coloured at maturity, awnless but shattering susceptible. Short and bold grains, kernel dull white, translucent, non-scented, and non-glutinous. Recommended for lower belts of the valley upto 1650 m; matures in 136-140 days, yields 5.0 - 5.5 t ha-1; photo period sensitive, lodging susceptible but resistant to cold.
- **K- 39 [WC].** A selection from a cross between China 1039 x IR 580; bred at Rice Research Station, Khudwani, Anantnag. Plants medium tall with profuse tillering. Grain bold and longer, translucent, non- glabrous and non- scented. Released by SVRC in 1978 for commercial cultivation in the lower belts of the valley up to 1659 m; matures in 140-145 days, yields 5.8 – 6.2 t ha-1; photoperiod sensitive, resistant to lodging, rice blast and cold.
- **K-60 [RC].** K-60 is a selection from a cross between China 47 x Rikku 132; bred at Rice Research Station, Khudwani, Anantnag and released by SVRC in 1962. Grains non-shattering short and bold, non-scented and non-glutinous. Suitable for cultivation, upto 1700 m; matures in 140-145 days; yields 5 to 5.5 t ha-1. Non- lodging. Cold tolerant at early stages of growth.

- **K-65** [NC]. A selection from a cross between Norin 8 x China 47; bred at Rice Research Station, Khudwani, Anantnag and released by SVRC in 1966. Variety semi tall, lodging susceptible but with easy threshability. Grains short and bold, non- scented, non- glutinous, with low head rice recovery. Suitable for cultivation in the valley up to 1650 m. Matures within 140-145 days, yields 5 – 5.5 t ha-1. Cold tolerant at seedling and grain filling stages.
- **K-78 (Barkat) [RC].** A selection from a cross between Shenei and China- 971; bred at Rice Research Station Khudwani, and released by SVRC in 1974 for cultivation in the valley especially in the higher belts of J& K State up to 1950 m altitude. Grains are long and bold; kernel dull white, translucent and non- scented. Matures in 130 – 140 days; yield, 3.8 to 4 t ha-1 with high head rice recovery. The plants are semi tall, photoperiod sensitive, lodging resistant, moderately resistant to blast and cold tolerant both at seedling and grain filling stages.
- **K-332 [NC].** A selection from a cross between Shenei and Norin II; bred at Rice Research Station, Khudwani, Anantnag. A medium tall variety. Grains long and bold, kernel dull white, translucent, non- scented and non-glutinous. Released by SVRC in 1982 for cultivation in the upper belts of J&K state up to 2000 m altitude; matures in 130-140 days; yields 4–4.5 t ha-1, photoperiod sensitive, lodging susceptible, moderately resistant to blast and cold.
- **K-429 (Kohsaar) [RC].** Selection from a cross between Shin-el and Ginmasari, bred at RR&RS, Khudwani Sub- Station, Larnoo, of SKUAST-K and released in 2001. Plants semi tall (75-80 cm) and non –lodging. Flag leaf angle erect, panicle compact, fully exerted and partly awned. Apiculus straw coloured at maturity. Grains short and bold, kernel dull white and translucent. Recommended for cultivation in high altitude areas of state, above 1700 m altitudes. Fertilizer responsive. Yields 4.0 to 4.5 t ha-1. Matures in 140-145 days. Tolerant to blast, *Helminthosporium* blight and tolerant to cold at seedling and grain filling stages.
- **Shenei [RC].** A selection from exotic material; bred at Rice Research Station, Khudwani, Anantnag, Kashmir and released by SVRC in 1967. A medium tall variety with short and bold grains, which are translucent, non- scented, and non-glutinous. Recommended for commercial cultivation in the higher belts of J&K state upto 1969 m altitudes; matures in 130-140 days; yields 3.8 to 4 t ha-1. Photo-period sensitive, non-lodging. Moderately resistant to blast and cold.

- **SKAU-23 (Chenab) [RC].** A selection from a cross between K 21-9-10 -1/IR 2053-521-1-1-2; bred at Rice Research and Regional Station, Khudwani, SKUAST-K & released by SVRC in 1993. Medium tall, erect and compact plant type, intermediate flag leaf angle and green apiculum. Panicle semi-compact, awnless; straw coloured lemma and palea. Medium bold grains; dull white kernel; translucent and non-scented. Suitable for lower belts of Kashmir valley and mid elevations of Jammu. Variety matures in 135-141 days with a productivity potential of 6.0 to 6.5 t ha -1.
- **SKAU- 27 (Jehlum) [WC].** A Selection from a cross between JAKUKU and IR 1444; bred at RR&RS, Khudwani, SKUAST-K & Released by SVRC in 1993. Medium tall, erect and compact plant type; green foliage, erect flag leaf and green apiculum which turns straw coloured at maturity; ligule white and colar green. Panicles straight and awnless Grains medium bolds, translucent, non-scented, whitish, non-glutinous with good cooking and milling qualities. Recommended for lower belts of Kashmir valley upto 1650 m altitudes. Matures in 136-141 days. Productivity potential of 6.0- 6.5 t ha-1 (Fig. O 18).
- **Shalimar Rice-1 [RC].** Developed by SKUAST-K and released by SVRC for general cultivation during 2004 in lower belts of the valley upto an altitude of 1650 m. The plant type is medium tall, extra vigorous and compact. Flag leaf is erect with purple basal leaf sheath colour and delayed leaf senescence. The grains are short, slender with non-glutinous endosperm and non-aromatic. Variety matures in 138-145 days and yields 6.5- 7.0 t /ha.
- **Shalimar Rice-2 [RC].** Rice variety developed and released by SKUAST of Kashmir in 2012 for cultivation in water loged low lying areas of Kashmir valley. The variety tested as SKUA-341 is a derivative of a cross between VL Dhan 221 and K-39. It posseses resistance to lodging, shattering and blast. Grains are medium bold, non-aromatic with good cooking quality, high milling and hulling recovery. Average yield ranges between 8.5 to 9.0 t / h (Fig. O 19).
- **Shalimar Rice-3 [RC].** Rice variety developed and released by SKUAST Kashmir in 2012 for cultivation in the plains of Kashmir and mountainous areas of Jammu.The variety is a derivative of the cross between IR 32429 -3-2-2 & K 438 and is suitable for early sown conditions under intense management conditions. Grains are medium bold, non-glutinous and non-aromatic. Matures within 132-135 days and yields 8.0 -8.5 t /h (Fig. O 20).

Osmanthus fragrans var. aurantiacus Makino (Oleaceae): Eng.- Fragrant Olive [NC]. An upright shrub or small tree with oblong to lance-shaped, leathery, toothed, glossy, dark green leaves. Flowers deliciously and strongly frangrant, yellowish orange, produced in late summer. Introduced in Kashmir as an ornamental, from China.

Oxalis acetosella L. (Oxalidaceae): Eng.- Wood Sorrel; Kash.-Tchouke - tchin [W]. A tender perennial rhizomatous shrub growing in wild on moist shady places. Leaves long stalked, trifoliate. Pearl white funnel shaped flowers appear from March to May. The leaves possess a refreshing sub-acid flavour, used as vegetable and in salads. Medicinally used to treat liver troubles and digestive disorders.

Oxalis corniculata L. (Oxalidaceae): Eng.- Creeping Sorrel; Hind- Khati booti; Kash.- Tchoke tchin [W]. Perennial herb,usually in shaded places and Himalayan grasslands. Stem creeping. Leaves trifoliate. Flowers yellow. Leaves eaten raw or cooked with other vegetables to give them acidic taste (Fig. O 21).

Oxyria digyna (L.) Hill (Polygonaceae): Eng.- Alpine Sorrell; Ladakhi-Chumcha [W]. Perennial tufted herb usually along high mountain glaciar streams at 3000- 3800 m. Leaves kidney shaped, fleshy. Flowers small, green and later redish, grouped in upright clusters.The shoots are kept in luke warm water and taken in the morning as an appetizer. Leaves used in preparation of chutnies (Fig. O 22).

Oxytropis japonica Maxim. (Fabaceae): Eng.- Japanese Locoweed; Ladakhi- Rashkun [W]. Short stemmed alpine/sub-alpine, hairy herb; leaflets 13-23, heads dense flowered. Powdered seeds used as diuretic and to decrease obesity. Flowers pinkish.

Oxytropis mollis Benth. (Fabaceae): Eng.- Soft Locoweed; Ladakhi-Shukshing [W]. An erect perennial herb of Himalayan grasslands above 2700 m. Rootstock woody; flowers pale-pink borne in long stalked condensed racemes. Extract of the powdered plant is used to cure cutaneous eruptions.

Oxytropis tatarica Baker. (Fabaceae): Ladakhi– Isarkash [W]. Short stemmed perennial herb of alpine grasslands. Roots thick and woody. Leaves stipulate, imparipinnately compound, densely pilose, hairs white. Calx densely pilose; corolla blue. Decoction of the whole plant used against ascites (Fig. O 23).

Fig. O 1. Oenothera glazioviana

Fig. O 2. Onosma hispida var kashmirica

Fig. O 3. Ophioglossum vulgatum

Fig. O 4. Origanum vulgare

Fig. O 5. Orchis latifolia

Fig. O 6. Orthosiphon rubicundus

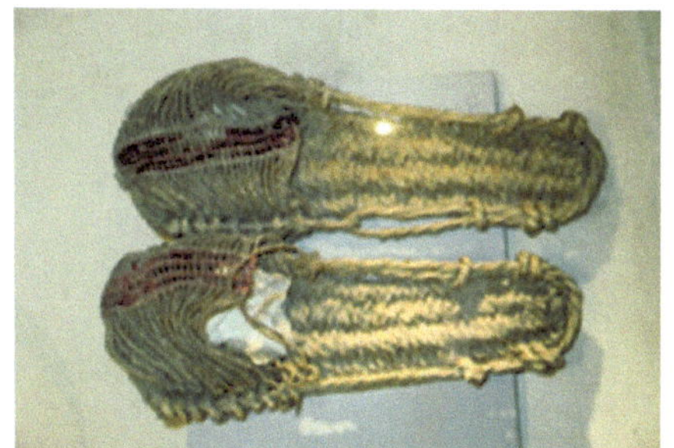

Fig. O 7. Shoes made from rice straw

Fig. O 8. var. Baber

Fig. O 9. var. Buta Baber

Fig. O 10. var. Kamad

Fig. O 11. var. Niver

Fig. O 12. var. Nun Buole

Fig. O 13. var. Poot Brear

Fig. O 14. var Qadir Ganai

Fig. O 15. var. Trel zagir

Fig. O 16. var Yamberzoule

Fig. O 17. var. China-988

Fig. O 18. var. Jhelum

Fig. O 19. var. Shalimar Rice-2

Fig. O 20. var. Shalimar Rice-3

Fig. O 21. Oxalis corniculata

Fig. O 22. Oxyria digyna

Fig. O 23. Oxytropis tatarica

P

Paeonia emodi Royle. (Paeoniaceae): Eng.- Himalayan Paeony; Hindi- Udsalap; Kash.- Mamekh [W]. A shurby perennial, found in W. Himalayas from Kashmir to Kumoan at 2400- 3300 m altitudes. Flowers yellowish white with a bunch of orange-yellow stamens. Leaves compound, 1-2 ft long with lance-shaped leaflets. Tubers used for uterine diseases, collics, bilious obstructions, dropsy, epilepsy, convulsions and hysteria. Seeds are emetic (Fig. P 1).

Paeonia lactiflora Pall. (Paeoniaceae): Eng.- Chinese Peony [NI]. A shrubby perennial grown for its handsome flowers and attractive foliage. Leaves compound, deep green. The scented, pure white single flowers have yellow stamens and appear from May to June.

Paeonia officinalis L. (Paeoniaceae): Eng.- Garden Peony [NI]. A herbaceous perennial ornamental with deeply cut light green leaves. Single crimson bowl shaped flowers appear from May to June.

Paeonia suffruticosa Andr. (Paeoniaceae): Eng.- Chinese Tree Peony [NI] An upright, sparsely branched, deciduous shrub. Leaves bi-pinnate. Flowers large, red, rose red to white, mostly solitary and fragrant.

Panicum miliaceum *L.* **(Poaceae): Eng. – Common Millet / Hog Millet; Hindi- Chin /Morha; Kash.- Pinga** [RC]. A grass species grown for its seed.Traditionally millets used to be grown under rainfed conditions and in lands where irrigation was rather impossible. As a food it was often relished by tribal people as a substitute to rice while as it was abused and denounced by Kashmiris as being cold. It used to be pounded in motars like rice but was difficult to cook. At present its cultivation is restricted to few pockets of Gurez and Kargil.

Papaver cambricum L. (syn. Meconopsis cambrica (L.) Vig. (Papaveraceae): Eng.- Welsh Poppy; Hindi- Kant Swanyus [NI]. Herbaceous perennial ornamental plant with poppy like flowers. The deep dissected mid-green leaves are slightly hairy and form basal tufts. Flowers showy, yellow or orange in colour.

Papaver kachroianum Tabinda, Dar & Naqshi. (Papaveraceae): Eng.- Wild Poppy. [W]. Annual herb inhabiting sub-alpine grasslands. Leaves hairy, lobed. Flowers rosy-pink.

Papaver nudicaule L. (Papaveraceae): Eng.- Iceland Poppy [W /UC]. A hairy perennial poppy growing on open alpine areas. Flowers bowl shaped, orange-yellow; sepals denely hairy. The smooth, soft green radical leaves are long petioled, deeply lobed and form a basal rosette.

Papaver pamporicum Tabinda, Dar & Naqshi. (Papaveraceae): Eng.- Wild Poppy. [W]. Annual herb of Karewa lands. Leaves light green, lobed and hairy. Flowers crimson, showy.

Papaver orientale L. (Papaveraceae): Eng.-Oriental Poppy. [NI]. A perennial poppy with large red to orange flowers. Leaves hairy, deep cut, deep green. Flowers scarlet with black blotch at the base of petals. Many garden varieties are under cultivation.

Papaver rhoeas L. (Papaveraceae): Eng. -Wild Poppy / Field Poppy; Hindi– Lalpost; Kash.- Gulala [W]. An annual herbaceous plant growing mostly wild as a weed in wheat/oat fields and orchards. Leaves pale- green and deeply lobed. Flower stems covered with coarse hairs. Flowers are deep red to rose-red with a dark centre and appear from May to June. Seeds have analgesic properties and are used as mild narcotic, and as good tonic useful in low fevers. The cooked roots are used for rheumatism (Yunani). In villages plants during their junnile phase are cooked as vegetable (Fig. P 2).

Papaver somniferum L. (Papaveraceae): Eng.- Opium Poppy; Hindi- Post; Kash.- Khashkhash [RC]. A hardy annual cultivated for seed as well as obtaining opium raw drug and for medicinal purposes. Plants attain a height of 60 to 80 cm, have deeply lobed, smooth, pale grey-green leaves. White, red, pink or purple flowers are produced from June to August followed by bulbous, flat- capsules. Dried exudates locally known as *Afin* is used traditional and ayurvedic system of medicine for treating various respiratory and throat related ailments. Crushed seed paste is used as a substitute to milk.Seeds also used in bakeries.

Parmelia perlata (Pameliaceae): Eng.- Stone Flower; Kash.- Paule maenze; Ladakhi- Dotak marpo/ Dotak magno [W]. A reddish-brown foliose lichen species growing on mountain rocks. In Ladakh they are used for treating chronic fever and as antidote for poison.

Parnassia nubicola Wall. ex Royle (Celastraceae): Eng.- Himalayan Bog Star; Ladakhi- Garsenarpo [W]. Glabrous perennial, scapigerous herb growing in alpine slopes at 2800 – 4000 m. Flowers white, solitary, borne on a slender stem. Leaves elliptic- heart shaped. Dried seed and root powder is used to treat senility and other nervous disorders.

Parnassia palustris L. (Celastraceae): Eng.- Marsh grass of Parnassus [W] A perennial herb found in moist woods of Kashmir. Flowers solitary, white with green venation, on terminating stems. Leaves basal, rosette, alternate. Decoction of the plant used as a sedative in nervous disorders.

Parnassia paxmanni Pall (Celastraceae): Ladakhi- Pangyan karpo [W] Perennial herb of wetlands of Himalayas. Plant used for stomach upsets and fevers.

Parrotiopsis jacquemontiana (Decne.) Rehder. (Hamamelidaceae): Kash. – Hatab/ Poh [W]. A large upright gregarious shrub or sometimes small tree with grey bark and rounded or broadly ovate, strongly veined, toothed leaves. Flowers borne on short lateral shoots usually with young leaves consisting of yellow anthered stamens and surrouned by 4-6 large conspicuous creamy petal-like obovate bracts. Branches used for making walking sticks and for handling agricultural implements. Young twigs used as cane frame of *Kangris* and baskets (Fig. P 3).

Parthenocissus semicordata (Wall.) Planch. (syn. Parthenocissus himalayana (Royle.) Planch. (Vitaceae) [NI]. A vigorous, self-clinging vine, almost as ubiquitous as the common ivy. Leaves long petioed, partly simple, broadly ovate, with 3 acuminate, coarsely serrate lobes, fall colour orange yellow and scarlet red, tendrils short but very well branched, with disc like hold fasts. Flowers yellow-green followed by dark-blue bloomy fruits in November.

Parthenocissus quinquefolia (L.) Planch. (Vitaceae): Eng.- Virginia Creeper [NI] A tall growing, more or less self-clinging vine, with leaves 5 parted, leaflets coarsely serrated, dull green above, bluish beneath, bright red and carmine in tendrils, terminating in disc like hold fasts.

Flowers golden-green in large terminal panicles. Fruits blue-black, slightly pruinose.

Paspalum conjugatum P.J. Bergius. (Poaceae): Eng.- Sour Paspalum [W]. Perennial pasture grass. Culms 20-60 cm long, creeping stoloniferous, rooting at nodes. Occurs throughout, in grasslands and on slopy mountain sides.

Passiflora caerullea L (Passifloraceae): Eng.- Passion Flower [NI]. A fast growing climber with moderately branching 4 angled, grooved stems bearing rich broad cordate green leaves. Bears solitary white to pinkish flowers, on long thin stalks in the leaf axils of the younger shoots, somewhat fragrant. Fruits ovoid, orange-yellow.

Paulownia tomentosa Steud. (syn. Paulownia imperialis Sieb. & Zucc.) (Paulowniaceae): Eng.- Foxglove Tree [NI]. A beautiful, broadly columnar tree with ovate, mid green, heart shaped leaves. Flowers fragrant, pale lilac with purple–speckled throats. Cultivated as an ornamental.

Pedicularis bicornuta Klotzsch. (syn. Pedicularis bicolour Diels.) (Orobanchaceae): Eng.- Horned Lousewort; Ladakhi- Langioulet [W]. A robust, erect perennial, upto 60 cm tall, bearing a large dense cluster of large globular yellow flowers inhabiting open alpine slopes of Ladakh. Leaves pinnatifid. The extract of the leaves is used against fevers, rheumatism and gout. Flowers have religious significance as they are specially offered to Godess Kali (Fig. P 4).

Pedicularis breviflora Regel. (Orobanchaceae) [W]. A perennial herb, upto 30 cm tall inhabiting sub-alpine grasslands, forest openings at 1500 to 1800 m altitudes. Stem erect, unbranched, finely pubescent. Basal leaves long petioled, stem leaves alternate, short petioled, pinnatisect, segments narrowly lanceolate. Inflorescence spicate, densely grey pubescent, bracts larger than calyx. Calyx campanulate– tubular. Corolla purple. Plant has been a source of scent.

Pedicularis cheilanthifolia var. albida (Pennel.) P C Tsoong (syn. Pedicularis albida Pennel.) (Orobanchaceae): Eng.- White Lousewort; Ladakhi- Langna [W]. Perennial herb (20-30 cm tall) of Kashmir Himalayas inhabiting alpine slopes at 2100 – 5200 m altitudes. Stem erect, unbranched with four lines of hairs. Flowers white borne in a spike, stalkless; sepals densely pillose along veins. Leaves linear, lance shaped, pinnately compound, cut into 8-12 pairs of segments. Powdered flowers and leaves, as poultice, are applied on swollen joint to relieve pain and oedema (Fig P 5).

Pedicularis longiflora var. tubiformis (Klotzsch) Tsoong (Orobanchaceae): Eng.- Long tube Lousewort [W]. A perennial herb with several erect and spreading stems. Flowers golden yellow in a cluster at the top of stem. Flower tube hairy. Leaves oblong blunt in outline, with many oblong toothed lobes. Plant inhabits alpine grasslands above 3000 m altitudes.

Pedicularis pectinata Wall. ex Benth. (Orobanchaceae): Eng.- Marsh Lousewort; Hindi- Mishran; Kash. Mishran/ Kousturi; Ladakhi- Lagtiat [W]. Perennial herb, usually along damp places in forests and alpine pastures. Leaves pinnatifid, toothed. Flowers rose, pink, in dense spikes. Powdered arial portion is used as astringent in the morning. Powedered leaves are used as antidote against poisoning in *Amchi* system of medicine (Fig. P 6).

Pedicularis punctata Decne. (Orobanchaceae): Eng.- Kashmiri Louswort; Ladakhi- Lungro-marpo [W]. Perennial herb, mostly along alpine water drains at 2700- 4500 m. Flowers purple with whitish blotch. Leaves elliptic with broad, ovate finely toothed lobes. Seedlings are consumed fresh in case of poisoning. Powdered flowers are used as a sedative.

Pedicularis rhinanthiodes Schrenk (Orobanchaceae): Eng.- Rattle Lousewort [W] A perennial herb inhabiting damp himalayam meadows at 3300- 4800 m altitudes. Plant has several stems. Flowers bright or pale-pink in a dense short cluster. Flower tube is slender, upper lip is sickle shaped; sepals 4 toothed and hairy.

Pedicularis siphonantha D. Don. (Orobanchaceae):Eng.- Tubewort; Ladakhi- Lungro-marpo [W]. Perennial herb, common along water drains in alpine meadows and swampy locations in Kashmir and also in Ladakh. Flowers, tubular, rose-red with white blotch. Plant used against uninary tract infections. In Ladakh the powdered plant used as sedative and as antidote for food poisoning (Fig. P 7).

Peganum harmala L. (Zygophyllaceae): Eng.- Wild Rue; Hindi- Gandhya /Ispanda; Kash. –Isband; Ladakhi-Techepak [W]. Perennial shrubby herb found in grasslands and Karewa slopes of Kashmir. Leaves pinnately divided into linear acute lobes. Fruit globular with dark brown seeds. Seeds used for treatment of asthma, hysteria, jaundice, colic pain and painful mensuration. Seeds of the plant have religious and social importance as they are put to burning charcoal on special occasions to avert evil eye. Dried powdered and roasted seeds used as a narcotic in Ladakh. Seeds have also been used to kill lice (Fig. P 8).

Pelargonium graveolens L' Herit (Geraniaceae): Eng.- Rose Geranium [NI]. A tender evergreen/deciduous plant suitably cultivated as pot plant. Leaves palmate, deeply lobed and aromatic. Rose- pink flowers appear on terminal umbels from June to October.

Pelargonium x hortorum L.H. Bailey (Geraniaceae): Eng.- Garden Geranium. [NI] A hybrid race of *Pelargonium* distinguished by their rounded, pale to mid-green leaves with conspicuous zones of bronze or maroon. The flowers are borne in dense rounded umnbels from upper leaf axils from May to October. Colors include white, pink, red and orange.

Perovskia abrotanoides Karel (Lamiaceae): Ladakhi- Burtche [W]. A sub-shrub, wide spreading, aromatic with thick woody rootstock. Leaves bipinnate with dendroid stellate, glandular hairs. Flowers blue in terminal spiks. In Ladakh dried plants are used as thatching material for houses.

Persicaria alpina (All.) H. Gross (syn. Polygonum alpinum All.) (Polygonaceae): Eng.- Prince's Feather / Alpine Knotweed; Hindi- Baqli hafiza; Kash.- Tchoke ladder *[W].* A tall perennial herb growing in Kashmir Himalayas at 1800 m - 3000 m as forest undergrowth. Inflorescence terminal, paniculate. Flowers white. Foliage and twigs are sour in taste and used as a vegetable by Gujars. Also given to cattle for curing dysentery. Roots used for rheumatic pain (Fig. P 9).

Persicaria amphibia (L.) Delarbre (syn. Polygonum amphibium L.) (Polygonaceae): Eng.- Water Knotweed [W]. Tall perennial, rhizomatous, acquatic cum land herb, common along water reservoirs and grasslands. Flowers pink, five lobed borne in dense terminal clusters. Leaves and rootstock used for tanning leather.

Persicaria amplexicaulis (D. Don.) Ronse Decr. (syn. Polygonum amplexicaule D. Don (Polygonaceae): Eng.- Mountain Fleece / Red Bistort; Kash.- Maichran [W]. A perennial rhizomatous herb very common in forests and on humus rich damp soils. Rhizomes horizontal, purple–brown. Flowers rose red in dense spikes. Leaves used as vegetables and also given to patients with bleeding piles and for checking profuse mensuration. Rhizomes were being used as substitute to tea (Fig. P 10).

Persicaria hydropiper (L.) Delarbre (syn. Polygonum hydropiper L.) (Polygonaceae): Eng.- Water Pepper; Kash.- Marche-wangan Gausse; Ladakhi- Chumerche [W]. Annual herb along rice fields and other moist places. Leaves alternate, sessile with blunt apex. Flowers pink.

The seeds are placed in water and boiled for 2-3 days. On cooling the extract is used as a diuretic and to decrease obesity. Leaves used for treating lung infection in animals.

Persicaria nepalensis (Meisn.)Miyabe (syn. Polygonum nepalense Meisn.) (Polygonaceae): Eng.- Nepalese Smartweed [W]. Erect or spreading annual herb, usually in damp situations. Leaves glabrous, alternate, elliptic often with a pair of dark blotches on each side of midrib. Leaves used in preparation of soups and stews. Flowers pink borne in clusters (Fig. P 11).

Persicaria vivipara (L.) Ronse Decr. (syn. Polygonum viviparum L.) (Polygonaceae):Eng.- Alpine Bistort; Kash.- Katur [W]. A perennial rhizomatous herb of mountainous areas. Stem erect, unbranched and hairless. Leaves hairless above and grey-green and hairy below. Flowers tiny white or pink borne on upper part of the spike. Lower flowers are replaced by bulbils. Whole plant used for diarrhoea and dysentery and lung diseases. Plant serves as an excellent fodder to wild life at high altitudes.Rhizomes were being used as substitute to tea (Fig. P 12).

Petunia hybrida Vilm. (Solanaceae): Eng.-Petunia; Kash.- Sornai Posh [NI] A half hardy perennial but grown as annual. Plant sticky with ovate mid to dark- green leaves. Flowers are trumpet shaped and showy. Colours varieties include cream, pink, white, red, mauve and purple, and blue. Several varieteies belonging to *Multiflora* and *Grandiflora* sub-groups are cultivated in gardens, parks for edging beds, and hanging baskets.

Phalaris minor Retz. (syn. Phalaris nepalensis Trin.): Eng.- Small Canary Grass [W] An important grass species distributed throughout North West Himalaya as an important element of grazing grounds.

Phaseolus lunatus L. (Fabaceae): Eng.– Lima Bean; Hindi – Bean; Kash.- Maz hema [WC]. Annual tall growing twiner. Leaves trifoliate; flowers white, pods flat, pubescent with a sharp beak; seeds rhomboidal flat and thin, white, brown, red or speckled. It is mostly consumed as shelled green beans.

Phaseolus vulgaris L. (Fabaceae): Eng.- Kidney Beans; Hindi– Rajmash; Kash. - Razmah. [WC]. A herbaceous annual grown for its seed or unripe fruit. Many kidney bean types, differing in habit and seed characters, are cultivated across Kashmir, mostly in Kandi rainfed areas (Fig. P 13). Dark to light red seed coloured types are mostly preferred by people as pulse. Types like *Sorekh Razmah, Narwan Razmah, Shopian Razmah, Baeeeru Razmah* are the most common types used as pulse.

Other cultivated varieties include:

- **Kashmiri Red [UC]:** A sub erect or twining annual. Leaves trifoliate, white to violet purple pods; slender straight or slightly curved, sides covex or rounded, surface glabrous or jointly pubescent, beak prominent. Seeds more or less kidney shaped elongated or nearly globular and dull red/dark red in colour.
- **Shalimar Rjamash-1 [RC]:** Developed by SKUAST-K during 2004 from a cross between Local Red and Canadian Red. It has erect bushy plant type with determinate growth habit and green stem colour. Pod shape is uniform with curved smooth surface and with slightly kidney shaped seeds. The pods on an average bear 4-6 seeds with test weight ranging from 36-37 g.

For tender pods both climbing as well as dwarf types are under cultivation. Fleshy tender pods are used fresh as well as after sundrying. Types used for tender pods include.

- **Shalimar French Bean-1 [RC]:** Developed by SKUAST-K through single plant selection form the open pollinated HOSE-1 population. Leaves are purple coloured where as pods are deep purplish green. The variety has medium maturity and pods are ready for picking after 55-60 days of sowing. Pods are stringless, plump, cylindrical without hollow cavity and with thick and tender flesh. The average green pod yield potential of the variety is 50 q ha -1 Other introduced cultivars include *Contender, Tender Green* and *Kentucky Wonder*.

Philadelphus coronarius L. (Hydrangeaceae): Eng.- English Dogwood [NI]. A broadly upright, deciduous ornamental shrub. Leaves ovate to ovate-elliptic, bright golden yelolow when young. Flowers creamy-white, very fragrant in racemes.

Philadelphus tomentosus Wallich ex G. Don. (Hydrangeaceae): Eng.- Mock Orange [NI]. A deciduous shrub with light grey bark. Leaves ovate lanceolate, long pointed, with prominent veins. Flowers cup shaped and fragrant.

Philadelphus x virginalis Rehder. (Hydrangeaceae): Eng.- Himalayan Mock Orange [W/UC]. A vigorous, upright, deciduous ornamental shrub. Leaves ovate and acuminate at base, dentate, nearly glabrous above and pubescent beneath. Flowers double or semi-double, very fragrant, pure white produced in loose racemes.

Phleum alpinum L. (Poaceae): Eng.- Alpine Timothy [W]. A perennial bunch grass, 20-40 cm tall. Inflorescence is cylindrical to oval mass of spikelets, upto 6 cm long. A widely distributed grass type of alpine pastures at 2500 -3000 m.

Phleum himalaicum Mez. (Poaceae): Eng.- Himalayan Timothy [W]. Annual herb growing on mountainous passes of Himalayas and constitutes an important fodder plant. Culms 8- 20 cm tall, usually erect. Leaf blades pale-green, glabrous.

Phlomoides bracteosa (Royle. ex Benth.) Kamelin & Makhm. (syn. Phlomis bracteosa Royle ex Benth.) (Lamiaceae): Eng.- Purple Jerusalem Sage; Kash.- Phaghurn; Ladakhi- Dandeait [W]. Perennial tomentose, herb on rocky slopes,near moist situations. Leaves petioled, ovate, obtuse, crenate. Flowers purple. Powdered leaves and shoots are used as blood purifier (Fig. P 14).

Phlox drummondii Hook (Polemoniaceae): Eng.- Drummond Phlox [NI]. A half hardy annual with lanceolate, light green leaves. The flowers are borne in dense heads from July to September and are in varying shades of pink, purple, lavender red and white.

Phormium tenax L. R. Forst. & G. Forst. (Xanthorrhoeaceae): Eng.- New Zealand Flax [NI]. A striking evergreen under-shrub grown for foliage effect, forming clumps of rigid, leathery somewhat glaucous, sword-like leaves. A superb architectural plant for creating contrasting and diverse effects. Flowers full red borne on branched panicles.

Phragmites australis (Cav.) Trin. ex Steud (syn. Phragmites communis Trin.) (Poaceae): Eng.– Common Reed [W]. Tall perennial herb,usually in shallow water ponds and damp places forming extensive stands. Flowers produced in dense dark purple panicles. Leaves were being used for thatching house boats, for making ropes and for plastering of walls. Also used as fodder for horses.

Phyllostachys nigra (Lodd. ex Lindl.) Munro (Poaceae): Eng.- Black Bamboo [NI]. A hardy plant, with culms and spikelets green but becoming black in second year, joints prominent. Leaves thin, pointed, finely serrate on one edge, closely tessellate. Grown as ornamental.

Physochlaina praealta (Decne.) Miers. (Solanaceae): Eng.- Tall Physochlaina; Ladakhi- Lungthung [W]. A tall poisnous herb growing wild on stony slopes in Ladakh region up to an altitude of 4000 m amsl. Flowers greenish-yellow with purple veining. Flowers funnel shaped, borne in lax, velvety corymbs. Leaves ovate, velvet hairy midrib prominent on underside. It has a great medicinal importance in Ladakhi system of medicine as its leaves are rich in hyocyamine and hysocine and is used in treating boils and as vermifuge and emetic. Powdered roots are taken as a sedative.

Phytolacca acinosa Roxb. (Phytolacaceae): Eng.- India Pokeweed; Hindi-Matazor; Kash.- Louber Hakh. [W]. A perennial glabrous herb growing in Kashmir above 2000 m on hill sides and forest margins. Stems erect, redish purple, terete, longitudinally grooved, fleshy. Flowers yellow-green. Berries purplish black at maturity. Its succulent leaves are cooked as vegetable by Gujars. Fruit with dark purple carpels, used as a flavouring agent (Fig. P 15).

Physalis alkekengi var. franchetii (Mast.) Makino (syn. Physalis franchetii Mast.) (Solanaceae): Eng.- Bladder Cherry/Chinese Lantern [NI]. A tender herbaceous ornamental grown for its bright orange to red papery fruits. Leaves mid- green deltoid to ovate, narrowing towards stalk.

Flowers white appear in June to July followed by globular orange fruits enclosed in an orange- red papery calyx.

Picea smithiana (Wall.) Boiss. (syn. Picea morinda Link.) (Pinaceae): Eng.- West Himalayan Spruce; Hindi- Bajur; Kash.- Katchul [W]. A large columnar and extremely beautiful evergreen tree distributed in silver fir forests at 2400-3000 m altitudes with branches upcurved at tips and long, pendulous branchlets, scaly grey bark, and pale brown shoots. Leaves mostly dark green, needle like, longest of any spruce; female cones cylindrical, green, later shining bright brown (Fig. P 16).

Picrorhiza kurroa Royle ex Benth (Plantaginaceae): Kash.- Chobi- kor; Hindi– Katki /Saiful mulook; Ladakhi– Honglan/Mikhe [W]. A low medicinal herb with underground long rhizomatous stolons, found in alpine Himalyas at 2700-4500 m altitudes. Leaves coarsely toothed, narrowed to a winged stalk. Flowers purplish-blue in compact spikes. Roots are used in treating liver and stomach aliments, rheumatism and cholera. Also used as tonic for horses (Fig. P 17).

Pimpinella acuminata (Edgew.) C B Clarke (Syn. Pimpinella hazariensis H. Wollf.) (Apiaceae) [W]. Biennial glabrous erect herb of alpine forests at 3000- 3700 m altitudes. Roots cylindrical.Stem little branched, often tinged at base. Leaf blade ternate. 2-pinnate; leaflets abaxially pubescent along veins,margins irregularly incised. Flowers white. Roots used medicinally.

Pinus canariensis C. Sm. (Pinaceae): Eng.- Canary Island Pine [NI]. A graceful pine with spreading branches and drooping branchlets with fissured, red brown bark, yellow shoots and large, ovate buds. Leaves in threes, yellow-green, 2-3 cm long and drooping. Male cones ellipsoid-ovoid; female cones borne solitary or in clusters. Grown as ornamental in public landscapes.

Pinus gerardiana Wallich ex D Don. (Pinaceae): Eng.- Neosia Pine; Kash.- Chilgoza [W]. A large tree with beautiful patchwork bark, thin smooth, greyish pink, flaking to reveal green, yellow and brown new bark. Leaves in threes, dark green; mature cones glaucous with very thick woody scales with stout recurved apices. The seeds, known as *Chilghoza* are edible. Grows in mountainous ranges at 2000- 3000 m in association with blue pine and cedars (Fig. P 18).

Pinus halepensis Miller (Pinaceae): Eng.- Aleppo Pine [NI]. A medium-sized ornamental tree with glaucous young shoots. Leaves in pairs, slightly twisted, bright fresh green; cones conic- ovoid, spreading or deflexed, short stalked, yellowish or reddish brown & glossy, persistant for several years.

Pinus pinea L. (Pinaceae): Eng.- Stone Pine. [NI]. A very distinct tree of small to medium size developing a characteristic dense, flat-topped or umbrella-shaped head. Leaves in pairs, glossy green, sharply pointed; solitary ovoid, shining brown female cones ripen in the third year and have wingless seeds; bark furrowed with grey-green plates.

Pinus radiata D. Don (Pinaceae): Eng.- Monterey Pine [W]. A narrow, conical tree, becoming broadly domed, with heavily ridged black grey green shoots, and cylindrical buds. Leaves in threes, slender, bright green, shining densely crowded on the branchlets; ovoid, glossy, yellowish female cones with 20 swollen outer scales, borne in whorls along the branches, often remaining intact for many years. A Himalayan forest conifer.

Pinus roxburghii Sarg. (syn. Pinus longifolia Roxb. ex Lamb.) (Pinaceae): Eng.- Long leaved Pine; Hindi– Chir; Kash.- Yaer / Snowber [W]. A large tree, 30- 50 m tall, of Himalaya forests, with very thick and deeply fissured rough bark; young shoots clothed with scale like leaves; leaves dark or bright green, needle like, borne in clusters of 3 on short shoots; cones solitary or clustered, ovoid-conical.Wood used for construction works.

Pinus strobus L. (Pinaceae): Eng.- Weymouth Pine [W]. A large dominant forest tree of conical habit when young, later developing a rounded head. Leaves in fives glaucous green, pendent on slender stalks, cylindrical, tapered green; female cones ripening to brown.

Pinus wallichiana Jackson (syn. Pinus excelsa Wall. ex D. Don.) (Pinaceae): Eng.- Indian Blue Pine; Hindi – Kail [W]. An elegant, symmetrical pyramidal coniferous tree growing in mountain valleys at altitudes of 1800 -4500 m; when young, developing a broad, domed crown, retaining its lowest branches when isolated. Olive green shoots with cylindrical conical buds. Leaves in fives, blue green, slender and drooping with age; cone stalked, solitary or in clusters of 2-3, banana shaped, ripening to brown.Wood used for construction and furniture works. The leaves of the plant were being used for washing clothes (Fig. P 19).

Piptatherum aequiglumis (Duthie ex Hook. f.) Roshev. (syn. Oryzopsis aequiglumis Duthie ex Hook.f) (Poaceae) [W]. A perennial grass species of North-West Himalayas with blackish brown fruiting bodies. Awns long without constriction.

Piptatherum gracile Mez. (Poaceae): Ladakhi- Chipkiang [W]. Perennial caespitose herb. Leaves pubescent, hairy adaxially. Its leaves are used for making large baskets in Ladakh known as *Tschekpo* for carrying vegetables.

Pistacia chinensis Bunge. subsp. Integerrima (Stewart ex Brandis)Rech.f (Anacardiaceae): Eng.- Pistacio [NI]. A deciduous ornamental tree, erect at first, then spreading. Pinnate leaves with oblong-elliptic, asymmetrical, toothed, leathery, glossy, dark green leaflets, with no terminal leaflets; fall foliage a beautiful carmine. Aromatic, red flowers are produced with the young leaves; the males are in crowded panicles, the female in loose panicles.

Pistacia vera L. (Anacardiaceae): Eng.- Green Almond; Hindi- Pista; Kash.-Pista [W]. A small ornamental tree with long petioled, odd-pinnate, leaflets usually 3-5, ovate to obovate, sessile, pubescent on both sides. Flowers in conspicuous panicles, followed by small, reddish fruits.

Pisum sativum subsp. arvense (L.) Asch. & Graebn. (syn. Pisum sativum L.) (Fabaceae): Eng. – Green Pea; Hindi– Mattar; Kashmiri- Mattar [UC]. Herbaceous annual ultivated for its pods and seeds. Commonly cultivated varieties include.

- **Local Peas:** A small round seeded variety being traditionally used as pulse, vegetable. Pods are smaller than introduced varieties.
- **Arkel:** An introduced wrinkled seeded, high yielding dwarf variety which gives fruit in early May under valley conditions.
- **Bonneville:** Wrinkled seeded, double podded, medium tall early variety. This is an exotic variety mostly grown on commercial lines.

Plantago asiatica L. (syn. Plantago major var. asiatica (L.) Decne.) (Plantaginaceae): Eng.- Common Plantain; Kash.- Woughte Gulla; Ladakhi- Karache [W]. A perennial scapigerous herb growing in grasslands and fruit orchards, especially in damp situations, as a weed. Leaves used as vegetable. Seeds used in treating uninary troubles. Cooked or boiled leaves used as blood purifier (Fig. P 20).

Plantago brachyphylla Edgew. ex Decne. (Plantaginaceae): Ladakhi- Tharam [W] A scapegerous herb of alpine grasslands. Plant used for intestinal troubles.

Plantago depressa Willd. (Plantaginaceae): Ladakhi- Ramboosuk [W]. A perennial herb of wet mountain slopes, wet fields upto 4000 m. Leaves basal, shortly petioled, lanceolate, densely white, pubescent. Spikes narrowly cylindrical densely flowered. Seeds of the plant used as stamachic and for intestinal troubles.

Plantago gentianoides subsp. grifithii (Decne.) Rech. f (Plantaginaceae): Ladakhi- Tharum [W]. Perennial herb with fibrous roots growing on stony pastures of Ladakh at 3000- 4300 m. Leaves basal, ovate. Seeds blackish-brown. Powdered seeds are used to cure dysetery and diarrhoea.

Plantago himalaica Pilger (Plantaginaceae): Eng.- Himalayan Plantain; Ladakhi- khichakarpo [W]. Perennial glabrous, stemless herb of alpine, sub-alpine grasslands at 3000 m. leaves ovate-lanceolate, petioled and basal. Scape stout, spike cylindric. In Ladakhi system of medicine seeds used as sedative against headache, lumber pain and renal colic. Leaves applied to wounds (Fig. P 21).

Plantago lanceolata L. (Plantaginaceae): Eng.- Ribwort Plantain; Hindi-Baltanga; Kash.- Gulla [W]. A nearly stemless, rosette forming perennial scapigerous herb, with lanceolate leaves commonly growing in grasslands, meadows and orchards. Inflorescence ovoid spikes of many small flowers.Tender leaves are cooked as vegetable in villages, in early spring (Fig. P 22).

Platanus occidentalis *L.* **(Platanaceae***):* **Eng.- American Plane Tree; Hindi-Chinar; Kash.-Boone** *[NI].* A broadly columnar, deciduous tree with flaking brown, grey, and creamy bark. Leaf margin coarsely sinuate, entire, base obtuse to cuneate, often tomentose; stipules very large, often conical or tubular; bright green leaves. Flowers inconspicuous, monocious, in globose heads. Fruit clusters mostly solitary, rounded at the apex and persist during autumn and winter. Planted in some gardens.

Platanus orientalis L. (Platanaceae): Eng.- Oriental Plane; Hindi-Chinar; Kash.-Boone. [WC]. A majestic deciduous shady tree with usually very broad and round head characterized by grey scaling bark, particularly on younger trunks and branches. The leaves are heart shaped in outline, broad, deeply cut into 5-7 narrow triangular-pointed, coarsely toothed lobes, leaf-stalk long. The leaves turn to striking bronze colour in October-November. Long-stalked pendulous globular clusters of numerous small greenish flowers appear with leaves in March-April. Fruiting heads globular. As a shade cum avenue tree the plant stands planted in the entire valley along road sides gardens and in sacred groves (Figs. P 23, 24).

Platycladus orientalis (L.) Franco. (syn. Thuja orientalis L.) (Cupressaceae): Eng.- Oriental Arbor vitae; Hindi-Morpankhi; Kash.- Sarauve [NI]. A large shrub or small evergreen tree of dense, conical or columnar habit with fibrous, red-brown bark and flat, vertical, irregularly arranged sprays of scale-like, blunt triangular, less-aromatic, mid green or yellow-green leaves, which often turn bronze in winter; upright, flask shaped grey bloomed, female cones. Makes fine avenue tree in gardens and parks.

Plectranthus oertendahlii T.C.E. Fr. (Lamiaceae) [NI]. A prostrate foliage plant, which roots from nodes and is generally grown in hanging baskets. Leaves are almost circular, bronze green with silver zones. The tubular flowers, white-purple in colour are borne in loose panicles.

Pleurotus ostreatus (Jacq. ex Fr.) P. Kumm. (Pleurotaceae): Eng.- Oyster fungi; Kash.- Kannae pappar [W]. A white-rot wood decay and edible mushroom with characteristic broad fan or oyster shaped dark-brown cap spanning 5-20 cm, with thich white and firm flesh. Stipe often absent or short. Found growing in damp wooded shaded areas.

Pleioblastus disticus (Mitford) Nakai (Poaceae): Eng.- Dwarf Fern leaf Bamboo [NI] A dwarf plant, with slender culms, zigzag, green or tinged purple; leaves tessellate on both edges, the sheaths overlapping the spreading blades rather close together and parallel, giving the plant a striking appearance, hence grown as ornamental.

Pleurospermum hookeri C B Clarke (Apiaceae) [W]. Annual glabrous herb growing in open alpine pastures, preferably by streams at 2700-5000 m. Stem ribbed. Leaf blade triangular-ovate, 3-4 ternate, pinnae 7-9 pairs. Umbells with white flowers. The decoction of the plant is taken to cure acidity.

Pleurospermum candollei (DC) Clarke (Apiaceae): Eng.- Paper Cupflower [W]. Thick stemed perennial. Stem covered with persistent leaf bases. Leaves pinnate, pinnae pinnatifid. Flowers creamish, borne in acompound umbel at the top of the stem. Umbels are covered in 10-15 papery bracteoles. Fruits medicinally used for treating flatulence, dyspepsia and renal problems.

Pleurotus sajor-caju (Fr.) Singer (Pleurotaceae): Eng.- Oyster Mushroom; Kash.- Kanna pappar [W]. An edible oyster mushroom, growing in moist shaded situations in coniferous forests. It has distinct light brown convex caps with margins rolled and shallow gills. Stipe short and offset from the centre of the cap (Fig. P 25).

Poa alpina L. (Poaceae): Eng.- Alpine Meadow Grass [W]. An alpine viviparous grass species common in Himalayan alpine grazing grounds above 2500 m altitude.

Poa angustifolia L. (Poaceae): Eng.-Narrow leaved Meadow Grass [W]. Perennial rhizomatous herb in wastelands and forest openings. Grows abundantly in grasslands across Kashmir Himalaya.

Poa bulbosa L. (Poaceae): Eng.- Bulbous Blue Grass [W]. Grows in dense clumps on high alpine meadows and forms a large part of the available pasture. Stem hollow with bulbous sections at their bases. Inflorescence a wide cluster of branches bearing leaf like spikelets containing bulbils.

Poa falconeri Hook. f (Poaceae) [W]. A perennial grass species having wide distribution in Himalayan grasslands.

Poa koelzii Bor. (Poaceae): Eng.- Koelz's Grass [W]. A perennial very glaucous, densely clumped species growing both in temperate conditions of Kashmir as well in cold arid conditions of Ladakh upto 4000 m altitudes. Panicles contracted, spikelets 2-5 flowered.

Poa lahulensis Bor. (Poaceae): [W]. A very common tufted perennial grass species growing under cold arid conditions of Ladakh. Leaf blade flat, abruptly contracted to a blunt tip.

Poa nemoralis L. (Poaceae): Eng.- Wood Blue Grass [W]. A shade tolerant or shade loving perennial grass found on shady rocks, having enough moisture. Panicle loose and branched.

Poa palustris L. (Poaceae): Eng.-Fowl blue Grass [W]. A loosely tufted grass species of the temperate regions of the world including Kashmir. It is a valuable fodder grass preferring marshy grasslands. Shoots extra vaginal.

Poa pratensis L. (Poaceae): Eng.- Smooth stalked Meadow Grass [W]. Perennial herb, usually common in wastelands and ravines. An excellent pasture grass, which can also be made in to hay. Leaves have boat shaped tips. Panicles conical, spikelets 2-5 flowered.

Poa sterilis M Bieb. (Poaceae) [W]. This densely tufted perennial is a common grass of the alpine pastures from 2000 to 3500 m. An importanct as fodder grass for wild life.

Poa stewartiana Bor. (Poaceae) [W]. Tall annual herb, common on humus rich soils among scrubs and tall herbs. A fairly common grass species in Himalayas above 2000 m. Leaf blade flat, tapering to a stout tip, scaberulous on margins. Panicles pyramidal, spikelets 3-4 flowered.

Polemonium caeruleum L. subsp. himalayanum (Baker) Hara (Polemoniaceae): Eng.- Green Valerian.; Kash.- Karipate [W]. A perennial herb of W Himalayan meadows, with ribbed and fleshy stem. Leaves alternate, pinnate. Flowers pale-blue borne on terminal branched clusters. Whole herb used as astringent; roots are sedative.

Polygala sibirica L. subsp. monopetala (Camb.) Chodat (Polygalaceae): Eng.- Chinese Senega, Milkwort [W]. Perennial herb, in dry rocky areas. Flowers purplish. Roots used medicinally. Plant is known to improve hearing, vision, mental activity and strengthen muscles and bones.

Polygonatum multiflorum (L.) All. (Asparagaceae): Eng.- Solomon's Seal; Kash.- Salem Misri [W]. Perennial rhizomatous herb on forest slopes. Flowers campanulate, greenish white. Roots used against suzak. Fruits bluish-black berries (Fig. P 26).

Polygonatum verticillatum (L.) All. (Asparagaceae): Eng.- Whorled Solomon's Seal; Kash.- Salemdhana [W]. Perennial rhizomatous herb on forest slopes near moist situations. Stem erect, glabrous. Leaves in whorls of 3, sub-sessile. Flowers, pendulous greenish yellow. Root used against suzak.

Polypogon fugax Nees ex Steud (syn. Polypogon littoralis var. higegaweri (Steud.) Hook. f) (Poaceae): Eng.- Asia minor Bluegrass [W]. It is a true annual hill grass growing in damp and swampy places all along North-Western Himalayas. Stems solitary, often decumbent at base and rooting from lower nodes. Leaf surface scaberulous. Spikelets solitary.

Polygonum affine D. Don (Polygonaceae): Eng.- Himalayan Fleece Flower; Ladakhi- Likatur [W]. Perennial matted herb, abundant on alpine slopes in Kashmir and Ladakh. Flowers purple, placed in dense spikes. Leaves elliptic, whitish on lower side, margins toothed. Extract of the roots used against fevers. Rhizomes were being used as substitute to tea.

Polygonum aviculare L. (Polygonaceae): Eng.- Common Knotgrass; Ladakhi – Kayakshua [W]. A glabrous much branched, pubescent annual. Stem semi erect, 10- 40 cm long. Leaves elliptic-oblong, subsessile, entire and short stalked. Flowers axillary, pink. Aerial parts used for treating boils. Leaves used as vegetable.

Polygonum hagei Royle. ex Bab. (Polygonaceae): Eng.- Himalayan Knotweed; Kash.- Maechran; Ladakhi- Chumerche [W]. A rhizomatous perennial herb, producing thick, hollow, erect stems. Decoction of seeds is used to cure jaundice and liver complaints.

Polygonum longisetum var. rotundatum A.J. Li (syn. Polygonum barbatum subsp. gracile Danser (Polygonaceae): Eng.- Jointweed; Kash.- Drouab [W]. Tall perennial rhizomatous herb in shaded and wet locations. Flowers greenish white. Leaves pubescent, base cuneate, apex acuminate. Plant used for lung diseases, diarrhoea and dysentery (Fig. P 27).

Polygonum plebeium R. Br. (Polygonaceae): Eng.- Small Knotweed; Kash.-Drouab [W]. A prostrate densely branched annual herb of Himalayan grasslands. Leaves lance like, elliptic, stalkless. Flowers axillary, pink. Plant used for lung disease and diarrhoea and dysentery.

Polystichum aculeatum (L.) Roth. ex Mert. (Dryopteridaceae): Eng.- Hard shield- Fern; Kash.- Kakai [W]. An evergreen fern on steep slopes in woodlands. Fronds dark-green, stiff, hard textured, bi-pinnate and drooping downslope. Young fronds are used against dysentery and diarrhoea (Fig. P 28).

Populus alba L. (Salicaceae): Eng.- White Poplar / Silver Poplar; Hindi– Safeda; Kash.- Kabuli Frast / Doude Frast [W/UC]. A large tree with light-grey bark, petioles and leaves covered beneath with dense white soft tomentum. Bark smooth, greenish-white with diamond shaped marks on young stem. Leaves shallowly lobed, persistently cottony -tomentose beneath, long stalked. Male catkins hairy, ciliate. Wood used for construction works.

Populus balsamifera L. (syn. Populus latifolia Moench.) (Salicaceae): Eng.- Balsam Poplar; Kash.- Frast. [NI]. A large tree with long flexous yellowish-brown or grey branchlets and sticky buds. Leaves are ovate, obtusely toothed, white and net veined beneath, give off a strong smell of balsam while unfolding. Male and female catekins sessile and drooping.Wood used for construction purposes.

Populus caspica (Bornm.) Bornm. (Salicaceae): Eng.– Poplar [NI]. A tall slender tree with branches rising irregularly. Leaves oval with cordate base and prominent dark green midribs. Bark fissured vertically. Introduced in 1960 in the valley for plantation.

Populus ciliata Wall. ex Royle. (Salicaceae): Eng.- Himalayan Poplar; Hindi –Tilaunja /Ban peepal [UC]. A large deciduous tree with greenish-grey bark and sticky buds. Main trunk straight and upright. Leaves broadly ovate, unequally toothed and glbrous. Branches angled, leaf buds viscidly resinous. The catkins long containing 40-45 capsules. The capsules are 3-4 valved and enclose 100-150 seeds with flecks of cotton. Wood used for construction works.

Populus deltoides Marshal (Salicaceae): Eng.– Eastern Cottonwood. [NI]. A large, 20-40 m tall, broad-headed, black poplar with rounded or angled twigs. Leaves large, deltoid and coarsely toothed and shake even to a slightest breeze. Bark dark-grey and fissured on old trees. Plant is dioecious. Grown extensively for its wood.

Populus euphratica Oliver (Salicaceae): Eng. Sindhi Poplar /Desert Poplar: Hindi – Ladakhi Safeda [W]. A large deciduous tree found along the Indus drainage from Sind to Ladakh at 2500 to 3500 m. Stem typically bent and forked. Leaves ovate to cordate, long petioled and usually dentate or variously lobed. Bark thick, fissured vertically. Flowers in lax catkins.Wood used for construction works and as fuel.

Populus nigra var. italica (Munchh) (syn. Populus nigra subsp. pyramidalis Celak) (Salicaceae): Eng.- Black Poplar; Ladakhi- Yulat/ Yerpa;

Kash.-Koushre Frast [W/UC]. A large tree, 20-30 m tall, with arching branches forming a narrow cylindric crown. Leaves 5-10 cm long, broadly ovate-rhomboid. Characterized by columnar habit. Wood used for construction purposes. Also planted along roadsides as avenue tree. In Ladakh the vessels used for preparation of *Chhangue* are made from its wood.

Populus yunnanensis Dode (Salicaceae): Eng.- Yunan Poplar [NI]. A fast-growing, medium-sized, upright balsam popular, with angular, glabrous shoots, new growth very reddish; leaves oval, reddish on new growth, bright green above whitish beneath, with a red midrib. Introduced and grown for plantation.

Portulaca grandiflora Hook (Protulacaceae): Eng.- Rose Moss/ Garden Portulaca [NI]. A succulent prostate annual, with semi-prostate red stems. The leaves are narrow, bright green and succulent. Flowers are saucer shaped with a central boss of bright yellow statmens. Variants range in flowers from red, orange, purple, yellow, pink and crimson. Both single and double varities are under cultivation.

Portulaca oleraceae L. (Portulacaceae): Eng. Common Purselane; Hindi-khurfa; Kash. -Nuner [W]. An annual succulent herb growing in temperate Himalaya up to 1750 m amsl. Leaves alternate, fleshy. Flowers yellow. The herb is used as minor vegetable. Seeds are used in curing scurvy, liver, kidney and bladder diseases. Leaves are now valued as refrigerant and as a article of diet in diseases of kidney and spleen (Fig. P 29).

Potamogeton lucens L. (Potamogetonaceae): Eng.- Shining Pond Weed; Kash.- Gur Hill [W]. Aquatic rhizomatous perennial plant growing in still and slow flowing streams. Leaves large, pale-green translucent, shining with distinctive netted veining. Fed to cattle as fodder.

Potentilla anserina L. (Rosaceae): Eng.- Silverweed /Cinquefoil; Kash.- Troma; Ladakhi- Shadiyaunk [W]. A stoloniferous perennial herb on alpine meadows and slopes. Leaves pinnate, leaflets crenate and covered with silky hairs. Flowers produced singly with 5 petals. Fresh plants are fed to pashmina goats to increase wool production. A thin paste of the powdered roots is used as a poultice to cure tumors and pimples.

Potentilla argentea L. (Rosaceae): Eng.- Silver Cinquefoil; Ladakhi- Shadum [W]. Erect perennial herb along roadsides and grasslands. Stem hairy. Leaves alternate, palmate, with three leaflets, dark-green above and densely white hairy underside. Flowers yellow. Extract of the whole plant is used as blood purifier and in treatment of pulmonary diseases.

Potentilla argyrophylla Wall. ex Lehman (Rosaceae): Eng.- Silver-leaved Cinquefoil; Ladakhi- Skial-daepo [W]. A hardy perennial herb on alpine slopes forming tufts. Roots terette. Flowering stems erect, densely silvery tomentose. Flowers yellow, purple at base. Used against heart problems.

Potentilla atrosanguinea G. Lodd. ex D.Don (Rosaceae): Eng.- Himalayan Cinquefoil A vigorous herbaceous perennial of Himalayan mountain slopes at lower altitudes. Leaves in clumps, long stalked, palmate; basal leaves divided into 3 finger like, long petioled, oblong- lanceolate, serrate, glossy and dark-green leaflets. Flowers 5 petalled, rose like, ruby red.

Potentilla kashmerica Hook.f (Rosaceae): Eng.- Silver Weed; Kash.- Rat-e dunde [W]. Erect branched herb, clothed with spreading hairs; Leaves shortly petioled, lower 5-foliate, obovate oblong, coarsely serrate. Flowers yellow.Leaves used as hemolitic.

Potentilla multifida L. (Rosaceae): Ladakhi- Thakto / zanskanpati [W]. A perennial herb on alpine meadows and banks of water channels. Leaves tomentose, 2-3 pinnate. Flowers yellow, sepals densely pillose. Seeds used for sleeplessness (Fig. P 30).

Potentilla nepalensis Hook. (Rosaceae): Eng.- Nepalese Cinquefoil; Kash.- Van chai [W]. A perennial herb, upto 30- 60 cm tall inhabiting alpine meadows at 2100 -2700 m. Leaves deep green, strawberry like, each composed of 5 broad, serrate leaflets. Flowers cup shaped, 5- petalled, red/deep pink with a darker centre. Leaves used for high fever and as beverage tea.

Potentilla reptans L. (Rosaceae): Eng.– Creeping Cinquefoil; Kash.- Marroach [W]. A creeping perennial herb growing in orchards, waste lands. Leaves long stalked. Flowers bright yellow. Root extract used as soothing agent.

Prangos pabularia Lindley (syn. Koelzella pabularia (Lindl.) Hiroe.) (Apiaceae): Eng.- Stony Herb; Ladakhi- Plans /Bishanzetsar [W]. A tall perennial herb, bushy, branched, leafy up to 1 m tall. Leaves long, linear, glabrous, 4-6 pinnate, segments filiform. Flowers yellow in compound umbels. Grows wild mostly around Drass under cold arid conditions of Ladakh up to 5500 m amsl. The plant after flowering is cut and fed to animals only after sundrying especially during winter. It serves as very valuable fodder for goat and sheep in Drass and neighbouring areas. It is highly nutritive and has the reputation of keeping the animals warm during severe winter months. Fruits/seeds are boiled in water and the extract is used as a sedative and pain killer during serious injuries (Fig. P 31).

Primula abconica L. (Primulaceae): Eng.- German Primrose [NI]. A perennial primrose grown as annual. The light green ovate leaves are slightly hairy. Clusters of pink, red, lilac or blue purple flowers are borne from March- May. Several varieities are under cultivation.

Primula clarkei Watt. (Primulaceae): Eng.- Clarkei Primrose. [W]. A dainty species with round or oblong, pale-green, serrated leaves. Flowers are rose-pink with yellow eyes, and borne in whorls. Occurs wild in alpine grasslands of Kashmir.

Primula denticulata Smith. (Primulaceae): Eng.- Drumstick Primrose. [W/UC] A vigorous wild primrose of Himalayan grasslands. Also grown as annual for spring bedding. The pale – green farinose leaves are broad and form a rosette. Flower are borne on dense globular heads with pale lilac to deep purple or deep carmine petals (Fig. P 32).

Primula macrophylla D. Don. (Primulaceae). A common alpine herb along snow banks, moist meadows from 3300 – 5400 m altitudes. Leaves erect, narrowly lanceolate or strap shaped. Flowers nodding, purple or lilac, often with darker eye (Fig. P 33).

Primula malacoides Franch (Primulaceae): Eng.- Fairy Primose [NI]. A perennial, winter green grown as annual. The hairy leaves are pale green and ovate with rounded teeth. Slender stems carry whorls of star like flowers. Colour variants are with brick-red, cherry–red, lilac, carmine red, snow white and light yellow flowers.

Primula minutissima Jacq & Duby (Primulaceae): Eng.- Heyde's Primrose [W] A tiny tufted perennial with rossete of leaves. Fowers stemless, purple. Plant grows along Himalayan mountain slopes at 3500 – 5200 m altitudes. Leaves spathulate, acute, shortly and acutely dentate, lanceolate and dark-green.

Primula obtusifolia Royle. (Primulaceae): Ladakhi- Khilchainakpo [W]. A perennial, with woody rootstock growing near rocky places near melting snow above 3500 m. Leaves ablanceolate, margins denticulate. Flowers heteromorphic, light pink, umbellate. Powdered rhizomes are given for bile complaints, as stomachic, for diarrhoea and dysentery.

Primula rosea Royle. (Primulaceae): Eng.- Himalayan Meadow Primrose [W] A compact tuft forming species with ovate leaves. Flowers appear in April in umbels of rose-pink colour. Mostly ooccurs wild in alpine grassland near moist places around Gulmarg and Sonamarg (Fig. P 34).

Prunella vulgaris L. (Lamiaceae): Eng.- Heal- all; Hindi- Usta qudoos; Kash.- Kala woeth; Ladakhi- Syangare [W]. A herb found in sub-alpine grasslands and shaded forests of Kashmir. Leaves lance-shaped, serrated. Flower heads club like, flowers two lipped, tubular with purple hood. Flower heads used in treating respiratory problems. In Ladakh a decoction of powdered flowers and leaves is used orally against cerebral and spinal disorders (Fig. P 35).

Prunus armeniaca L. (Rosaceae): Eng.- Apricot; Hindi- Khobani; Kash.- Tcher; Ladakhi-Phating [W/UC]. A multipurpose deciduous economic plant, fairly common in ravines, alpine slopes and woodlands in wild state. Also cultivated from 850 m to about 4000m amsl in Kashmir and Inner Himalayas. Bark reddish-brown bark and brownish on branchlets. Leaves broad ovate or orbicular-ovate, sub-cordate or rounded base, apex acuminate, margins crenate, stipules lanceolate. Flowers pinkish –white appearing before leaves (Fig. P 36).

Apricot culture is one of the very old cultures in Ladakh and Kashmir Himalayas as such its diversity is represented by a good number of varieties and types. In Kashmir local apricots are mostly grown in waste places, Karewa slopes and orchards, primarily for its wood, foliage and stones. Fruits are greenish yellow at maturity, edible but kernel is bitter and yield oil which has medicinal and industrial uses. The stony mesocarp is used as a excellent fuel for cooking purposes since centuries. In Ladakh oil from its kernels is given to women after delivery for energy, to stimulate growth of long and healthy hair. In Gomphas it is used for illumination near the image of Budha, especially during worship.Commercially cultivated varieties include *Australian, Amba, Frogmore, Early Gilgati Sweet, Round, Turky, Quetta, New Castle and Charmagz*. These types have sweet and juicy flesh with sweet kernel, larger fruits than wild types and are collectively known as *Bota Tchera*.

Ladakh has the greatest diversity in apricots and the types vary with respect to their shape, leaf characteristic, date of bloom, flower colour, fruit colour, size and shape, and pubsence. Most of the apricots types of Ladakh are sweet and juicy, possess capacity to resist very low moisture and fluctuating temperatures, and can easily be sundried, thus becoming the most popular, widespread and number one commercial fruit of Ladakh. Some most popular apricot types of Ladakh include:

- **Rakchecarpo:** The variety grows mostly in Indus and Nubra valleys, above 4000 m amsl. Fruits are very sweet, highly juicy, aromatic with sweet white kernels.

- **Halman:** An introduction from Helmand river basin area of Afganistant. Fruits sweet, juicy, thick with edible kernel. Large quantities of its fruit are sundried/machine dried for commercial purposes (Fig. P 37).
- **Chuli: A** wild apricot with cling stone growing mostly around Chulichum area of Ladakh bordering Gilggit district. Kernels are sweet as well as bitter and are often used for the extraction of oil. Trees have the ability to stand extremities of climate better than other varieties. Large quantities of its fruit are sundried for sale.

  Other lesser known apricot varieties of Ladakh include - *Amba, Afgan, Australian, Aam, Adzo Arghoun, Bongti, Brochuli, Badamchuli, Chulikarpo, Charmagz, Chharap, Deychuli, Germoy Tago, Ghoun Chuli, Gotsos, Gilgit Sweet, Jeembochuli, Qoban, Qoban Khantey, Rdongmer, Rogan, Regivespachan, Shaqanda, Shambear, Shepachesmo,Ston Chuli, Serkiangmo tilli, Sultan Chuli, Suffaxida, Shakarpara, Tripa, Toqpopa, Toqpopakhantey, Thul chuli, Urgen Chuli, Wartsiokhangrigyaop, Zgoaq Chuli and Kaisha.*

Prunus avium *(L.)* **L. (Rosaceae): Eng.- Sweet Cherry; Kash.- Gillas [WC].** A deciduous tree with open habit, few suckers and smooth reddish-brown bark. Leaves broadly-obovate, acuminate and crenate-serrate, appearing crimson in Autumn. Flowers whitish and appearing on pendulous clusters on naked branches by end of March.

Popular sweet cherry desert and commercial varieties under cultivation include:

- **Awal Number Gilas (Early Purple Heart) [UC]:** Tree much branched, average sized (1.5 to 4.0m) with semi-erect crown and greyish white bark. Leaves obovate or lanceolate with long pinkish petioles. Flowers by Ist week of April. Flowers white and arranged in fascicled and simple racemes. Fruits oblate, fat, broader than long; skin light to fast carmine, flesh light carmine and sweet. Matures by 3rd week of May.
- **Double Glass [UC]:** Tree medium sized with thick and scarfy white bark, much branched with semi-erect crown, average sized (1.8 – 3.5 m). Flowers in second week of April; flowers dull white, larger as compared to other varieties and arranged in fascicled racemes born on short peduncles. Fruits ovoid, heart shaped, skin semi-amber, speckled light deep rose, flesh whitish, and sweet.
- **Gogji Makhmali Glass** [UC]: Trees 1.5 to 4 m tall with semi- erect crown and rough and scarfy bark. Leaves dentate, pointed and broadly elliptical. Flowers white and arranged in simple racemes. Fruits spherical, peduncles long, shining deep red, turning to black red on ripening; pulp

pome-granate and full of colour marks, juicy and sub- acid to sweet. Plants bloom in second week of April and mature by 2nd week of June.

- **Mishri Glass** [UC]: Trees much branched with semi erect crown and smooth, shining, greyish white, bark. Leaves dark green, lanceolate and dentate. Flowers are creamy white and borne in simple racemes and comparatively smaller than other sweet cherry varieties. Fruits ovoid / heart shaped, skin intense carmine and with light violet rose flesh and sweet acid. Plants bloom in 2nd week of April and mature by 3rd week of June. Stone small and round (Fig. P 38).
- **Siyah Glass (Black Heart**) [RC]: Trees big tall, pyramidal in spread with scaly, greyish brown bark. Leaves big, lanceolate, dentate, dark green and drooping. Fruits big, heart shaped, black purple, sap light black with agreeable sweet taste.
- **Tontal Glass [UC]:** A semi tall tree with greyish brown bark. Leaves dark green bidentate and lanceolate. Flowers white, arranged as simple racemes. Trees bloom in Ist week of April and fruits mature by last week of May. Fruits are big, unequal lobed, pointed, suture line distinct, skin shining, black purple, flesh violet, mild sub- acid to sweet.

Prunus cerasifera Ehrh. (Rosaceae): Eng.- Cherry Plum; Kash.- Gurdoule [W]. A sub- alpine small fruit tree of Kashmir growing wild on forest slopes and ravines. Trees medium, strong trunked. Lenticels are elongated, redish brown and raised. Bark shining redishes brown on old wood, olive green on new wood. Flowering time is from last week of March to 2nd week of April. Bears annually. Leaves are thin, light green glabrous, short ovate, finely serrate. Flowers solitary, white, interior somewhat pink, appearing before the foliage in early April. Fruits small, lobose, cherry sized, juicy, yellow with slight red cheek; flesh soft juicy, sub – acid, cling stone and edible (Figs. P 39, 40).

Prunus cerasifera var. atropurpuria (Rosaceae):Eng.- Ornamental Plum /Cherry Plum [W/UC]. Handsome ornamental tree with dark purple foliage. Bark purplish, smooth with few sunken lenticels. Leaves medium, oval, dark purple, serrate. Flowering time is during first fortnight of April. Fruits mature by late August. Fruits are small oblong, dark purple, flesh yellowish green, sub- acid, cling stone.

Prunus cerasus L. (Rosaceae): Eng.- Sour Cherry; Hindi- Alubalu glauss; Kash.-Aelchi [W/UC]. A tree with slender, spreading and pendulous branches and roots with many suckers. Leaves elliptic-obovate, erect, firm, shining and crenate-serrate. Flowers appear in pendulous clusters, white with calyx tube campanulate. Fruit globular, dark red, acidic; stone globular, smooth. Fruits used in salads and chetnies and also cooked. Commonly cultivated and wild sour cherry varieties of plains, foothills and uplands of Kashmir are:

- **Aelchei** [W]: A cherry cultivar growing in fruit archards and in forest areas.A medium sized tree, taller than sweet cherry cultivars, gaining a height of 2.5 to 5.0 m. Crown erect, branches mostly horizontal or drooping downwards. Leaves ovate or obovate with double serrate margins. Flowers white and appear in simple racemes. The plants bloom in Ist to second week of April. Fruits small (2.3 to 2.8 gm in weight), obovate, broader than long; skin thin and shiny, intense carmine to blackish red; flesh clinging to stone, carmine, acidic and juicy; stone small and smooth (Fig. P 41, 42).
- **Hachhi Gillas (Esperen) [RC]:** Trees big, stout and spreading. Leaves are smaller, greenish yellow, oval and thicker. Maturity falls in Ist week of July. Fruits heart shaped, yellowish white with crimson rosy shades on exposed cheeks, skin hard, pulp yellowish red delicious and with longer shelf life.
- **Kashmiri Gillas [RC]:** An indigenous cherry cultivar having strong trunk with shining silvery chocolate bark. Fruits big, cordate, rose red on sunny side, yellowish white base, pulp slightly sweet.
- **Turasha Tetha Gilas (Gobet Grande) [RC]: S**Trees medium tall with scaly olive greenish grey bark. Leaves small, short, oval, olive green and oval. Fruits mature by end of June and are spherical, yellow amber to rose marble skinned with yellowish pulp with acid-sweet taste.

Prunus cornuta (Wall. ex Royle) Steud. (syn. Padus cornuta (Wall ex Royle) Carri Sre.; Prunus padus Brandis) (Rosaceac): Eng.- Himalayan Bird cherry; Kash.-Zomb, Wotil [W]. A medium-sized tree with rough grey-brown bark; leaves lanceolate, serrate and pointed. Flowers white, in terminal and axillary, pendulous racemes; fruits glossy, red then dark purple to black, drooping in grape-like clusters. Growing commonly in forests with other deciduous elements. Wood used for making agricultural tools.

Prunus domestica L. (Rosaceae): Eng.- Common Plum; Kash.- Alu-Bukhara; Kash.- Aure [WC]. A small tree with young branchlets glabrous. Leaves elliptic or obovate, coarsely crenate-serrate. Flowers appearing with the leaves, petals white. Fruit ovoid, drooping, yellow purple or blue black, with greenish flesh.The main commercial plum cultivars under cultivation are grouped as under:

A: Japanese Group of Plums

- **Amba Auer (Wickson) [RC]:** Trees medium tall, with upright branches. Lenticels are raised, oblong with redish centre. Bark is dark-brown. Leaves lanceolate, green, crenate, glabrous, held out, petiole redish-brown. Flowering time is from last week of March to Ist week of April. Periodicity of fruiting is annual. Fruits mature by late July. Fruits are heart- shaped, conical, yellowish red, smooth, waxy; flesh sweet, aromatic, yellowish, tender and juicy.
- **Awal Number (Transparent Gage) [RC]:** Trees short statured, with angular trunk. Bark is dark- brown to redish green; lenticels brownish, oblong raised and pointed. Leaves obovate, serrate, curved with thick and pubescent petioles. From 2nd to 4th week of April trees flower. Fruits mature by end of July. Periodicity of fruiting is biennial. Fruits are roundish, pale yellowish skin, with and whitish bloom all around; flesh, yellowish red and sub- acid (Fig. P 43).
- **Chogandari Auer (Satsuma) [RC]:** Tree medium tall, trunk strong and stout having dark – grey bark and small sized conspicuous and irregularly shaped lenticels. Leaves are lanceolate, bicrenate margins, dull-green stipules and mid rib and veins crimson. Flowers appear from last week of March to 2nd week of April. Fruits mature by end of July and are round, heart shaped, skin dark- crimson with yellowish dots, bluish bloom predominates on fruits; flesh dark red or beet red, juicy, firm melting, sweet when ripe; cling stone (Fig. P 44).
- **Rupa Auer (Burbank) [RC]:** Medium to dwarf tree but with vigorous and prolific growth. Old wood chocolate coloured with elongated and raised lenticels. Leaves are lanceolate, crenate, (midrib and venation pink on dorsal side), glabrous, and dark green. Flower appear from last week of March to 2nd week of April. Fruits mature by late July- August. Bearing is annual. Fruits are conic with whitish bloom; skin smooth and shining; flesh golden yellow, crisp, juicy, sweet and aromatic; cling stone.
- **Wazul Rupa Auer [RC]:** Trees medium with vigorous growth. Bark is redish brown to olive green. Lenticels oblong, raised and redish brown. Leaves lanceolate, bicrenate, undulating, light green. Flowering occurs between 2nd to 4th week of April. Fruits mature by August and bearing is biennial. Fruits are medium to large, heart shaped with suture lines, skin crimson with silver bloom all over; flesh pale- orange, crisp, juicy, acidic cling stone.

- **Wazul Thula Auer [RC]:** Trees medium, spreading and dense branching. Bark brownish red and scaly. Lenticels are ovate, greyish brown. Leaves obovate, serrate, yellowish green. Flowering time is from 2nd week to last week April. Bearing is biennial. Fruits mature by late July to August. Fruits are large, oval, suture line distinct, skin red covered with white bloom; flesh yellowish, juicy sweet; cling stone.

B: European Group of Plums

- **Bunafsha Auer (Kirks Blue) [UC]:** Tree medium tall, upright and spreading habit. Trunk stout, scaly with greenish- brown bark and oval, slightly raised lenticels. Leaves are obovate, crenate, pale- green and short petioles. Flowering time is from last week of March to 3rd week of April. Fruits are large, round to oval, skin dark purple, speckled with bluish bloom and russet markings; flesh greenish yellow, firm, juicy, sweet; free stone.
- **Damson Surakh (Clande Red) [RC]:** Trees large, stout, with drooping branches. Bark chocolate and rough. Lenticels irregular in outline and bark- brown. Leaves big oblong, crenate, green and hairy. Flowering time is from Ist to 3rd week of April. Bears biennially. Fruits mature during August. Fruits are smaller, round, smooth skinned, light crimson with purplish white bloom; flesh pleasant and free stone.
- **Khazir Auer**: Dwarf trees but with spreading and straggly branching. Trunk week covered with dark brown bark having few elongated and raised lenticels. Leaves smaller, oblong, thick, crenate, erect, up- folded, dark- green and with pink redish petioles. Flowering time is from Ist 3rd week of April. Periodicity of fruiting is alternate. Fruits mature by August. Fruits are elongated, pear shaped with a narrow neck, skin pinkish red with bluish bloom and russet scars; flesh olive- green, pleasant but acidic; free stone. Fruits are also sundried.
- **Misri Surakh Auer (Santarosa) [UC]:** Tree medium strong, spreading with upright and stout branches. Bark is scaly and dark-brown. Lenticels are oblong and greyish brown. Leaves are ovate, crenate, thick, green with small petioles. Flowering time is second fortnight of April. Fruits mature by late July and periodicity of bearing is annual. Fruits are medium, round, skin yellowish with red bloom; flesh juicy, sweet. Stone skin breaks in humid and hot weather.
- **Saubni Auer [RC]:** Trees medium, vigorous with dense, spreading habit. Trunk has greyish bark. Lenticels are round and slightly raised. Leaves are ovate, crenate, dark- green, with stout petioles. Flowering period is from Ist to 3rd week of April. Fruits mature by August. Bears

alternatively by heavy and light crops. Fruits are large sized, greenish yellow with pale- bloom; flesh yellow, juicy, sweet; free stone.

- **Siraj Auer (Jafferson's Gags) [RC]:** Trees medium tall with strong, stout trunk covered with olive- green to dark- brown bark. Lenticels are oblong to oval, greenish brown. Leaves are obovate serrate, with pink coloured petioles. Flowers from last week of March to middle of April. Fruits mature by July-August. Bears biennially. Fruits are large, roundish - oval, yellow spotted red and deep crimson; flesh slightly orange, juicy, and sweet.
- **Subza (Bryanstons Gage) [NC]:** Medium tall tree with strong, upright trunk and branches. Lenticels are redish brown, round and few. Bark greyish black, smooth, shining on old wood and pinkish green on new wood. Leaves are smaller, ovate glabrous, serrate, grey, down curved with small pinkish petioles. Flowering time is from last week of March to middle of April. Fruits mature by July-August. Fruits are small, yellowish skinned with slight red cheek; flesh soft, juicy, sub-acid, cling stone.
- **Thulla Auer (Yellow Egg) [RC]:** Trees large, vigorous, upright, spreading. Bark greyish brown on old wood and brownish green on new wood. Lenticels are few, scattered, oval, greyish brown. Leaves ovate, serrate, yellowish green, pubescent with short petioles. Flowering time is from Ist to 3rd week of April. Fruits mature by Early September. Bearing is biennial. Fruits are large, golden yellow with whitish bloom; flesh yellow, juicy, acidic but sweeter when ripe, free stone.
- **Wangan Auer (Grand Duke) [RC]:** Trees medium tall, upright. Trunk ribbed, with dark-brown bark, and oblong dark brown lenticels. Leaves oblong, crenate, upcupped but drooping. Flowering occurs during 2nd to 4th week of April. Periodicity of bearing is biennial. Fruits mature by September. Fruits are large, oval, doubles and malformations common; skin smooth deep purple with bluish bloom; flesh yellowish, firm, sub-acid to sweet; cling stone.
- **Zarad Auer (Mirabellee Hative) [RC]:** Tree small sized, stout and spreading. Branches small, weak bearing the slender twigs. Bark is dark brown on old wood and redish brown on new wood. Lenticels oval, raised and greyish. Leaves small club shaped and pale- green. Fruits mature by late August. Periodicity of fruiting is biennial. Fruits are small, oval in bunches like grapes, yellowish green with whitish bloom; flesh yellowish green acidic; cling stone.

- **Zarad Bor Auer (Coes Golden Drop) [RC]:** Trees medium tall, spreading habit stout branches but small, scarfy twigs. Lenticels are round, greyish brown. Bark greyish brown on old wood, scarfy olive-green on new wood. Flowering period is from 2nd week to 4th week of April. Fruits mature by September. Bears alternatively light and heavy crops. Leaves are oblong, crenate, dark- green, thick, pubescent with short petioles. Fruits are large, oval to oblong, golden yellow, dotted red; flesh yellow delicious.

C. Indigenous Varieties of Plums

- **Jangle Kala Auer (Wild Goose Plum)** [W]: Tree small to medium, spreading; upright and spreading branches. Bark is dark redish brown with oval, greyish brown lenticels. Leaves obovate, thick, dark- green, crenate, up- cupped with short petioles. Bears annually. Flowering time is from Ist to 3rd week of April. Fruits are small, roundish oval, dark-green with bluish bloom. Grow wild in foothill orchards.
- **Jangli Auer [W]:** Trees medium tall, strong and stout. Trunk scaly, scarfy with redish brown bark and red brown and raised lenticels. Leaves medium, obovate, green, serrate, glabrous, petiole hairy and short. Flowering time is from Ist to 3rd week of April. Fruits mature by August. Fruits are small, dark pinkish red covered with bluish bloom; flesh greenish yellow, acidic, juicy; cling stone. Fruits used for culinary purposes.
- **Jangle Ancher Auer [W]:** Trees stout, strong, with stout, thick branching. Bark greyish brown on old wood and grayish white on new wood. Lenticels are round, grey and scattered all over trunk and branches. Leaves are obovate, green, apex macronate, with serrate margins. Flowering time is from Ist to 3rd week of April. Fruits mature by August. Periodicity of bearing is annual. Fruits are oval/ roundish, dark purple with whitish bloom; flesh greenish, crisp, sub- acidic. Grows in wild state mostly in sub- alpine forests.

Prunus dulcis (Miller.) D A Webb.) (syn. Prunus amygdalus Batsch.) (Rosaceae): Eng.- Almond; Kash.- Baadum [WC]. A deciduous tree with grey bark and glabrous branchlets. Leaves mid- green, ovate-Ianceolate, serrulate, short stalked, stipules fimbriate. Flowers are almond rich pink appearing on the naked branches in March. Fruit a drupe.

Almonds are growing in Kashmir and Inner Himalayas from 1600 to 3100 m amsl across Kashmir and Ladakh regions however its main cultivation has remained concentrated on the *Karewa b*elts of Kashmir valley (Fig.

P 45). All the types of almonds - sweet and bitter, thin and thick shelled types, have been cultivated in the region but most of the types grown have been wild types producing thick shelled and bitter fruits. However at present the diversity has been enriched with improved thin shelled and high yielding types also. Popuplar varieties under cultivation are:

A: Traditional Almond Varieties

- **Dade Badam [NC]**: Trees large, strong trunk. Fruits medium sized with small kernel and a hard shell.
- **Khar Badam** [NC]: Trees large with upright branches with thick dark brown bark. Flowers profusely during Ist week of March. Fruits large sized, very thick shelled and edible.
- **Tchheghta Badam [NC]:** Trees large, with dark brown scaly bark. Flowers in the Ist week of March. Fruits average sized, thick shelled with bitter kernels. Nuts were traditionally used for oil extraction and for fuel. The hard shells were being used for making famous Kashmir ink (Sharma, 1986).

B. Almond Varieties Released by SKUAST-K

- **Mukhdoom** [UC]: Variety released by SKUAST-K on the merits of its high yield and quality shells. Trees large in size and spreading. Fruits are borne mostly on one year shoots, but some are borne on spurs. Nut size medium, shell colour moderate yellowish with sparsely pored surface. Shell semi-soft, shell seal good, kernel medium sized and plump. Shelling percentage around 38-40%. Shell yield on an average 7 Q/h.
- **Parbat [UC]:** High yielding sweet almond variety released by SKUAST-K. Trees of medium size and upright to spreading. Blossoms by the end of March. Fruits are borne on one year shoots, some on spurs, nut size small, shell colour whitish yellow with better surface; matures at 140 days after full bloom. Kernels are smooth having good colour and taste; shelling percentage on an average 49%. Mid season variety and ready for harvest from 1-2nd week of August; performs well under rainfed conditions. Cultivar yields (in shell) around 5Q/ha under normal conditions.
- **Waris [UC]:** Thin shelled almond variety developed and released by SKUAST-K during 2001 for commercial cultivation. Trees is upright, medium is size and a regular bearer; bears mostly on one year old wood; nut size is medium and has soft shell; shell colour moderately yellow with densely pored surface; shell seal good, medium sized and plump kernel. Mid to late variety, ready to harvest in third to last week of August. Shell yield is around 6-7 Q/ha.

- **Shalimar** [UC]: A quality and thin shelled almond variety developed by SKUAST-K during 2001 for commercial cultivation. Tree is medium in size with spreading behavior. It is mid bloomer. High shelling percentage (40%); shell colour light yellow with moderately pored surface, mid to late maturing group. Kernels are smooth having good colour and very good appearance and taste. The cultivars yields (in shell) on an average 5Q/ha.

Prunus glandulosa Thunb. (Rosaceae) Eng.- Dwarf flowering Almond: [W]. A glabrous bushy shrub, of neat habit. Leaves ovate– lanceolate, and mid-green. Flowers white to light pink, solitary or paired in the leaf axils.

Prunus 'kanzan' (Rosaceae): Eng.- Japanese Cherry [NI]. An upright deciduous ornamental tree, vase-shaped when young, spreading wider with age, new growth brown; having ovate, dark green leaves, to mature leaves some what reddish, bronze when young. Double deep pink flowers.

Prunus laurocerasus L. (syn. Laurocerasus officinalis M. Roem.) (Rosaceae): Eng.- Common Laurel/ Cherry Laurel [NI]. A vigorous, bushy, evergreen shrub or small tree. Leaves oblong to obovate or elliptic, stiff and leathery and dark green. The small white flowers, are produced in upright racemes, from leaf-axils (mid April to late April) followed by conical, cherry-like red fruit ripening to black

Prunus lusitanica L. (Rosaceae): Eng.- Portugal Laurel [NI]. A large evergreen shrub or small to medium sized tree with red stalked, oblong-ovate, acuminate, serrate leaves, dark green and glossy above, lighter beneath. Flowers cupshaped, small, white, hawthorn- scented, borne in slender, ascending, spreading or pendent racemes. Fruits ovoid, cherry-like red.

Prunus mume (Siebold) Siebold & Zucc. (Rosaceae): Eng.- Japanese Apricot [NI] A round crowned tree or a tall shrub, with bark of the twigs green, thin. Leaves ovate to elliptic, long acuminate, finely and scabrous serrate, bright green above, lighter beneath. Flowers solitary or paired, sessile, white to pale pink, very fragrant, appearing before the foliage. Fruit globose, yellow to greenish, slightly pubescent, flesh sour to bitter.

Prunus persica (L.) Batsch (Rosaceae): Eng.- Flowering Peach. [WC] An ornamental tree with glabrous branchlets and pubescent buds. Leaves elliptic lanceolate or oblong-lanceolate, long-acuminate, cuneate and serrate. Flowers usually solitary *on* short pedicels, calyx-tube campanulate, lobes with rounded apex, pubescent outside, petals pink. Fruit sub-globose, tomentose; stone deeply pitted and furrowed.

Prunus persica *L.* **Batsch. (Rosaceae): Eng.- Peach; Kash.- Tchunun [UC].** Medium sized trees with spreading habit grown for fruit. It has also run wild in some forest areas and its fruit relished by wild animals. Wild types have bitter as well as sweet types. Large, sweet and juicy varieties, after having been introduced are now growing in many orchards and home gardens. Some of varieties under cultivation are: *Alexander, Elberta, Elton, Peshawari and Quetta.* Varieties recorded from Ladakh include *Yarken, Shanbear, Papa and Moro.*

Prunus prostrata Labill. (Rosaceae): Eng.- Mountain Cherry / Prostrate Cherry [W/UC]. *A* dwarf, spreading ornamental shrub. Leaves broadly ovate, scabrous serrate, deeply incised, dark green above, white tomentose beneath. Flowers pink, solitary or paired, usually sessile, borne along the wiry stems in April. Fruit dark red, ovate with thin flesh (Figs. P 46,47).

Prunus 'taihaku' (Rosaceae): Eng.- Great white Cherry [NI]. A most beautiful, strong, growing, white flowering tree, with attractively copper-red new growth. Flowers single, snow white, sepals long and narrow.

Prunus tomentosa Thunb. (Rosaceae):Eng.- Nanking Cherry /Mountain Cherry; Kash.- Ushkan Glass [W]. Upright large trees with long drooping branches growing in sub-alpine forests. Bark shining chocolate coloured. Leaves yellowish green, thicker, long and lanceolate. Fruits medium, oblong, cordate, skin red, flesh yellowish pale, juicy and excellent in taste, and mature by end of May (Fig. P 48).

Pseudomertensia nemorosa (A. DC.) Stewart & Kazmi (Boraginaceae) [W] A perennial herb, inhabiting shady alpine slopes. Branches slender; leaves long petioled, clasping at the base, ovate, acuminate. Inflorescence lax; flowers bluish purple (Fig. P 49).

Pseudosasa disticha (Mitford.) Nakai (syn. Pleioblastus distichus (Mitford) Nakai) (Poaceae): Eng.- Dwarf fern leaf Bamboo [NI]. A dwarf plant, with culms slender, zigzag, green or tinged purple, the branches borne singly; leaves tessellate, serrate on both edges, the sheats over-lapping, the spreading blades rather close together and parallel, giving the plant a striking appearance.

Psychrogeton amorphoglossus (Boiss.) Novopokr. (syn.Psychrogeton andryloides DC. Novopokr ex Krash (Asteraceae): Eng.- Ladakhi Fleabane; Ladakhi- Luchchugba [W]. A densely grey woolly haired tufted, perennial with a woody rootstock inhabiting mountain slopes of

Ladakh at 5500 m altitudes. Leaves crowded at base, radical spathulate, cauline few and linear. Rays cream,disc cream or yellow. Decoction of fresh plant used for expulsion of worms.Also used as fuel in Ladakh.

Pteris vittata L. (syn. Pteris ensifolia Poir.) (Pteridaceae): Eng.- Chinese Brake; Kash.- Lousa Gasse [W]. A perennial fern locally common on forest openings. Fronds tufted, pinnate with 20-40 pinnae. Rachis glabrous, pale brown. Plant used for post delivery bath of ladies.

Pueraria montana var. lobata (Willd.) Sanjappa & Pradeep (syn. Pueraria hirsuta (Thunb.) Matsum. (Fabaceae): Eng.- Thumbard Kudzu Vine; Hindi– Bidari kand [NI]. A very vigorous, deciduous, twining ornamental climber, woody at base, with tough thickened or tuberous roots, bears 3-palmate leaves composed of ovate to diamond shaped lobed leaflets. Fragrant pea-like purple flowers are produced in erect racemes, in July.

Punica granatum L. (Lythraceae): Eng.- Pomegranate; Hindi– Anar; Kash.- Daen [WC]. An upright, sometimes spiny shrub or small, rounded tree, with short-petioled, opposite, narrowly oblong, glossy, bright green leaves. Bears funnel shaped, 5-petalled, scarlet or orange-red flowers with crumpled petals, singly or in clusters of up to 5, followed by spherical, yellow-brown, edible fruit.

Pomegranates are found growing throughout the valley and have been traditionally used for medicinal purposes by local *Hakims*. Six sweet varietieis have been recorded in the valley.

- **Anar Sheereen:** Thick skinned, pale fruit with sweet and juicy pulp.
- **Anar Turash:** Bearing thin skinned pale coloured fruits. Pulp flesh and juicy and blood red (Fig. P 50).
- **Sheerin Bedana:** Thin skinned fruits attaining crimson colour at maturity.
- **Sheerin Danadar:** Thick skinned fruits becoming light crimson at maturity.

Pyracantha crenulata (Roxb. ex D. Don) M Roem (Rosaceae): Eng.- Himalayan Firethorn [W/UC]. A spiny evergreen ornamental shrub, with usually crowded narrow-oblong blunt shining leathery leaves. Flowers white appear in June. Fruit globular orange-red, nuts protruding from the persistent calyx. Grows wild on forest slopes.

Pyracantha cv. Orange Glow (Rosaceae) [NI]. Loosely branched shrub with broadly elliptic to obovate glossy, dark green leaves. Small white flowers are produced in corymbs in mid May followed by a profusion of persistent, orange-red fruits.

Pyrethrum pyrethroides (Kar. & Kir.) Fedtsch. Ex. Krasch (syn. Chrysanthemum pyrethroides (Kar & kir.) Fedtsch. (Asteraceae): Eng.- Himalayan Chrysanthemum; Ladakhi- Sigtik [W]. Short, densely woolly perennial herb on dry alpine slopes above 2000 m altitude. Leaves cauline, 2-3 pinnatisect. Heads solitary, with white-pinkish ray florets and yellow central disc. Powdered roots used to cure itching (Fig. P 51).

Pyrus calleryana Decne. (Rosaceae): Eng.- Callery Pear; Kash.- Chakajal weight Taunge [W]. Tree tall, vigorous and spreading tree inhabiting forest woodlands. Lenticels scattered, raised and light brown. Bark is furrowed, scaly and brown on main trunk. Flowers appear by last fortnight of April while fruits mature by late October. Bearing is annual. Leaves are dark-green, crenate, down curved. Fruits are small, flatish round, green overspread with raised, conspicuous russet dots; flesh hard, crisp, sweet with bitter taste.

Pyrus communis L. (syn. Pyrus communis var. sativa (DC) DC. (Rosaceae): Eng.- Pear; Hindi- Nashpati; Kash.- Taunge [WC]. A broadly pyramidal tree with young branchlets glabrous or slightly pubecent. Leaves orbicular-ovate to elliptic, shortly acuminate, crenate-serrulate, glabrous. Flowers in umbels, calyx-lobes spreading acute villous, petals orbicular, clawed, white. Fruit turbinate, on slender stalks green or yellow.

Commercial cultivars of Pear include

- **Awal Number / Bagu Gosh (Bartlet- Williams)** [UC]: Trees spreading habit and medium tall. Branches drooping and slender. Bark chocolate on trunk and brownish red on shoots. Flowers from 2nd to 3rd week of April and fruits mature by end of July to August. Periodicity of bearing is annual but partial and full bearing alternate. Leaf ovate, entire, finely serrate, green and ex-stipulate. Fruits larger, oval, pyriform, bright yellowish-green with russet dots all over, exposed sides attain red cheeks with faint stripes; flesh white, clear, very juicy, sweet with musky flavour.
- **Bihi Taung [RC]:** Medium tall with strong and stout trunk and branches. Bark chocolate coloured. Lenticels oval, redish-brown. Leaves green ovate, serrate and down hanging. Flowering time is Ist to 3rd week of April and fruits mature by September. Bears annually; fruits are round, quince faced, light green with reddish brown flush; flesh whitish yellow, juicy and sweet.

- **Bor Farans Taunge (NC):** Trees large, upright, spreading vigorous and prolific. Bark greyish, furrowed and scaly. Lenticels round, big, raised. Leaves round/ ovate, finely serrate, green and up –cupped. Flowering from 2nd to 4th week of April. Fruits mature by October-November. Bears heavily and annually. Fruits are large pyriform, skin rough, pale-yellow with brown russet patches and dots, slightly red cheek on exposed side; flesh whitish, melting, slightly gritty, very juicy, wine flavoured. Commercially important.
- **Brare Taunge [W]:** Trees with vigorous growth and medium tall. Lenticels and bark redish brown. Flowering in first fortnight of April and fruits mature by October. Leaves lanceolate, serrate, down hanging, green with white grooved long petioles. Fruits are medium, round/ conical, tapering to eye, ribbed, ripening to pale-yellow. The fruit assumes a cat face appearance, hence the name.
- **Chenni Taunge [NC]:** Tree medium tall. Branches thin, drooping. Bark greyish brown. Lenticels oval/ elongated, greenish brown. Leaves ovate, almost entire, down curved, pale-green. Flowering from Ist to 3rd week of April. Bearing is alternate. Fruits mature by September. Fruits are round, yellowish- green, russeted skin; flesh yellowish white, melting, sweet, slightly acidic.
- **Dobas Taunge [NC]:** Medium tall, spreading habit with upright and stout branches. Bark redish brown on main trunk having few scattered, sunken yellowish brown lenticels. Trees bloom during first fortnight of April and fruits mature by October. Bearing is biennial but profolic. Leaves are lanceolate/ovate, pale-green, finely serrate. Fruits are large, round, tapering towards stem, flattened at the base, light yellowish green with brown red flush and covered with russet dots; flesh white, crisp, melting, juicy, delicious flavour, with little gritty core.
- **Faraeshe Taunge [UC]:** Trees medium tall with heavy bearing capacity. Bark scaly and reddish brown. Lenticels, raised, oblong and brownish. Leaves small obovate down hanging, slightly serrate, pale-green petiolate. Flowering time is from Ist to 4th week of April and fruits mature by July. Bearing is alternate. Fruits are medium, pyriform, tapering towards stem, greenish-yellow, with reddish brown flush and russet markings. Flesh whitish, melting, with sweet flavour.
- **Gew Taunge [NC]:** Medium tall the bark scaly and scarfy. Lenticels scattered all over trunk. Bark rough, scaly and scarf grayish on old wood. Leaves ovate, yellowish green and crenate. Flowering time is first fortnight of April. Fruits mature by October. Fruits are medium to large,

conic, tapering towards stem, greyish brown, russet on yellow skin; flesh pale-yellow, sweet, aromatic, and melting.

- **Gole Taunge (Summer Rose) [NC]:** Vigorously growing trees with greyish and scaly bark. Lenticels oblong, raised and greyish brown. Leaves ovate, serrate, deep-green. Flowering from 2nd to 4th week of April and fruits mature by August. Periodicity of bearing is annual, but light and heavy crops alternate. Fruits are small to medium, round like apple, pale-yellow with red flushes on exposed sides; flesh crisp, whitish, gritty and sweet.
- **Goshbugo (Citron Citron Des Carmes) [UC]:** Trees medium to big, spreading, scaly. Bark greyish on old wood and olive green on new wood, scurfy. Lenticels conspicuous and sunken. Leaves ovate, serrated, green with long petioles. Flowering period is Ist fortnight of April and fruits mature by end of July. Bearing is annual. Fruits medium, conical, greenish white with russet dots; flesh greenish white with russet dots; juicy with sweet flavour.
- **Gulab Taunge [RC]:** Medium tall. Bark is scaly, scarfy, redish brown on main trunk. Lenticels sunken, grey, round is outline. Leaves ovate, yellowish green, serrate, held out. Flowering time is 2nd to 4th week of April. Fruits mature by late September. Bears light and heavy crops alternatively. Fruits are medium and pyriform. Skin smooth, greenish yellow covered with patches of orange russet, red cheek on exposed sides; flesh yellowish white, sweet and aromatic.
- **Hara Nakh (Kings Pear) [UC]:** Trees large sized, stout trunks, spreading and upright branching. Bark brown with greenish hue all over. Lenticels round, raised, greyish white. Leaves large, ovate, light green, glabrous, serrate and exstipulate. Flowering commences in the 2nd to 3rd week of April and fruits mature by August. Bearing is annual but heavy and light crops alternate. Fruits are medium, conical, tapering towards stem, yellowish green, with russet dots, flesh juicy, crisp, but slightly sweet.
- **Hengal Taunge [NC]:** Tree medium tall and a prolific bearer. Branches stout, with slender drooping twigs. Bark scaly and redish brown. Leaves ovate, down curved, finely serrate, green with long petioles. Flowering period Ist to 3rd week of April. Fruits mature by November. Bears almost annually. Fruits are medium, pyriform, unequal and uneven, smooth, shining, greenish yellow with russet dots/ patches; flesh whitish, good flavoured. Fruits have good keeping quality.
- **Jafiri Taunge [NC]:** Trees dwarf with spreading habit; branches drooping and selender. Bark greyish brown on old wood. Lenticels many,

conspicuous, raised and oval. Leaves large, narrow, down-hanging, thinly serrate, green. Flowering time 2nd to 3rd week of April. Bearing is annual and fruits mature by November. Fruits pyriform, ridged with deep crimson cheek and bronze russet patches; flesh white, juicy, sweet with musky flavour.

- **Kagzi Nakh (Claps Favourite) [UC]:** Medium to large sized tree. Prolific bearer. Bark scaly, greyish. Lenticels whitish green, raised conspicuous. Leaves ovate, regular, serrate, green, glabrous, down curved; flowering time 2nd to 4th week of April. Bearing is annual. Fruits mature by September. Fruits medium, tapering towards eye, smooth, russet dots all over, pale-milky yellow, with scarlet red flushes/ stripes; flesh pale yellowish white, very juicy, sweet and slightly gritty (Fig. P 52).
- **Khar Paddar [RC]:** Medium tall, trees with greyish green bark. Lenticels oblong/round, raised and conspicuous. Leaves ovate, crenate, upfolded and green. Flowering time is Ist to 3rd week of April. Fruits mature by October. Bearing annual. Fruits are bigger, pyriform, yellowish-green, russeted with dots/ patches; flesh white, slightly gritty and sour in taste.
- **Khari Taunge [NC]:** Medium tall tree with scaly, scarfy and brownish-red bark. Lenticels are numerous, round to oblong, sunken on bark and brownish. Leaves ovate, finely serrate, yellowish green, up-folding. Flowering time is from 2nd to 4th week of April but bearing is biennial and fruits mature by late October. Fruits are round, conic, *bergomottee*, skin pale-yellowish at maturity with thick grey russet patches; flesh whitish, melting, juicy, acid sweet, good flavoured.
- **Kirmiz Taunge [W]:** Strong trunked, medium tall tree with greyish bark and raised conspicuous lenticels. Leaves ovate, yellowish green, entire, upfolding and down- curved. Flowering time is Ist to 3rd week of April. Fruiting is annual. Fruits are round conic, uneven, skin rough with numerous dots and patches and crimson cheeks, yellowish- green, flesh whitish, juicy, crisp sweet and aromatic.
- **Kottarnal / Misri (Jar gonelle) [RC]:** Large tree, strong and with vigorous spreading branching; prolfic bearer. Bark scaly reddish brown to greyish. Lenticels are round, whitish-brown. Leaves ovate, yellowish green, serrate with long petioles. Flowers during 2nd to 4th week of April and fruits mature by August. Periodicity is annual. Fruits are conical, greenish yellow with brownish red blushes, covered with greenish russet dots; flesh pale- yellow, sweet and aromatic.
- **Madur Taunge [NC]:** Trees medium, spreading, upright and a good

cropper. Twigs thin slender and drooping. Bark rough, greyish. Flowering time Ist to 3rd week of April. Fruits mature by October. Periodicity of bearing is annual. Leaves ovate, serrate, greenish yellow, shining; fruits medium, conical, even, yellowish green with small russet dots and red cheek on exposed side; flesh yellow white, melting and very sweet.

- **Misri Taunge (Fertility) [RC]:** Medium tall tree, spreading habit with strong and stout trunk covered with furrowed and greenish brown bark. Lenticels few, elongated, greenish brown and raised. Leaves are ovate/ oblong, greenish yellow, fine serrate. Flowering time is 2nd to 3rd week of April. Fruits mature by October. Bearing abundantly but alternatively. Fruits are smaller, conical, yellowish-green, russeted, giving a dull golden loot; flesh whitish and sweet (Fig. P. 53).
- **Nabed Taunge (Charles Congnese**) [RC]: Trees medium tall, strong and vigorous. Bark redish brown with scarf on old wood, olive green–redish brown on new wood. Lenticels conspicuous raised and greyish. Leaves medium, ovate, light- green, entire, undulating down curved. Flowering time Ist to 3rd week of April. Fruits mature by October. Bears annually with light heavy fruiting alternating. Fruits conic, golden yellow, with red cheeks, rust dots and markings all over; flesh whitish, melting, very sweet.
- **Nakhi Kashmiri [UC]:** Trees large, upright and spreading and densely branching. Bark is scaly grayish; lenticels raised, oval. Leaves ovate, finely serrate, yellowish green, with exstipulate petioles. Trees flower from Ist to 3rd week of April. Fruits mature by October. Bears light and heavy fruits crops alternatively. Fruits are conic, round, smooth, yellowish- green, scarlet cheek on exposed side. Flesh white, very juicy, sweet and gritty.
- **Panze Taunge [W]:** Trees medium tall, strong trunked, with upright branches. Bark rough, whitish brown. Lenticels redish brown, scattered all over. Leaves greenish–yellow, greatly up-folded. Periodicity is annual. Flowering time is first fortnight of April and fruits mature by late November. Fruits big, pyriform, with ridges; skin smooth, yellowish-green with russet dots; flesh sweet, delicious, aromatic without any grit or sand.
- **Safed Taunge [W]:** Trees upright and medium tall. Bark redish brown with scarf on trunk. Lenticels greenish white and oval. Leaves ovate, serrate, up-held and with greenish white long petioles. Flowering time early April but fruits mature by October. Fruits large, pyriform, tapering to stem smooth, ribbed and almost pentagonal; flesh tender, yellowish white, sweet and good flavoured.

- **Saharanpuri (Colmar-De-Ete) [UC]:** Medium tall, spreading and upright. Bark greenish brown, scaly, both on trunk and branches. Leaves lanceolate, slightly serrate, pale-yellow-green, down hanging. Flowers from first fortnight of April and fruits mature by July. Periodicity is alternate. Fruits medium, conical, straw green with brownish red flush; flesh yellowish white, melting and juicy.
- **Satarwatti (Souvenier De Congress) [UC]:** Stout and shiny trunked tree with short slightly drooping branches. Bark sooty grey. Lenticels oval, brownish red and raised. Flowers during Ist to 3rd week of April. Fruits mature by late September. Bears almost annually. Fruits are large, *calebebasse* form, yellowish green, scarlet cheeks with russet lines; flesh yellowish white, juicy, sweet with musky flavour (Fig. P 54).
- **Saterwatti Kalan (Vicar of Wink Field) [UC]:** Tree medium tall and a good copper. Branches upright and strong. Lenticels very prominent, greyish brown. Bark also greyish brown. Flowering time is fist fortnight of April and fruits mature by November. Fruits very large, uneven, skin green to yellow, smooth, soft, rough due to russet dots and patches; flesh yellowish white, juicy, sweet and flavoured.
- **Seber Taunge [NC]:** Trees are big, vigorous upright with scarfy scaly greyish bark. Lenticels oval, greyish brown and raised. Leaves ovate, down curved, entire, petioles greenish white. Flowering time is last fortnight of April. Fruits mature by November. Fruits are shortened, pyriform, even, skin smooth, lemon yellow, russet dots spread all over resembling the bold spots of trout fish, flesh white and sweet.
- **Seki Seber Taunge [NC]:** Large tree, spreading, upright with scarfy, scaly, greyish white bark. Lenticels are round to oval, raised and greyish. Leaves smaller finely serrate, down curved and yellowish green. Flowering commences from 2nd to 4th week of April. Fruiting is annual and fruits mature by October. Fruits are small to medium, pale–yellowish at maturity with brown russet dots; flesh whitish, sandy and gritty, juicy and slightly sweet.
- **Sirka Taunge [NC]:** Trees medium sized, upright on the thick, dense, scaly and scarfy branches. Lenticels very few, round/ oblong, greyish. Leaves are ovate, thick, crenate, serrate, yellowish green with long yellow green, thick petioles. Flowering time is 2nd to 4th week of April and maturity period is October. Fruits very large, massive, pyriform, tapering to stem, greenish yellow flushed with red check on exposed sides, skin rough with russet dots; flesh whitish, less juicy and sub-acidic in taste. Fruits being acidic, were used in fermentation, hence the name.

- **Sona Taunge (Nouveau Poiteau) [RC]:** Medium tall tree with drooping branches. Bark greyish brown. Lenticels oval to oblong, brownish and raised on trunk/branches. Leaves large, finely serrate, green and with long petioles. Flowers in 2nd to 3rd week of April and fruits mature by August. Bearing is annual. Fruits large, pyriform, greenish yellow covered with reddish russet, presenting a golden hue; flesh whitish, melting and sweet.
- **Sourakh Taunge [RC]:** Dwarf, upright think branching which droop down. Lenticels round, brownish. Bark chocolate brown. Flowering time 2nd to 3rd week of April. Fruits mature by end of September. Fruiting periodicity is annual. Fruits oval to pyriform, skin rough, pale-yellowish with red flush; flesh yellowish white sweet and aromatic.
- **Suraj Bali (Doyenne Bussoch) [RC]:** Trees medium, spreading, trunk scaly. Bark furrowed and greyish. Lenticels are, conspicuous, greyish-brown. Leaves ovate, green, down curved, serrate with long petioles. Flowers in second fortnight of April and fruits mature by September. Bearing is alternate. Fruits are medium to large *Bergomottee* type, yellowish-green with patches and dots of russet, red cheeks on exposed sides; flesh white, juicy, aromatic and sweet.
- **Surakh Nika Nakh [NC]:** Tree large, upright with spreading habit. Bark on old wood is redish brown, scarfy and scaly. Flowering period is from Ist to 3rd week of April. Bears annually and fruits mature by October-November. Leaves are ovate, narrow, crenate, yellow-green.

  Fruits are conical, to oval, smaller, skin smooth, yellowish green with red flush and prominent russet dots at maturity, flesh whitish, melting and delicious.
- **Tetha Taunge [W]:** Tree tall and large, spreading with thick, scaly branches. Bark greyish white. Lenticels oval, greyish-white. Leaves lanceolate with entire margins, light green. Flowering time Ist to 2nd week of April; fruits mature by September. Bearing is annual with light and heavy crops alternatively. Fruits are conical, uneven, pale – greenish with dull red flush, skin smooth with russet dots; flesh white, juicy sweet with bitterness.
- **Zieth Sabaz Taunge [NC]:** Medium tall tree, with scaly bark. Lenticels round, brownish and round. Leaves larger, ovate, irregularly serrate, green and down hanging. Flowering occurs in second fortnight of April and fruits mature by October. Fruits are longer, *calebasse*, tappering to stem and eye, skin rough, dark green which changes to pale- yellow on full ripening, covered with much russet dots. Flesh pale yellowish with a pinkish tinge around core, melting, very juicy and sweet.

- **Zirhama (Beurre- de- Amanlis) [UC]:** Trees are large, tall, spreading, with stout and long branches. Bark is chocolate on old wood, scaly, rough and greenish brown on new wood. Lenticels are round, raised, conspicuous. Leaves long, ovate, entire or slightly serrate, green, up – folded. Flowering time is from last week of March to middle of April. Fruits bearing is annual. Fruits are medium, conical, pale to yellow-green with brownish red flush on exposed sides; flesh white, juicy sweet with good flavour.

Pyrus kumaensis Koidz. (Rosaceae): Kash.- Gochal Weigh Taunge [W]. Large sized tree with curved, scarfy and drooping branches. Bark scaly, scarfy and greyish. Lenticels oval/ round. Leaves lanceolate, green, shining, crenate. Flowering time is the second fortnight of April. Bearing is annual; fruits are small (walnut sized), greenish yellow covered with numerous russet dots; flesh white, less juicy, crisp, gritty and of bitter taste, less edible, mature by late October- November.

Pyrus lanata Miq. (Rosaceae): Eng.- Beam Tree; Kash.- Mael Tange [W] Trees medium tall with stout trunk growing wild in foot hills, though their frequency has drastically reduced. Fruits small, gritty and sweet bitter in taste.

Pyrus ovoidea Rehdr. (Rosaceae): Kash.- Hapat Taunge [W]. Trees tall, with vigorous and spreading branching. Lenticels are few greyish white and round. Bark is scaly scarfy, furrowed and greyish black. Flowers appear in just fortnight of April while fruits mature by end of November. Leaves arc oblong to ovate, a shallowly serrate. Fruits are small (walnut sized), round at base but tapering from middle towards truncate apex giving fruit an ovoid shape, skin greenish yellow russeted with dots; flesh white, crispy and not edible.

Pyrus pashia Buch.- Ham. ex D. Don (Rosaceae): Eng.- Himalayan Pear; Hindi – Mehal [W]. A deciduous tree, 6-8 m tall, usually thorny, young shoots usually woolly at first later glabrous, red-brown. Leaves ovate to oblong, acuminate, crenate or obtusely serrate. Flowers white. Fruits nearly globose, brown, dense, warty, punctuate. Used as rootstock for pear. Grows wild in hillside forests in mid hill zone upto 2000 m.

Pyrus phaeocarpa Rehder. (Rosaceae): Eng.- Dusky Pear; Kash.- Taungi Gulabi [NC]. Trees large, strong, upright and spreading. Bark is greyish white, scaly and scarfy and furrowed on main trunk. Lenticels are raised, whitish grey. Flowering time is from 2nd to 4th week of April and fruits mature by end of August. Bearing is annual. Fruits are pyriform, small and greenish yellow with rosy flush on exposed side; flesh is whitish yellow, crisp, juicy sweet, core gritty. Grows wild in hillside forests.

Pyrus ussuriensis Maxim. ex Rupr. (Rosaceae): Eng.- Ussurian Pear; Kash.- Madur Gol Tange [NC] · Tree large, upright, spreading and vigorous. Bark, furrowed, greyish, scaly, lenticels few and raised. Flowering period extends from Ist to 4th week of April. Bears annually. Fruits mature by October. Fruits are small, globose tapering to stem, greenish yellow at maturity; flesh crisp, juicy, gritty and sweet. Grow wild in hillside hardwood forests of Kashmir.

The pear varieties recorded from Ladakh region include **Hosourneoty, Naqponeoty and Karponeoty.**

Fig. P 1: Paeonia emodi

Fig. P 2: Papaver rhoeas

Fig. P 3: Parratiopsis jacquemontiana

Fig. P 4: Pedicularis biocornuta

Fig. P 5: Pedicularis cheilanthifolia

Fig. P 6: Pedicularis pectinata

Fig. P 7: Pedicularis siphonantha

Fig. P 8: Peganum harmala

Fig. P 9: Persicaria alpina

Fig. P 10: Persicaria amplexicaulis

Fig. P 11: Persicaria nepalensis

Fig. P 12: Persicaria vivipara

Fig. P 13: Common bean types of Kashmir

Fig. P 14: Phlomoides bracteosa

Fig. P 15: Phytolacea acinosa

Fig. P 16: Picea smithiana

Fig. P 17: Picrorhiza kurroa

Fig. P 18: Pinus gerardiana

Fig. P 19: Pinus wallichiana

Fig. P 20: Plantago asiatica

Fig. P 21: Plantago himalaica

Fig. P 22: Plantago lanceolata

Fig. P 23: Platanus orientalis

Fig. P 24: Platanus orientalis

Fig. P 25: Pleurotus sajar-caju

Fig. P 29: Portulaca oleraceae

Fig. P 26: Polygonatum multiflorum

Fig. P 30: Potentilla multifida

Fig. P 27: Polygonum longisetum

Fig. P 31: Prangos pabularia

Fig. P 28: Polysticum aculeatum

Fig. P 32: Primula denticulata

Fig. P 33: Primula macrophylla

Fig. P 37: var. Halman

Fig. P 34: Primula rosea

Fig. P 38: var. Mishri

Fig. P 35: Prunella vulgaris

Fig. P 39: Prunus cerasifera

Fig. P 36: Prunus armeniaca

Fig. P 40: Prunus cerasifera

Fig. P 41: Prunus cerasus

Fig. P 42: Prunus cerasus

Fig. P 43: var. Awal number

Fig. P 44: var. Chougandari

Fig. P 45: Prunus dulcis

Fig. P 46: Prunus prostrata

Fig. P 47: Prunus prostrata

Fig. P 48: Prunus tomentosa

Fig. P 49: Pseudomertensia nemorosa

Fig. P 52: var. Kagzi Nakhe

Fig. P 50: var. Anar Tursh

Fig. P 53: var. Mishri Tanje

Fig. P 51: Pyrethrum pyrethroides

Fig. P 54: var. Sattar watti

Q

Quercus baloot Griffith (Fagaceae): Eng.- Holy Oak; Kash.- Jamal Gota/ Baloote Kashmir [W/UC]. A small rounded evergreen tree. Leaves are coriaceous, oblong-ovate to lance shaped, entire or with spiny -toothed margins. Acorns oblong-ovoid to nearly rounded, solitary or in groups of 2 or 3. Grows as a hard wood element of forests. Also grown as ornamental.

Quercus oblongata D.Don (syn. Quercus leucotrichophora A. Camus (Fagaceae): Eng.- Banjh Oak [NI]. A large or medium sized evergreen ornamental tree, distinguished by its leaves which have a dense white-woolly hairs beneath, and its white woolly young shoots. Leaves leathery,dull-green, sharply toothed followed by ovoid acorns.

Quercus incana Bartram (Fagaceae): Eng.- White Oak/ Upland Willow oak; Hindi- Baunje; Kash.- Baloote Kashmir/ Jamal Gota [W]. A large evergreen tree of temperate Himalaya. Bark dark brown. Leaves oval, glossy green above and woolly haired beneath. Acorns provide food for wild life.Wood used as fire wood and for handling of tools. Bark used for tanning.

Quercus robur L. (Fagaceae): Eng.- British Oak; Kash.- Hum [W]. A large spreading, deciduous tree with fissured, grey-brown bark. Leaves short-petioled, oblong-obovate, dark green. Ovoid acorns are borne singly or in clusters of 2 or 3. Common cultivar is *fastigiata*. Growing within Dachigam Sanctury in being important for wild life.

Quercus suber L. (Fagaceae): Eng.- Cork Oak [NI]. A short-stemmed, rounded evergreen tree, with rugged and corky bark. Leaves oval or oblong, broadly-toothed, leathery, lustrous green above, greenish- grey-felted beneath; bears ovoid-oblong acorns.

R

Rabdosia rugosa (Wallich ex Benth.) Hara (Lamiaceae): [W]. An erect branched leafy shrub or under-shrub growing as high altitude shrub above 2000 m altitudes on alpine slopes.Leaves ovate, wrinkled, hairy, coarsely toothed, conspicuously grey-woolly beneath, in contrast to green wrinkled upper surface. Flowers tiny, fragrant, white, spotted with purple. Shrub is of great significance for honey bee colonies.

Ranunculus hirtellus Royle. (Ranunculaceae): Eng.- Softly hairy Buttercup [W]. Erect, small pubescent alpine herb inhabiting shady situations. Basal leaves rounded, deeply cut into branched lobes; stem leaves divided into 3-5 lanceshaped lobes. Flowers yellow. Achenes on long receptacles, beaked. Aerial parts used medicinally.

Ranunculus sceleratus L. (Ranunculaceae): Eng.- Celery-leaved Buttercup / Crowfoot Buttercup; Kash.- Kawe khoure [W]. An erect herb, found in moist grasslands. Leaves dark-green, lobed like a crow foot. Flowers yellow. Plant cooked as vegetable in juvenile phase (Fig. R 1).

Ranunculus hyperboreus Rottb. (syn. Ranunculus radicans var. multifidus (G.Don.) Regel.) (Ranunculaceae): Eng.- Arctic Buttercup; Ladakhi- Chechah [W]. Small perennial high altitude herb on crevices of moist rocks in W- Himalayas. Stem prostrate, glabrous, rooting nodally. Leaves deeply 3- lobed. Flowers yellow; sepals and petals 3-4; nectar on petal surface. Powdered seeds taken for vitality and fertility.

Raphanus raphanistrum subsp. sativus (L.) Domin (syn. Raphanus sativus L.) (Brassicaceae): Eng.- Radish; Hindi – Mooli; Kash. - Mujhe [WC]. Biennial herb with a white or brightly colured tuberous tap root. Stems simple or branched, erect, 20-100 cm; basal leaves long, lyrate, pinnate or pinnate sect, coarsely toothed; cauline leaves simple, linear. Flowers in long terminal racemes usually white or lilac with purple viens; fruits inflated with long tapering beak; seed globose, yellow to brown. Local as well as introduced types are grown. The local types are with crimson red/ scarlet and white napiform root which are eaten raw, cooked as vegetable and also used in pickling. Tender leaves are also used as vegetable.

Common cultivated radish types are:-

- **Dal Muje:** Roots with whitish skin globose to napiform in shape, flesh hard and white. Cultivated mostly in Dal lake areas and around Srinagar in having local preference for its taste and flavour.
- **Gaume Muje:** Roots with dull white skin, globose to napiform in shape, flesh hard and white. Cultivated mostly in the villages of Anantnag and Pulwama Districts due to its local preference. Roots have long shelf life and are more pungent.
- **Islambadi Muje:** Cultivated mostly in and around District Anantnag and Pulwama by vegetable growers. Its roots have appealing crimson coloured skin, large in size, with white sweeter flesh.
- **Japanese White:** *An* introduced table radish variety under cultivation. Roots tapering towards soil side and with white skin and sweet flesh.
- **Scarlet Globe:** Introduced and cultivated mostly by vegetable growers especially in Spring for its small, scarlet coloured roots which are preferred for table purposes.

Rhamnus purpureus Edgew (Rhamnaceae): Eng.- Purple Buckthorn [W]. A small tree, inhabiting open sub-alpine slopes.Branches purplish with white spots.Young leaves pubescent beneath, ovate-lanceolate, shortly acumunate, minutely serrulate. Fruit shortly pedicelled.Fruit is used as purgative. Wood used for handling tools.

Rhaponticum repens (L.) Hidalgo (syn. Centaurea picris (Pall) M. Bieb (Asteraceae): Eng.- Russian Knapweed; Ladakhi- Bushaka [W]. A bushy rhizomatous perennial of cultivated fields, pastures and roadsides. Stems and leaves are arachnoid –tomentose. The rosettre leaves oblanceolate, entire, lower cauline leaves smaller, pinnately lobed, upper leaves sessile, serrate to entire. Heads many terminating the branches. Flowers pink to purplish. In Ladakh whole plant used for skin diseases and as blood purifier (fig. R 2).

Rheum australe D. Don (syn. Rheum emodi Wallich ex Meisn.) (Polygonaceae); Eng.- Himalayan Rhurbarb; Hindi – Dolu; Kash.-. Pamb-e-Chalan; Ladakhi- Chercheat [W]. A perennial rhizomatous herb found in alpine Himalayas from 3000-4000 m amsl at open slopes, alpine grasslands and forest clearings. Flowers small, pinkish-white borne in bunches. Leaves large, triangular with fleshy petioles. Roots and rhizomes are source of drug used as laxative, toxic and purgative and for treatment of arthritis and swollen ulerus. Dried roots are commonly used for treating wounds and ulcers. Fresh leaves used as vegetable and for pickling. In Ladakhi system of medicine powdered roots, in the form tablets with honey, are used as purgative and in curing dyspepsia (Fig. R 3).

Rheum tibeticum Maxim ex Hook.f. (Polygonaceae): Ladakhi- Chrucha [W]. A perennial stemless, rhizomatous herb growing on drier slopes of Ladakh at 4000- 5000 m. Leaves leathery, orbicular, cordate, entire or crenate. Flowers light purple to yellowish-green. Extract of the plant used against rheumatism.

Rheum webbianum Royle (Polygonaceae): Eng.- Small Himalayan Rhubarb; Hindi- Dolu; Kash.- Pumbe Hakh / Lachoo [W]. A stout perennial herb, found in central and western Himalayas from 3000 m to 4200 m. Leaves large, long stalked, circular, heart shaped. Flowers yellowish white. Leaves are sour in taste and are cooked fresh as well as after sundrying by local *Gujars* and *Bakerwalls.* Sundried leaves are usually blended with fish during winter season (Fig. R 4).

Rhodiola fastigata (Hook.f. & Thomson) S H. Fu (Crassulaceae): Eng.- Clustered Rhodiola [W]. A perennial rhizomatous, clustered herb on rock surfaces on alpine slopes, grassy slopes at 3600- 5500 m altitudes. Flowering stems many, parallel to each other. Stem leaves entire, stalkless, linear-ovate. Flowers on top of stems, yellow or red in compact corymb like cymes (Fig. R 5).

Rhodiola himalensis (D Don) S H Fu (Crassulaceae): Eng.- Himalayan Rhodiola [W]. A large showy plant with many erect stems covered with narrow elliptic fleshy leaves. Flowers are borne in clusters on top of stems and are dark-red to dark-pink. Plant inhabits alpine forest scrubs and slopes at 3300- 4800 m.

Rhodiola tibetica (Hook. f. & Thomson) S.H Fu (syn. Sedum tibeticum Hook. f. & Thomson.) (Crassulaceae): Eng.- Tibetian Rhodiola;

Ladakhi- Sholo [W]. Perennial rhizomatous herb on rocky alpine slopes of Ladakh. Leaves alternate, sessile. Flowers borne in many flowered cyme. Semi crushed leaves used as diuretic and against obesity.

Rhododendron anthopogon D.Don. (Ericaceae): Eng.- Dwarf Rhododendron; Hindi- Cherailu; Kash.- Yurmi [NI]. Dwarf hardy evergreen shrub, forming patches in association with *Juniperus recurva* at 3300-4000m. Leaves dark-green, oval, aromatic and scaly underside. Flowers yellow, tinged with pink, in clusters of 4 to 6. Leaves used for skin ailments.

Rhododendron campanulatum D. Don (Ericaceae): Eng.- Bell Rhododendron; Hindi- Cherailu; Kash.-Ponda nass/Younge patre; **Ladakhi- Tama** [W]. A widely branched, spreading shrub growing in association with *Betula,* in forest thickets, mountain slopes, at 3100-3500 m. Leaves oblong-elliptic, dark green with brown feted whooly hairs on lower surface. Flowers bell-shaped in a lax cluster varying in colour from pale rose to lavender– blue with darker spots inside. Leaves are used in chronic rheumatism, against syphilis and sciatica. Powdered leaves are used as snuff for headache (Fig. R 6).

Rhododendron lepidotum Wall ex G Don (Ericaceae): Eng.- Pink Scaly Rhododendron; Hindi- Atarasu, Sumral; Kash.- Tazak tsun. [W]. A small alpine shrublet. Leaves elliptic, obtuse, scabrid with brown tomentose / fleshy scales on lower surface. Flowers pink or purple borne in clusters of 2- 4 on slender stalks. Fruit a capsule, densely scaly and also with persistent sepals. Leaves used medicinally.

Rhus succedanea L. (Anacardiaceae): Eng.- Wild varnish Tree; Kash.- Arkhol [W] A deciduous tree common in ravines. Fresh juice from the plant causes blisters and intense skin itching. Galls are astringent and expectorant. Wood used for agricultural implements and other tools. The plant is antidote, antivinous, cholagogue, febrifuge and ophthalmic.

Rhus cotinus L. (syn. Cotinus coggygria Scop.) (Anacardiaceae): Eng.- Smoke Tree [NI]. A multiple branching ornamental shrub, 5-7 m tall, with an open,spreading irregular habit. Leaves rounded oval with waxy glacous sheen. Flowers with 5 yellow petals.

Ribes nigrum L. (Grossulariaceae): Eng.- European Black Currant; Hindi– Nabar; Kash.- Hargil [W]. A medium sized deciduous shrub growing on rocky surfaces. Leaves palmate with serrate margins. Fruits dark purple, edible. Calyx hairy with yellow glands. Fruit dark-purple with persistant calyx.

Ribes orientale Desf. (Grossulariaceae): Eng.- Flowering Currant / Oriental Gooseberry [W/UC]. A polygamous dioecious, sticky glandular and minutely hairy shrub growing in cold arid regions of Himalayas. Leaves rounded reniform, crenate, obscurely 3-5-lobed. Flowers in erect racemes and small. Berries, deep red, spherical, glandular(Fig. R 7).

Ribes rubrum L. (Grossulariaceae): Eng. -Red Currant; Hindi –Dak [W]. Deciduous, upright shrub growing on rocky surfaces. Berries red, edible. Leaves 5 lobed, arranged spirally on stem. Flowers small, yellow green, in pendulous racemes.

Ribes uva-crispa L. (syn. Ribes grossularia L.) (Grossulariaceae): Eng.- Wild Gooseberry; Hindi- Amlanch [W]. Deciduous straggling shrub, mostly on rocky cliffs. Berries purple, edible. Branches with sharp spines. Flowers bell shaped.

Robinia ambigua Poir var. Bella- rosea (Fabaceae): Eng.- Purple Robe Locust [NI] A medium sized tree having glutinous branches with small thorns. Leaflets 15-21, flowers light pink. Branches slightly glutinous, thorns stouter; flowers larger and somewhat darker. Grown as ornamental.

Robinia pseudoacacia L. (Fabaceae): Eng.- Black Locust; Hindi – Kikar; Kash.- kekur. [W/UC]. A large deciduous tree, with deeply furrowed brown bark, branchlets and foliage glabrous, leaflets elliptic or ovate, rounded or truncate and mucronate at apex, glabrous and pale green beneath and dark green above. Flowers in long drooping pendulous racemes which are cream white, with pinkish tinge, fragrant. Grown as a specimen shaddy tree in parks.Wood used for furniture works, making agricultural implements and as fire wood. Leaves have fodder value (Fig. R 8).

Roripa indica (Ceder.) Barbea (Brassicaceae): Eng.- Indian Field Cress, Yellow Cress [W]. Erect herb of ruderal habitats. Basal leaves stalked, lance shaped, irregularly toothed, upper leaves short stalked. Fruit a seeded elongated capsul. Flowers yellow.Whole plant used medicinally.

Rosa banksiae R Br. (Rosaceae): Eng.- Bank's Rose [NI]. A tall-growing, vigorous, semi-evergreen ornamental climber with few or no prickles. Twigs slender and glabrous. Leaflets 3 or 5, oblong-lanceolate, sharp-serrulate. Flowers small, few to many in umbel-like clusters, slightly fragrant.

Rosa brunonii Lindl. (Rosaceae): Kash.- Krichi. [W]. Deciduous armed and arching shrub common on sub-alpine woodlands and slopes. Flowers white (Fig. R.9).

Rosa x damascena Herrm. (Rosaceae): Eng.- Damask Rose; Kash.- Koushur Gulab [UC]. A small shrub with densely thorny stems and branches armed with many hook-like, regular thorns. Leaflets 5-7, ovate, serrate, glabrous and more grey-green above, underside softy pubescent. Flowers usually many in groups, pink to rose-red, double, fragrant. One of its cultivars *Trigintipetala* has red pink semi double flowers which are highly fragrant. Grown for its fragrant flowers in home gardens. Flowers used traditionally in the preparation of *Gule-Kande*, used medicinally for the treatment of common cold and rhinitis (Fig. R 10).

Rosa foetida Herrm. (Rosaceae): Eng.- Austrian Briar [NI]. An erect ornamental shrub with slender stems and branches with straight prickles. Leaflets 5-9, elliplic-obovate or orbicular. Flowers conspicuously, solitary, butter cup-type, yellow.

Rosa macrophylla Lindl. (Rosaceae): Eng.- Himalayan Rose [W]. A large leaved Himalayan rose species inhabiting open forest slopes at 2400-3600 m. A tall bush, moderately prickly with long bluish-green leaflets. Flowers medium sized rosy pink (Fig. R 11).

Rosa moschata Herrm. (syn. Rosa brunonii Lindl.) (Rosaceae): Eng.- Himalayan Musk Rose; Hindi- Kunja; Kash.- Krichi / Kreer [W]. A rampant climbing species with hooked prickles common in forests and on humus rich soils. Leaves compound with 5-7 elliptic to oblong -lanceolate, finely toothed leaflets. Flowers fragrant, white, cup shaped, borne on large trusses during June- July. Fruits purple brown, ovoid, smooth, shining and edible (Fig. R 12,13).

Rosa multiflora Thunb. (Rosaceae): Eng.- Polyantha Rose/ Multiflora Rose. [NI]. A vigorous, large ornamental shrub or rambler with long stems, stipules prominent. Flowers fragrant, white, borne in long panicles.

Rosa pedunculosis D. Don. (syn. Rosa pedunculata Kult.) (Rosaceae) [W]. Scandent shrub, deciduous, on dry slopes. Flowers pink.

Rosa rugosa Thunb. (Rosaceae): Eng.- Rugosa Rose/ Japanese Rose {NI]. A strong-growing, perpetual-flowering shrub with stout densely prickly and bristly stems. Leaves dark green, each composed of 7-9 oblong leaflets. Flowers fragrant, violet-carmine-red showing yellow stamens, borne singly or in small clusters. Two varieties are cultivated namely *alba* (flowerss white) and *rubra* (flowers wine crimson)

Rosa sericea Wall. ex Lindl. (Rosaceae): Eng.- Silky Rose [NI]. An erect spineless shrub with leaves having 7-17 narrow leaflets, with serrated margins. Flowers four petalled, cream white appearing single during June followed by small red or yellow pear shaped hips with persistent sepals.

Rosa webbiana Wall. ex Royle. (Rosaceae): Eng.- Webb's Rose; Kash.- Aura-wal; Ladakhi- Sebiamaituk/ Bangsertze [W]. A slender native rose shrub with copious prickles on branches common on dry open forest slopes. Flowers mostly solitary on short pedicels and light pink in colour. Fruits bottle shaped to globular, red. In *amche* system of medicine in Ladakh the extract of the fresh ripened thalamus is used to relieve toothache and throat infection. In Kashmir petals of the plant are used to cure nasal bleeding and liver problems.The plant has a name in the Kashmiri Sufi poetic songs (Fig. R. 14 & 15).

Rosmarinus officinalis L. (Lamiaceae): Eng.- Rosemarry; Hindi- Rusmari; Kash.- Rosmari [NI]. An upright to rounded, dense, bushy, aromatic, evergreen ornamental shrub, with linear, leathery, dark green leaves. Whorls of tubular, 2 lipped, purple-blue to white flowers, are produced in numerous axillary clusters along the branches of the previous year.

Rosularia alpestris (Kar. & Kir.) Boriss (syn. Sempervivella accuminata (Dcne.) Berger) (Crassulaceae): Eng.- Lower mountain Sedum; Ladakhi- Chungchange [W]. A perennial succulent herb of lower Himalayas inhabiting rock crevices at 1500 to 5000 m altitudes. Rootstock thick and leafy rosettes form a dense carpet. Stem sessile. Rosette leaves oblong, glabrous. Inflorescence cymose– corymbiform, flowers white. Decoction of the leaves and flowers is given with *Changue* as antiflatulant, antispasmodic and sedative.

Rotala densiflora (Roth.) Koehne (Lythraceae): Eng.- Dense Flowered Rotala [W]. Annual erect herb, usually along rice fields. Leaves ovate, lance shaped. Flowers purplish borne in leaf axils; hypanthium bell shaped. Leaves, flowers used medicinally.

Rubia cordifolia L (Rubiaceae): Eng.- Indian Madder; Kash.- Mazait [W] Perennial climbing herb, climbing by scabrid hairs, among scrubs. Flowers pale-white in dense racemes. Decoction of leaves used as vermifuge. Berries red to black. Roots a source of red pigment used traditionally for colouring clothes.

Rudbeckia bicolor Nutt. (Asteraceae):Eng.- Cone Flower [NI]. An upright, branching annual having brightly stems and foliage. The leaves are oblong and mid-green. The flower heads are yellow with a black central disc. Several garden varieties are grown with larger flowers.

Rubus niveus Thunb. (syn. Rubus lasiocarpus Sm.) (Rosaceae): Eng.- Hill Raspberry; Hindi- Kalianchi; Kash.- Chhanche/ Ras [W]. A large rambling perennial shrub usually near forest clearings. Stem arching, terete, tomentose when young or glabrous and bloomy purple with hooked and compressed prickles extending to petiole and rachis. Leaves pinnate, leaflets ovate-elliptic, dark-green above and white tomentose below. Flowers rose – purple, appear in many flowered axillary corymbs. Drupes in compact heads, numerous, black with white tomentose bloom, edible.

Rubus saxatilis L. (Rosaceae): Eng. -Yellow Raspberry; Kash.- Popai / Hapat phall [W]. A large deciduous armed shrub growing around forest clearings and hill orchards. Leaves compound with 3 leaflets with serrate margins. Fruit red, containing large pips. Fruits edible with acid flavour.

Rubus ulmifolius Schott. (syn. Rubus furticosus var. discolor (Weihe & Nees.) Syme) (Rosaceae): Eng.- Black berry; Hindi- Anchoua; Kash.- Daen chaunch [W]. A rambling armed shrub, near forest openings and slopes; stem arching with broad-based prickles straight on the stem and hooked on the branchlets. Leaves compound, leaflets orbicular, obovate or elliptic, sharply doubly serrate towards the apex, dark-green and glabrous above white tomentose beneath. Flowers white to light pink in thyrsoid panicles. Fruits purple–black,edible (Figs. R 16, 17).

Rumex acetosa *L.* **(Polygonaceae): Eng.- Sheep Sorrel: Hindi – Khata Palekh; Kash.- Abouje** *[W].* A tall rhizomatous perennial herb found commonly growing in Kashmir, in shaddy and moist places. Flowers pinkish. Leaves are acidic and used as vegetable and specially fed to people with liver affections and other digestive problems (Fig. R 18).

Rumex dentatus L. (Polygonaceae): Eng.- Toothed Dock; Ladakhi-Sekephala [W] Annual herb common in deep situations in sub-alpine grasslands. Leaves oval with wavy margins. Flowers in clusters of 10-20. Decoction of whole plant is used against acidity (Fig. R 19).

Rumex hastatus D. Don (Polygonaceae): Eng.- Arrow Leaf Dock; Kash.-Jungle Obuje [W]. An erect, branched, shrubby perennial growing from a woody rootstock on dry alpine slopes. Leaves rhombic, deltoid or hastately 3-lobed. Flowers appear in few flowered whorls forming slender, panicled racemes. Fruit pinkish. Leaf extract applied to wounds and cuts to check bleeding. Root is laxative, tonic, anti- rheumatic.

Rumex nepalensis Spreng. (Polygonaceae): Eng.- Dock; Kash.- Obuje [W]. Tall perennial inhabiting moist sub-alpine slopes. Radical leaves oblong -ovate, petioled, upper sessile, winged valves toothed. Leaves used as herbal vegetable.

Rumex patientia subsp. orientalis (Bernh. ex Schult. & Schult. f.) Danser. (Polygonaceae): Eng. – Patience Dock; Kash. - Abuje; Ladakhi-Sheneumo [W] A tall perrennial herb distributed in Kashmir Himalaya, common on deep shaded paces, from 1550 -3600 m. Leaves are sour and mucilagenous and cooked as vegetable at tender stage. Young leaves are used in pickles and chutnies (Fig. R 20).

Rumex paulsenianus Rech. f (Polygonaceae): Ladakhi- Khinpo [W]. A perennial stout herb of Ladakh. Extract of the leaves and young shoots used as digestive tonic. Flowers in many flowered whorls.

Rumex vesicarius L. (Polygonaceae): Eng.- Bladder Dock; Hindi-Hamaze; Ladakhi- Tnankijariatz [W]. Annual herb, on stony slopes of Ladakh. Branching, succulent and fragile. Leaves triangular, arrow shaped, entire. Flowers in racemes, flowering stalks, solitary bearing twin flowers. Nuts greyish-brown. Decoction of the dried roots is used orally in treatment of fevers, pulmonary disorders and throat infections (Fig. R 21).

Ruscus aculeatus L. (Asparagaceae): Eng.- Butchers Broom [NI]. Clump-forming, rhizomatous ornamental evergreen sub-shrub with upright branched shoots bearing ovate, spine-tipped glossy, dark green cladophylls; berries resembling bright, sealing-wax-red cherries, sometimes abundantly produced when plants of both sexes are present.

Ruscus hypoglossum L. (Asparagaceae): Eng.- Horse Tongue Lily [NI]. Clump-forming rhizomatous ornamental sub-shrub producing arching, unbranched shoots and obovate to broadly ovate, glossy, mid green cladophylls which carry a tiny, green flowers on the upper surface and on female plants. Large, red, cherry- like fruits.

Fig. R 1. Ranunculus sceleratus

Fig. R 2. Rhaponticum repens

Fig. R 3. Rheum australe

Fig. R 4. Rheum webbianum

Fig. R 5. Rhodiola fastigata

Fig. R 6. Rhododendron campanulatum

Fig. R 7. Ribes orientale

Fig. R 10. Rosa damascena

Fig. R 8. Robinia pseudo-acacia

Fig. R 11. Rosa macrophylla

Fig. R 9. Rosa brunonii

Fig. R 12. Rosa moschata

Fig. R 13. Rosa moschata

Fig. R 14. Rosa webbiana (plant)

Fig. R 15. Rosa webbiana

Fig. R 16. Rubus ulmifolius

Fig. R 17. Rubus ulmifolius

Fig. R 18. Rumex acetosa

Fig. R 19. Rumex dentatus

Fig. R 20. Rumex patientia

Fig. R 21. Rumex vesicarius

S

Sagittaria sagittifolia L. (Alismataceae):Eng.- Old World Arrow Head; Hindi- Chotakut; Kash.- Kyou [W]. Perennial herb inhabiting shallow water ponds, swampy rice fields. Leaves arrow shaped. Flowers have small sepals and three white petals, numerous purple stamens.Tubers edible (Figs. S 1, 2).

Sagina procumbens L. (Caryophyllaceae): Eng.- Birdeye Pearlwort, Procumbent Pearlwort [W]. Perennial clump forming herb, usually along stream banks, moist shaded places. Leaves opposite, stalkless with entire margins. Flowers solitary, with small white petals. Whole plant used medicinally.

Sagina saginoides (L.) H. Karst. (Caryophyllaceae): Ladakhi- Kanganchhoo [W] Perennial herb, usually on moist alpine slopes. Flowers greenish white. Whole plant used for headaches.

Salix acmophylla Boiss. (Salicaceae): Eng.- Willow; Hindi- Bed; Kash.- Vir [W/UC]. A medium sized male tree, preferring damp situations, with young twigs pubescent, later glabrous and reddish or olive brown. Leaves 8- 12 cm long, narrow, linear-lanceolate, acuminate, tapering at base, entire, olive green above, glaucous beneath; catkins appearing with the leaves, dense, oblong-cylindrical, on glabrous leafy peduncles. Leaves lopped for fodder.Wood used as fuel.

Salix alba L. (Salicaceae): Eng.- White Willow; Hindi- Bed/ Bis; Kash.- Buteh vir [W/UC]. A fairly large male tree with young twigs silky, later glabrous. Leaves lanceolate silky greyish-green above and silvery underneath. Male catkins appearing with the leaves, dense, oblong cylindrical on leafy pubescent peduncles. Four white willow varieties are found growing in the valley namely, *Chermesina* (male catkin with remarkable form and conspicuous in winter), *Coerulea* (spreading branches and sea- green lanceolate leaves), *Vitelina* (smaller tree, egg-yolk coloured young shoots) and *Vitellina pendulla* (golden yellow weeping branchlets).Wood is commonly used as fuelwood, manufacture of cricket bats, packing cases, fruit boxes and plywood. In Ladakhi system of traditional medicine its leaf extract is used for fevers.

Salix babylonica L. (Salicaceae): Eng.- Babylon Willow / Weeping Willow; Hindi – Bedi Majnun [NI]. A medium sized tree with pendulous branchlets. Leaves linear-Ianceolate, glabrous above and sparsely pubescent beneath; catkins yellow green appearing with the leaves, male catkins 1-3 cm long, on short leafy peduncles. Planted in parks and gardens as ornamental.Wood also used as fuelwood.

Salix caprea L. (Salicaceae): Eng. –Great Sallow / Great Willow; Hindi - Bed Mushak; Kash.- Gure– vir / Mooth vir [NI]. A small tree usually grown as a shrub for its sweet scented male catkins appearing before the leaves. Catkins appear before leaves and mature yellow at pollen release. During bloom they are collected, sundried and used in many ayurvedic medicines. Decoction of leaves used as stimulant and aphrodisiac. Extract of the bark is used as astringent application for piles. Catkins of the plant have also been the source of scents (Fig. S 3).

Salix daphnoides Vill. (Salicaceae): Eng.- Black Willow/ European Voilet Willow; Kash.- Kaeni Vir [UC]. A large deciduous shrub or small tree growing in Kashmir Himalayan Zone at 2000 – 3000m. Leaves purplish, acumuniate, serrate. Bark smooth, greyish- green. Twigs used for making baskets and *kangris.*

Salix denticulata Anderson (Salicaceae): Kash.- Gujure Vir; Ladakhi- Lomakarpo [UC]. A medium sized tree growing under cold -arid region of Ladakh. Leaves dull green with toothed margins. Its branches are used for construction of roofs.

Salix disperma Roxb. ex D. Don (syn. Salix julacea Andreson) (Salicaceae) : Kash.- Geur [UC]. A small tree or shrub, with shoots and leaves silky, leaves sub- sessile long, ovoate- lanceolate, obscurely serrulate, silky on both surfaces when young, nearly glabrous above and glaucous beneath. Male catkins appearing before the leaves, female catkins 8-10 cm long, drooping with an elongated conical apex.

Salix fragilis L. (Salicaceae): Eng.- Crack Willow / Kashmir Willow; Kash.- Tetha Vir [UC]. A small to large tree with glabrous and fragile twigs. Leaves long stalked, lanceolate, acute or caudate-acuminate, glabrous or somewhat hairy, glandular-serrate. Catkins appearing after the leaves on leafy peduncles. Wood used as fuel and for industrial purposes (Fig. S 4). Leaves immersed in hot water used to foment feet to remove fatigue.

Salix flabellaris Anderson (Salicaceae): Kash.-Pahale Vir [W]. Plant a shrub, with prostrate habit with stout, glabrous, red-brown branches. Leaves broadly obovate or elliptic, cuneate, often toothed or crenulated, shiny green above, paler below. Catkins cylenrical, male yellow. Plant inhabits alpine slopes and moraines, often spreading over rocks at 3500 – 4000 m altitudes.

Salix mutsudana var. tortuosa Vilm. (Salicaceae): Eng.- Chinese Willow; Kash.- Peche Vir [UC]. A fast growing and upright tree with curiously twisted and contorted yellow green shoots, that are particularly striking in winter. Leaves narrow-Ianceolate, twisted acuminate, serrate and mid- green; male catkins green yellow borne on leafy stalks, appearing with the leaves. Wood used for making bats and as fuel wood.

Salix nigra Marshall. (Salicaceae): Eng.- Black Willow; Hindi – Bed; Kashmiri -Vir. [W]. Small to medium tree grown in North Western Himalaya for wood, twigs and for fodder. Plant prefers rich swamps, river banks and wet meadow dykes.

Salix pentandra L. (Salicaceae): Eng.- Bay Willow; Kash.- Kaykas Vir [W]. A small tree or shrub growing in rich swamps.Young shoots shining. Leaves lanceolate, finely toothed. Catkins appear after leaves. Twigs used for making baskets and many other items.

Salix purpurea L. (Salicaceae): Eng.- Purple Willow; Kash.- Kala Vir [NI] A spreading shrub to upright tree of wetlands, with arching, frequently red-tinged shoots. Leaves oblanceolate dark green to blue green, glabrous, paler and finely serrate. Catkins are silver green, slender and produced all along the shoots before the leaves. Male catkins have purple anthers that turn yellow. The twigs used for manufacture of baskets, tables, chairs, trays,etc.

Salix tetrasperma Roxb. (Salicaceae): Eng.- Indian Willow; Hindi- Baishi; Kash.- Chu Mukhi Vir / Mushke vir [NI]. A moderate sized tree of sub- Himalayan zone inhabiting wet and swampy places. Shoots silky, leaves lanceolate, acuminate finely serrulate. Catkins appear after leaves. Bark rough with vertical fissures. Wood used as fuel and for match works. Twigs used for making baskets, etc. Bark medicinal, used for treating fevers.

Salix viminalis L. (Salicaceae): Eng.- Basket Willow; Kash.- Kreer / Louri Vir [NI] A slender shrub or small tree, inhabiting wet areas.Twigs flexible. Leaves linear-Ianceolate, hoary above find tomentose beneath. Male catkins 2-3 cm long; sessile, ovoid, silky, rachis silky.Twigs used for making baskets and chairs (Fig. S 5).

Salix wallichiana Anderson. (Salicaceae): Eng.- Willow; Kash.- Dantan Vir/ Geur [W]. A small deciduous tree or large shrub inhabiting mountain slopes, margins of woods, riversides. Leaves ovate to laceolate, acuminate, serrate, silvery tomentose on both surfaces when young. Catkins appear with leaves. Branches used for basket making and as tooth brushes.Wood used as fuel and in industries (Fig. S 6).

Salsola collina Pall. (Amaranthaceae): Eng.- Tumble Weed; Ladakhi- Cpapochechi [W]. Annual herb, growing in waste places and as a weed of cultivation. Stem erect, densely papillose or hispid. Leaves alternate, blade filiform to narrowly linear. Decoction of the aerial portions of the plant is administered orally as an antidote for food poisoning.

Salvia hians Royle ex Benth. (Lamiaceae) [W]. A mound forming perennial, native to Himalayas,growing at 2400 to 4000 m altitudes on open slopes and forests. Leaves hairy, broadly ovate, basaly cordate to hastate with margins having mixture of blunt and sharp teeth. Flowers are deep blue purple to violet, tubiform and borne on flowering stems in 4 -6 flowered whorls. Root used as stimulant.

Salvia moorcroftiana Wall. ex Benth (Lamiaceae): Eng.- Kashmir Salvia; Kash.- Sholar; Ladakhi- Shrematus [W]. Perennial hispid herb, along sub-alpine slopes and uplands at 1700-1900 m. Flowers pale- blue to dull white, two liped and borne in spikes.Root used for high fevers (Fig. S 7).

Salvia lanata Roxb. (Lamiaceae) [W]. Perennial herb of temperate Himalayas, on forest openings. Basal stem woody.Stem several, unbranched, hairy. Leaves hairy, clustered at the bases of stems, lanate to pannose below. Inflorescence 4- 8 flowered distant verticillaster. Flowers dark violet. Seeds and root used medicinally.

Salvia nubicola Wall ex Sweet. (Lamiaceae): Eng.- Himalayan Yellow Sage/ Himalaya Cloud Sage [W]. An erect herbaceous perennial, up to 80 cm tall, inhabiting alpine woodlands. Leaves triangular, radical leaves larger at the base of the plant, petiolate. Flowers pale-yellow that are spotted maroon on upper lips; calx green, hairy,sticky, galandular.

Traditionally used in treating oral problems.

Salvia pratensis L. (Lamiaceae): Eng.- Meadow Sage [NI]. Half hardy tender annual used for summer bedding. The mid green leaves are ovate and pointed; flowers are clear blue, tubular and hooded.

Salvia splendens Sellow ex Schult. (Salvia colorans Benth.) (Lamiaceae): Eng.- Scarlet Sage [NI]. A half hardy perennial- grown as annual. The bright green leaves are ovate with toothed edges. Scarlet flowers are surrounded with coloured bracts. Many varieties are grown in gardens like *Fire ball, Violet purple,* etc.

Salvinia natans (L) All. (Salviniaceae): Eng.- Floating Fern; Kash.- Mongole [W]. A small floating acquatic fern inhabiting all water bodies and rice fields as a weed. Stem creeping, branched bearing trimerous leaf whorls; each whorl with two green, entire, short petioled and floating leaves while the third is root like and pendant. Sporocarps are basifixed. Used as poultry feed.

Sambucus nigra L. (Adoxaceae): Eng.- European Elder; Hindi- Mushkiara [NI] An upright, bushy ornamental shrub with stout shoots and a rugged fissured bark. Leaves dark-green, pinnate with ovate, sharply toothed mid- green leaflets. Bears musk-scented white flowers in flattened panicles, followed by spherical glossy black berries.

Sambucus wightiana Wall. ex Wight & Arnott. (Adoxaceae): Eng.- Kashmir Elder; Hindi – Mushkiara; Kash. - Phokland; Ladakhi- Kown [W]. A small shrub with herbaceous stem growing in Kashmir Himalayas from 2200-3000 m amsl. Corymbs many rayed with creamy white flowers. Fruits drupe, globose, orange coloured. Fruits are diuretic and used in the preparation of diuretic drinks which promote free urination. Leaves are expectorant and strongly purgative & mostly used in dropsy (Fig. S 8).

Santolina chamaecyparissus var. corsicafiori Nana (Asteraceae): Eng.- Cotton Lavender [NI]. A much branched evergreen ornamental sub-shrub 15-20 cm high. The silver, woolly leaves are finely dissected and carried on white fetted branches. Bright lemon- yellow flowers are borne during July.

Sapindus mukorossi Gaertn. (syn. Sapindus detergens Roxb.) (Sapindaceae): Eng.- Soap Nut Tree; Hindi- Ritha; Kash.- Raenta [W]. A deciduous tree growing in lower foothills of Himalaya upto 4000 m; leaves pinnate, leaflets 8-13, oval-oblong to oblong-Ianceolate, glabrous beneath. Flowers white, in large, terminal panicles. Fruit a fleshy 2 cm thick drupe, yellow-brown. Fruits were being used for washing hair, silken clothes and shawls.

Sapium sebiferum *(L.)* **Roxb. (Euphorbiaceae): Eng.- Chinese Vegetable Tallow; Hindi- Viliati Shisham [NI].** A small tree, poplar-like in appearance, grown as ornamental. Leaves ovate-rhombic abruptly acuminate, attractively yellow and red in fall. Flowers yellowish green, in terminal panicles.

Saponaria officinalis L. (Caryophyllaceae): Eng.- Soapwort, Crow Soap; Kash.- Kown [W]. A herbaceous hardy perennial with woody base inhabiting shady sub-alpine areas along roadways and gardens. Flowers in shades of pink. Leaf poultice is used for rheumatic pain.

Saussurea atkinsonii C. B Clarke (Asteraceae): Ladakhi- Psangipakchan [W]. A perennial herb of Ladakh Himalayan zone. Leaves obovate to elliptic, membranous, toothed. Heads bluish purple, solitary. Decoction of roots and leaves is given against whooping cough and cold.

Saussurea bracteata Decne (Asteraceae): Eng.- Narrow Leaved Saw-Wort; Ladakhi- Phansi [W]. A high altitude Himalayan plant distinguished by its pinkish pale papery boat shaped bracts which surround the dense flower heads borne on densely woolly stalks. Aerial parts used for cough and bad cold (Fig. S 9).

Saussurea ceratocarpa (C B Clarke ex Hook. f.) Lipsch. (syn. Jurinea ceratocarpa var. depressa Clarke ex Hook.f) (Asteraceae): Ladakhi-Leturzil [W]. Thick stemless perennial growing on alpine slopes. Powdered seeds with milk are taken as purgative.

Saussurea costus (Falc.) Lipsch. (syn. Saussurea lappa (Decne.) Sch. Bip.) (Asteraceae): Eng. Costus /Snow Lotus; Hindi- Kunt; Kash.-Kought [W]. A tall perennial herb found in Kashmir Himalaya from 3000 -3800 m amsl,on alpine slopes.Leaves large, triangular, irregularly toothed, pubescent beneath. Flower heads purplish. Roots are aromatic, used medicinally as tonic, carminative and as stimulant, and are effective against bronchial asthama, lucoderma, epilepsy, and hysteria. They are also having antifungal and insecticidal properties. Roots have been used as a source of scents and cosmetics (Fig. S 10).

Saussurea gnaphalodes (Royle ex Royle) Sch. Bip (Asteraceae) [W]. Dwarf, densely tufted herb inhabiting high altitude moist scree slopes at 2700-5800 m altitudes. Leaves oblong obovate, obscurely toothed, densely woolly on both sides. Heads cylindrical, florets are rose –purple and embedded in densely woolly hairs.

Saussurea roylei C.B. Clarke (Asteraceae): [W]. A perennial herb growing on open slopes above 3000 m. Flower heads solitary with dark-purple florets. Paste of the plant used as to relieve aching joints (Fig. S 11).

Saussurea simpsoniana (Fielding & Gaedner) Lipsch. (syn. Saussurea sacra Edgew.) (Asteraceae): Eng.- Phen Kamal; Kash.- Jogi Padshah. [W]. A high altitude curious looking (looking like a hairy, woolly mound) woolly haired perennial herb of alpine slopes at 3800 – 5500 m. Leaves linear, lanceolate, coarse toothed, covered with woolly hairs. Whole herb used for healing boils, against nervine debility (Fig. S 12).

Saussurea stoliczkae C.B. Clarke. (Asteraceae): Ladakhi-Lupha [W]. Densely tufted,stemless perennial herb on alpine slopes above 4500 m altitude. Leaves obovate, lyrate, cottony on both surfaces. Heads sessile. Receptacle densely bristly. Powdered leaves and flowers used externally to treat fevers.

Saussurea taraxacifolia (Lindl.) Wall. ex DC (Asteraceae): Eng.- Dandelion Saw wort; Ladakhi- Psangijarpachan [W]. A perennial herb with erect and solitary stem. Flowers purple. Leaves marrowly elliptical, greyish-green, pinnately lobbed and pointing downwards like dandelion leaves. The sundried rhizomes are powdered and added to preboiled milk, kept for one day and used against fever.

Saussurea thomsoni C.B. Clarke. (Asteraceae): Ladakhi- Eripakchan [W]. Perennial, dwarf herb of alpine slopes, growing besides boulders at 5600 – 6000 m. Leaves obovate, coriacious, with toothed margins. Heads bluish purple, subglobose, densely clustered. Decoction of the dried roots is used for treating asthama,, chronic bronchitis and whooping cough.

Saxifraga asarifolia Stenb. (Saxifragaceae): Eng.- Rockfoil [W]. Annual dainty herb 15-30 cm tall on shady locations in alpine forests. Leaves glabrous sparsely ciliate. Bulbills in the axils of basal leaves. Flowers white.

Saxifraga flagellaris Willd. (Saxifragaceae): Eng.- Whiplash Saxifrage; Ladakhi-Teetaserzing [W]. Pernnial stoloniferous herb, on rocky and open high, moist alpine slopes. Stem single, erect, leafy.The basal leaves in dense rosette. Each stem bears a terminal flower with golden yellow petals.Crushed leaves used as poultice to heal open wounds.

Saxifraga jacquemontiana Decne. (Saxifragaceae): Eng.- Jacquemont's Saxifrage; Ladakhi- Teetasariat [W]. Perennial, caespitose herb below moist rocks, on alpine ranges of Kashmir and Ladakh. Leaves sessile, glandular, pubescent, forming rosette at the tip of branches. Flowers yellow. Powdered shoots used as laxative (Fig. S 13).

Saxifraga kashmeriana Dhar & Kachroo (Saxifragaceae): Eng.- Mountain Saxifraga [W]. Perennial herb inhabiting alpine scrubs of Kashmir. Stem slender, loosely tufted, glandular pubescent throughout. Radical leaves entire or sometimes lobed, obovate-lanceolate, shortly petioled. Flowers white.

Scabiosa speciosa Royle (Caprifoliaceae) [W]. A pubescent handsome tufted herb common on alpine slopes at 2700- 4300 m altitudes. Leaves lanceolate to blong-lanceolate, some basally dissected into lanceolate segments. Flower heads radiant, terminal on peduncles, light mauve coloured.

Scandix pecten-veneris L. (Apiaceae): Eng.- Shepherd's Needle [W]. An annual herb of grasslands and orchards.Leaves bi to tri- pinnate. Umbells bearing white flowers. Fruit cylindrical, compressed laterally, beaked dorsally looking like a needle. Stem and foliage edible, and used as vegetable.

Scripus dialgamensis G Singh (Cyperaceae) [W]. An aquatic herb on marshy situations. Flowers brownish. An important fodder plant for horses.

Scripus lacustris L. (Cyperaceae): Eng.- Great Bulrush [W]. A large perennial herb inhabiting alpine/ sub-apline moist soils. Serves as fodder for cattle. Culms used for thatching of roofs in Ladakh.

Schoenoplectiella articulata (L.) Lye (syn. Scirpus articulatus L.) (Cyperaceae): Eng.- River Bulrush [W]. A rhizomatous perennial growing on moist soils. Culms densely tufted, terete, clothed at base. Leaf sheath terete obliquely truncate at mucronate mouth. Spikelets sessile. Nutletts yellowish grey at maturity.

Scorzonera divaricata Turcz. (Asteraceae): Ladakhi- Tharnoo [W]. Erect, branched, glabrous perennial on alpine slopes/near base of mountains of Ladakh. Stem erect, basally woody. Basal leaves linear, stem leaves linear to filiform. Capitula terminal with 4-5 yellow florets.Whole plant used for rheumatism.

Scorzonera virgata DC. (Asteraceae): Ladakhi- Tharnoo [W]. Erect, branched, glabrous perennial herb on dry alpine slopes above 3000m. Leaves long, linear, recurved. Achenes linear, ribbed, pappus feathery. Whole plant used for rheumatism.

Scrophularia decomposita Royle. ex Benth. (Scrophulariaceae): Ladakhi- Zimpontziamt [W]. Tall, robust, glabrous perennial herb,on open alpine /sub-alpine slopes. Leaves 2 pinnatisect. Powdered inflorescence used for treatment of dropsy and chronic bronchitis.

Scrophularia koelzii Penn. (Scrophulariaceae): Ladakhi- Zimbatiq / yarma [W] Perennial herb, usually on dry rocky and savana slopes. Flowers yellowish white. Powdered roots are used for treating gastric disorders. In Kashmiri ethnomedicinal practice it is a herbal remedy for rheumatic pain.

Scrophularia lucida L. (Scrophulariaceae): Ladakhi- Neeremarpo [W]. A perennial herb,usually on dry rocky and savana slopes from 1800-2300m. The extract of the powdered plant is used to cure epidemic fevers and migraines.

Scutellaria galericulata L. (Lamiaceae): Eng.- Common Skullcap, Marsh Skullcap[W]. Erect rhizomatous perennial inhabiting damp situations.

Leaves ovate- lance shaped. Flowers axillary on apical part of stem. Coralla purple to blue. Nutlets yellow, ovoid.Used as nerve tonic, sedative and as diuretic.

Scutellaria heydei Hook.f. (Lamiaceae)): Ladakhi- Jimthiglae [W]. Perennial, suffruticose herb with thick woody rootstock. Stem prostrate, ascending, purplish. Leaves thick textured, rugose. Flowers dull creamy white with pink, orange or purplish markings. Nutlets balck, covered with hairs. Aerial parts are dried near fire and powdered and added to water which is used against eye troubles. The powder is also added to curds and used as diuretic (Fig. S 14).

Scutellaria prostrata Jacq. ex Benth. (Lamiaceae) [W]. Dwarf prostrate much branched alpine herb. Leaves ovate,small, petioled and toothed. Flowers yellow tipped with violet. Whole plant used medicinally.

Secale cereael L. (Poaceae): Eng.- Rye Grass; Kash.- Vilayati Knakh. [W]. A perennial grass with short awns and tough rachis. Generally grows as a weed of cultivation in wheat fields. It is used as an important fodder plant in fresh as well as hay. It used to be cultivated as a grain crop in Ladakh, till recently.

Sedum crassipes Wall. ex Hook.f & Thomson (Crassulaceae): Ladakhi-Koad [W] A herbaceous perennial having a thick caudex covered with old dead growth. Grows on open slopes among rocks. Stem erect upto 25 cm tall. Leaves narrow, distantly dentate. Flowers yellowish white. Plant extract has soothing effect on human body.

Sedum ewersii Ledeb. (Crassulaceae): Eng.- Stone Crop; Ladakhi-Churupa [W]. Perennial glabrous herb forming patches on moist rocks. Leaves flat, fleshy, bluish green, opposite, sessile. Flowers bluish purple. Decoction of young leaves used as appetizer.

Sedum quadrifidum Pall. (Crassulaceae): Ladakhi- Sholomarpo [W]. Perennial herb of alpine areas, forming mats on rocks. Stem and leaves glabrous or puberulous. Flowers red. Decoction of roots in milk is given for constipation.

Sedum oreades (Decne.) Reym.- Hamet (Crussulaceae): [W]. Low annual hairless herb growing in alpine rock crevices at 3000- 4000 m altitudes. Stem sub-erect, branched, glabrous. Leaves dense rosulate, alternate, sessile, fleshy, entire. Flowers white or yellowish, solitary.

Sedum spectabile Boreau (Crussulaceae): Eng.- Rose Root [NI]. A half hardy tender ever green succulent mainly grown as plants under green house conditions. Leaves oval, white- green. Pink flower heads appear from September- October.

Selinum vaginatum C.B. Clarke (Apiaceae): Eng.- Bhut-keshi [W]. A perennial stout glabrous herb of alpine meadowlands. Flowers white in large clusters. Leaves larger, bi-pinnate with long stalks. Roots posses sweet and musky odour and used as an incense and nerve sedative.

Selinum wallichianum (DC) Raizada & H. O Saxena (syn. Selinum tenuifolium Salisb.) (Apiaceae): Eng.- Wallich Milk Parsley [W]. A perennial tall, glabrous herb of Alpine zone. Leaves fern like. Flowers white. Rootstock fibrous. Roots used in the preparation of incenses.

Sempervivum heuffelii Schott. (Crussulaceae): Eng.- House Leek [NI]. A rosette forming succulent, suitably placed in pots and for rock gardening. Leaves are fleshy, yellow green, glacous and curved to form a rosette.

Sempervivum montanum L. (Crussulaceae): Eng.- Live Fore Ever [NI]. Succulent leaves form rosettes. Pale-red purple flowers are borne on stem from June to August.

Sempervivella accuminata (Decne.) Berger (Crassulaceae): Eng.- Lower Mountain Sedum [W]. A succulent annual. Decoction of the leaves and flowers is given with *Chanuge* as anti-flatualant, antispasmodic and sedative.

Senecio chrysanthemoides DC (Asteraceae): Ladakhi- Churupa [W]. Perennial erect herb growing along borders of rice fields, in forests and on shady alpine slopes. Leaves toothed. Flower heads yellow.Whole plant has soothing effect.

Senccio graciliflorus (Wall.) DC (Asteraceae): Eng.- Graceful Senecio; Kash.- Phagga [W]. Erect perennial herb. The long narrow lance-like leaves are pinnately lobbed with 6-8 pairs of lobes which are toothed. Flower heads yellow, graceful borne in flat topped clusters. Water extract used for skin eruptions (Fig. S 15).

Senecio jacquemontianus (Decne.) Benth. ex Hook (Asteraceae): Eng.- Ragwort; Kash.- Khalar [W]. A tall robust glabrous herb on shady alpine slopes. Heads yellow. Root powder used for rheumatic pain.

Senecio krascheninnikovii Schischk. (syn. Senecio pedunculatus Edgew.) (Asteraceae): Eng.- Hinduksh Senecio; Ladakhi- Unarswah / Sianthi [W]. Annual, glabrous dwarf herb on alpine/sub-alpine sandy slopes. Leaves dissected into linear segments. Flower heads yellow.The paste of fresh leaves applied to forehead to relieve headache and on inflamed parts to reduce pain.

Senecio vulgaris L. (Asteraceae): Eng.- Groundsel; Kash.-Shawla Loute [W]. A tenacious annual or biennial herb, a weed of cultivation in orchards. Flower heads yellowish. Leaves pinnately lobbed covered with soft, smooth hairs. Used in many medicines as diaphoretic, diuretic and tonic (Fig. S 18).

Serissa japonica (Thunb.) Thunb. (syn. Serissa foetida (L. f.) Lam. var. variegata) (Rubiaceae): Eng.- Snow Rose *[NI]*. A densely branched, evergreen ornamental shrub, upright, shoots glabrous or pubescent, dark unpleasantly scented when crushed. Leaves opposite deep green above with white margins lighter beneath. Flowers solitary or few, axillary and terminal, corolla salver shaped. Fruit small berry with 2 stony seeds.

Sesamum indicum L. (Pedaliaceae): Eng.– Sesame; Hindi- Til; Kash.- Taile [RC] A herb, grown as an oil bearing crop, however its cultivation has almost been abandoned. Flowers yellow, tubular. Fruit a capsule. In Hindu religious functions its oil was used as an offering to God; the oil in a lamp is being used in obesequial rites for deceased father or other relations.

Setaria italica (L.) Beauv. (syn. Panicum italicum L.) (Poaceae): Fox-tail Millets; Hindi- Kangni; Kash.- Shoule [RC]. Annual grass with thin, vertical, leafy stem. Seed head is dense hairy panicle. A second grade cereal, once considered an extremely useful grain yielding plant (as substitute to rice) in Kashmir is now rarely cultivated and its cultivation restricted to few pockets of upper hilly regions where rice cultivation is rather impossible. The grain, which is husked like rice, has never been esteemed by Kashmiris as food, as it was considered to have heating properties and thus its use was restricted to some tribal belts only. There were two varieties of millets traditional grown in Kashmir namely, **Shol/Kangni kalan** - grains longer bold and white; and **Shol/ Kangni Khuard**- grains reddish and smaller (Fig. S 17).

Sibbaldia cuneata Schouw ex. Kunze (Rosaceae): Eng.- Wedgeleaf Sibbaldia; Kash.-Sanjoo [W]. Low perennial herb of Himalayan slopes at 3000-4000m altitudes, with branched woody rootstock. Leaves with 3 leaflets. Common in alpine and sub-alpine meadows. Flowers tiny, yellow arise in compact, hairy clusters at the top of the stem.

Sibbaldia parviflora Willd. (syn. Potentilla cuneata Wall ex. Lehm.) (Rosaceae): Eng.- Five Finger Cinquifoil; Kash.- Sanjoo [W]. A small spreading perennial herb of alpine meadowlands at 2500 – 4500 m altitudes. Leaves ternate with 3 small obovate leaflets, coarsely toothed at the tip. Flowers rich yellow, solitary.

Sibbaldianthe bifurca (L.) Kurtto & T. Erikss. (syn. Potentilla bifurca L.) (Rosaceae): Ladakhi- Psangijarpa [W]. A low, thick rootstocked perennial herb, growing on moist shady places in alpine grasslands. The powdered plant, mixed with honey, is used as heart stimulant and nervous depressant.

Siegesbeckia orientalis L. (Asteraceae): Eng.- Common St. Paul'swort [W]. A small shrub common under forest shade, also inhabits ruderal habitats. Stem terete, with white pubescence. Heads yellow enclosed in five involcral bracts, having sticky glandular hairs. Plant used medicinally for treating wounds and inflammations.

Silene flos- jovis (L.) Desr. (syn. Lychnis flos-jovis (L.) Desr.) (Caryophyllaceae): Eng.-Campion [NI]. A half hardy annual, ornamental. The stems and the lanceolate thick leaves are silvery grey. Purple or red flowers are borne in a loose rounded inflorescences.

Silene latifolia subsp. alba (Mill.) Greuter & Burdet (syn. Lychnis alba Mill. (Caryophyllaceae): Eng. –White Alpine Campion; Kash- Watte-Krum [W]. A perennial tomentose herb growing on rocky surfaces. Stem with densely leafy shoots. Its tender leaves are used as vegetable. Flowers white (Fig. S 18).

Silene moorcroftiana Wall.ex Benth. (Caryophyllaceae): Eng.- Moorcroft Campion; Ladakhi- Luksukpa [W]. Perennial clustered, glandular, pubescent herb, with woody rootstock, on drier alpine slopes above 3000 m. Flowers yellowish brown. Ash of the plant used for itching and cutaneous eruptions.

Silene pendula L. (Caryophyllaceae): Eng.- Campion Sweet William [NI]. An annual herbaceous plant with hairy green leaves. Pale pink axillary flowers are carried in loose cluster from May to July. Several garden varieties with salmon pink or crimson flowers are grown in gardens.

Silene uralensis subsp. apetala (L.) Bocquet (syn. Lychnis apetala L.; Lychnis himalayansis (Rohrb.) Edgew & Hook. f.) (Caryophyllaceae): Ladakhi – Khizingtse [W]. Perennial tomentose herb of Ladakh Himalayas. Flowers nodding, petals rose lavender to brownish purple. Leaves basally congested. Extract of the plant used as an anthemintic.

Sinopodophyllum hexandrum (Royle) T.S. Ying (syn. Podophyllum hexandrum Royle.; Podophyllum emodi Wall ex Hook. f. & Thomson) (Berberidaceae): Eng.- Indian Podophyllum / Himalayan May Apple; Hindi– Paora; Kash.- Wanwangune; Ladakhi-Bengpheloutche [W]. A perennial high altitude herb found in the interior ranges of Himalayas from 2400-4000 m amsl. Plant is having with unbranched stem and flourishes as forest undergrowth. Leaves glossy green, drooping and cut into 3 ovate toothed lobes. Flowers cup shaped,white, with six petals.

Fruit a large scarlet or reddish berry. Rhizomes and fruits are source of drug which is used as stimulant and purgative. The drug is also reported to have anticancerous properties (Fig. S 19).

Sisymbrium brassiciforme C A Mey (Brassicaceae): Eng.- Himalayan Tumble-Mustard; Ladakhi- Paleyoung [W]. Tall, annual, glabrous alpine/ sub-alpine herb inhabiting alpine grasslands. Lower leaves dandelion like, cut into many lobes; upper ones linear sessile. Leaves fleshy. Flowers yellow. Whole plant used as diuretic.

Sisymbrium irio L. (Brassicaceae): Eng.- London Rocket; Hindi- Khoob-Kalan; Kash.- Chaeri Latchije [W]. Annual herb,upto 90 cm tall, inhabiting mud walls, waste places, orchards.Branches open; basal leaves broad and lobed, upper linear.Fruit cylindrical siliqua;seeds redish yellow, oblong. Flowers yellow. Seeds are expectorant and stimulant. Leaves used for treatment of throat and chest infections.

Sisymbrium loeselii L. (Brassicaceae): Eng.- Tumble Weed Mustard; Hindi – Khoobe Kalan; Kash.- Dande Hakh [W]. An erect biennial herb, common in wastelands and orchards. Stem coarsely hairy. Flowers yellow. Seeds used for scurvy and scrofula (Fig. S 20).

Sisymbrium orientale L. (Brassicaceae): Eng.- Indian Hedge Mustard; Ladakhi- Staga [W]. Biennial herb growing on rocky alpine slopes. Stem branched, hairy. Stem produces racemes of yellow florwers. Lower leaves lobed, upper leaves lance-shaped. The powdered seeds are rolled into small tablets with butter or milk and used as an appetizer and carminative.

Skimmia anquetilia N. P. Taylor & Airy Shaw (Rutaceae): Kash.- Neer [W]. An erect, dome-shaped, diocious ornamental shrub of Kashmir Himalaya, producing inversely lance shaped to oblong elliptic, leathery, strongly aromatic, dark or yellowish green leaves. Bears small, yellow-green flowers, in terminal panicles. Berries persistent on female plants. Leaves contain strong anti-oxidants and used for treating swollen joints and inflammations. Bark used for healing wounds and skin infections (Fig. S 21).

Skimmia laureola Franch. (Rutaceae): Kash.- Shingli Mingli / Kani Wathir [W] An aromatic shrub of sub- alpine forests of West Himalayan range, preferring shaded lacations. Leaves opposite, oblong-lanceolate, shiny dark-green adaxially. Flowers white, berries round and red. Leaves are being used as incense by Hindus, especially while worshiping lord Shiva. Leaves used for intestinal disturbances.

Smilax vaginata Decne. (Smilacaceae): Eng.- Catbriers; Kash.-Thir/ Chob Chini [W/UC]. Deciduous under-shrub, among scrubs and in forests on humus rich soils. Branches terete, unarmed. Flowers minute, purple. Berries globose, blue-black. Decoction of root used for treating veneral diseases.

Solanum americanum Mill. (syn. Solanum nigrum L.) (Solanaceae): Eng.-Black Nightshade; Hindi- Makoi; Kash.- Kambai. [W]. Annual herb, grows in waste places and also as weed of cultivation. An infusion of plant is given to patients with abdominal upsets and as remedy for anthrax pustules. Fruits useful in cirrhosis of liver (Fig. S 22).

Solanum melongena L. (Solanaceae): Eng.– Egg Plant; Hind- Baingan; Kash. – Wangun [UC]. A herbaceous, prickly annual, cultivated for its edible fruit. Fruits are consumed in large quantities and cooked in fresh as well as sundried forms. Presently many exotic varieties, besides the local types, are cultivated extensively, however the local types are preferred to over the exotic types because of their taste, cooking qualities, longer shelf life and in being resistant to fruit borer.

Commonly cultivated brinjal types are:

- **Dilruba:** Variety developed by State Department of Agriculture in 1982 by crossing, *Kashmiri Round* brinjal variety and one exotic variety *Osaka Honaga*. Plants are medium tall (80-110 cm) with characteristic oval, deep purple coloured fruits. It yields 150-200 q ha-1. Plants are resistant to brinjal diseases and pests.
- **Kashmiri Long:** A local brinjal variety of Kashmir. Plants tall. Leaves oval, sinuate. Flowers white, mostly appearing solitary; calyx thick with prickles. Fruits cylindrical and long, light pink when edible. The variety yield as 200-240 q ha -1 (Fig. S 23).
- **Kashmiri Round:** A local brinjal cultivar, with characteristic roundish fruits. Plants tall (80-120 cm); leaves oval, sinuate; flowers white, mostly appearing solitary and also in clusters. Fruits light pink at edible stage and round to oval in shape. Variety yields 180-220 q ha -1.
- **Shalimar Brinjal Hybrid -1:** A single cross brinjal hybrid between two lines-SH-B-4 and SH-B-12 developed and released by SKUAST-K during 2010. A light pink fruited hybrid with very high yield potential of 875 q/ha with 122% yield superiority over Local Long. It matures within 55-60 days. It is tolerant to wilt, blight and fruit rot.
- **Shalimar Brinjal Hybrid-2:** Developed by SKUAST-K as a single cross

hybrid involving SH-B-11 and SH-B-12. It is a purple fruited brinjal hybrid with average yield of 653 q/ha with 101% yield superiority over Local Long. It matures within 55-57 days and is tolerant to wilt, blight and fruit rot.

Other recently introduced and popular brinjal varieties indlude **Pusa Purple Long, Pusa Purple Cluster, Punjab Chamkeela** and some hybrids.

Solanum nigrum L. (Solanaceae): Hindi- Makowe; Kash.- Kambai [W]. Annual herb, usually a weed of cultivation in orchards, often along pathways. Flowers white. Fruits dark blue at maturity and edible. Fruit used for treating liver ailments.

Solanum pseudocapsicum var. diflorum (Vell.) Bitter. (syn. Solanum capsicastrum Link. ex Schauer) (Solanaceae): Eng.- Winter Cherry [NI]. A popular house plant, grown principally for its attractive berries. Leaves dark- green, lanceolate. Bears axillary cymes, of star shaped white flowers, followed by oblong ellipsoid to ovoid, pointed, red or orange-red fruit which are persistent.

Solanum tuberosum L. (Solanaceae): Eng. Potato; Hindi - Auloo; Kashmiri - Aeulowe. [UC]. Potato thrives well in Kashmir especially in well drained soils of hilly areas in the Distts. of Pulwama and Anantnag, Shopian, Baramulla. and Srinagar. Local types and other high yielding varieties of potatoes under cultivaton are:-

- **Hirpura Auloo:**Tubers roundish, skin smooth/rough scaly or russeted light pinkish with white flesh. The type is specifically cultivated in Hirpura locality of Shopian hence the name.
- **Gulmarg Special:** Tuber oval, skin colour light red with purple or violet or blue eyes. Cultivated in the higher belts of Tangmarg and Gulmarg areas of Kashmir (2200 - 2600 m altitude).
- **Sedow Auloo:** A famous potato cultivar of Shopian area known for its cooking qualities and long shelf life. Flesh white/pale yellow, reddish around the vascular ring. Yields -170-220 q ha -1.
- **Gurez Auloo:** A potato cultivars of Gurez. Tubers with light skin, dull white flesh and few eyes.
- **Shalimar Potato -1:** A high yielding, white skinned potato variety, developed by SKUAST-K through clonal selection; suitable for cultivation under valley conditions. Variety matures in 120 days and has a average tuber yield of 420 q/ha showing 31% yield superiority over *Kufri Jyoti*. Tolerant to early blight, tuber rot and moderately tolerant to late blight.

- **Shalimar Potato -2:** A high yielding red skinned potato variety developed by SKUAST-K during 2010 through clonal selection. Matures in 115 days and has average tuber yield of 238 q/ha, having 22% yield superiority over *Hirpura Local*. Tuber firm fleshed, pinkish red skined with excellent cooking and keeping qualities. Tolerant to early blight, tuber rot and moderately tolerant to late blight. Tubers good for chips and French fries.

Other introduced and popularly cultivated potato varieties include *Kufri jyoti* and *Kufri Chandra Mukhi.*

Solanum villosum Mill. (syn. Solanum miniatum Bernh. ex. Willd.) (Solanaceae): Kash.- Kambai [W]. Annual herb inhabiting orchards and waste places. Whole plant used for skin infections.

Solanum virginianum L. (syn. Solanum xanthocarpum Schrad. & Wendl.) (Solanaceae): Eng.- Yellow Berried Nightshade; Ladakhi- Batkatel [W]. A spreading armed herb, stem prickly, growing in wastelands and river banks. Flowers purplish produced in bunches. Berries globular, drooping, yellow or pale. Flowers used against indigestion.

Solidago virga-aurea L. (Asteraceae): Eng.- Woundwort [W]. A graceful, glabrous perennial herb of alpine meadowlands and open slopes. Heads in cymes and yellow. Plant has diuretic and lithotropic properties. Also applied on skin to heal wounds.

Sonchus asper (L.) Hill. (Asteraceae): Eng.- Roughmilk Thistle; Ladakhi- Kahlaserpo [W]. Annual herb of Kashmir and Ladakh, commonly in waste lands. Leaves bluish-green, simple, lanceolate with lobed margins and covered with spines on margins and beneath. Flower heads yellow. Leaves and stem emit a milky sap when cut. Decoction of the leaves, in Ladakhi system of medicine, is used against thyroid swelling and sore throat.

Sonchus oleraceus L. (Asteraceae): Eng.- Hare's Lettuce [W]. Annual coarse herb, usually in wastelands. Heads yellow. Roots, stem and leaves used medicinally.

Sophora japonica L. (Fabaceae): Eng.- Japanese Pagoda Tree; Kash.- Chatre Kul [NI]. An ornamental tree planted in gardens. Flowers yellow, in racemes, showy.

Sophora mollis (Royle) Baker: Eng.- Soft Sophora / Himalayan Laburnum (Fabaceae) [W]. A spineless shrub or a small tree with finely grey-downy branches. Leaves pinnate, pale grey-green. Flowers borne on long racemes,in axils, are dull yellow and showy. Pods like a string of beads. Flowers and seeds used medicinally.

Sophora moorcroftiana (Benth.) Baker (syn. Caragana moorcroftiana Benth. (Fabaceae): Ladakhi- Takay Vonpo [W]. A deciduous shrub of cold-arid areas of Ladakh with densely hairy branches. Leaves hairy, spinescent with 11-15 leaflets. Flowers blue-purple. Fresh leaves are boiled in milk and taken regularly as blood purifier. Decoction of the plant is used against sinusitis, gastric ulcer.

Sorbaria tomentosa (Lindl.) Rehder (syn. Sorbaria tomentosa var. angustifolia (Wenz.) Rahn.; Sorbaria aitchisonii (Hemsl.) Hemsl. ex. Rehder) (Rosaceae): Eng.- Kashmir False Spireal [NI]. A large deciduous, ornamental shrub of Kashmir Himalaya, with wide spreading branches, totally glabrous and reddish when young. Leaves pinnate, slender, pointed, ash like, mid green on red stems. Leaflets 15-21 lanceolate to narrowly lanceolate, fern like. Panicles 25-30 cm long, out spread, loose, bearing minute yellowish white flowers during August (Fig. S 24).

Sorbus aucuparia L. (Rosaceae): Eng.- Rowan Mountain Ash (Rosaceae); Hindi - Battal [W]. A deciduous tree, with rough bark and glabrous branches, winter buds glutinous, glossy. Leaves compound, leaflets narrowly oblong, scabrous serrate, glabrous above. Flowers appear on wide heads followed by large branches of globular orange red berries. Grows around damp coniferous forest margins.

Sorbus cashmiriana Hedl. (Rosaceae): Eng.- Himalayan White Bean/ Kashmiri Rowan [W]. A short lived deciduous ornamental tree native to Himalaya.The leaves are mid- green with grey green undersides. Flowers pale- pink borne in wide pendulous clusters. Fruits pink tinged globular, persistent even after leaf fall. Grows wild is alpine and sub-alpine forest slopes of Kashmir.

Sorbus lanata (D. Don) Schauer. (Rosaceae): Eng.- Hairy Rowan; Kash.- Charin [W] Tree or tree-like shrub, often multi-stemmed, crown broadly conical, branches grey tomentose at first, later olive-brown, much furrowed. Leaves larger, elliptic, scabrous and densely serrate. Grows wild on alpine forest slopes. Plant has white woolly, aromatic, flower clusters. Fruits yellowish-red, edible (Fig. S 25).

Sorghum bicolor (L.) Moench. (syn. Sorghum vulgare Pers.; Andropogon sorghum (L.) Brot.: (Poaceae): Eng. – Sorghum /Broom Corn; Hindi- Joar; Kash.- Jowar [RC]. Annual, erect, tillering annual with erect and solid stem attaining 2 to 5m height. Sorghum, though a grain crop in other parts of the country, after being introduced in the valley of Kashmir, is however grown exclusively as a kharif forage crop. *M.P.*

Charry is the most common variety being cultivated for forage purposes. It is harvested after flowering and used for stall-feeding.

Sorghum halepense (L.) Pers. (syn. Andropogon halepensis (L.) Brot. (Poaceae): Eng. Johnson Grass; Kash.- Drehma / Haess Gausse [W]. Perrennial grass with creeping rhizomes inhabiting pastures, abandoned fields and crop lands. Plants with narrow leaf blades, slender culms and small pedicelled spikelets. It is poisonous in vegetative phase and forms one of the best cattle feeds after flowering phase.

Spinacia oleracea L. (Amaranthaceae): Eng.- Spinach; Hindi– Palak; Kash.- Palakh [WC]. An biannial herb grown for its foliage and young shoots which are used as vegetable. Flowers small, yellow-green.

Spiraea cantoniensis var. florepleno Maxim. (Rosaceae): Eng.- Bridal Wreath [NI] A spreading shrub, totally glabrous, shoots cylindrical, slender. Leaves rhombic -lanceolate, coarsely serrate, dark-green above, blue-green beneath. Flowers many in numerous hemispherical corymbs, pure white borne at the tips of young shoots.

Spiraea japonica L. f (Rosaceae): Eng.- Japanese Spiraea [NI]. A shrub with stuffy erect shoots, slightly branched, glabrous or only pubescent when young, striated or nearly cylindrical. Leaves oval-oblong sharply toothed. Flowers lighter or darker pink-red, borne on terminal corymbs.

Spiraea prunifolia Siebold & Zucc. (syn. Spiraea prunifolia var. plena Schneid.) (Rosaceae): Eng.- Bridal Wreath [NI]. An upright shrub, shoots thin, long rod like, nodding, finely pubescent at first. Leaves oblong-elliptic, finely dentate, bright green above, soft gray pubescent beneath, orange to red-brown in fall. Flowers pure white, 3 - 6 in sessile umbels.

Spiraea x vanhouttei (Briot) Zabel. (Rosaceae) [NI]. An ornamental shrub upto 180 cm high; long, rod like branches gracefully nodding, glabrous. Leaves ovate-rhombic, crenate, dark-green above, bluish beneath, glabrous. Flowers pure white borne in flat corymbs.

Sporobolus indicus (L.) R. Br. (Poaceae): Eng.- Smut Grass [W]. A robust, tufted perennial grass with slender culms up to 120 cm tall growing across grasslands up to 3000 m amsl. Inflorescence a dense spike like panicle of light-brown spikelets. Leaves often coated with black smut fungus.

Stachys sericea Cav. (Lamiaceae): Eng.– Silky Woundwort; Ladakhi– Sianthi [W]. Perennial, densely villous erect herb on open slopes and forest openings in Kasmir and Ladakh regions at 2500-3300m. Flowers bluish purple. In Ladakhi system of medicine aerial portion, other than inflorescences, is dried on fire, powdered and used as laxative, febrifuge, antispasmodic, astringent, anodyne and stomachic (Fig. S 26).

Stachys tibetica Vatke. (Lamiaceae): Eng.– Tibetian Wound Wort; Ladakhi– Jatuk-Napo [W]. A perennial woody clump forming herb of high altitude alpine ranges of Ladakh. Flowers pink. Leaves glandular and hairy. Fresh roots and young shoots edible. Leaves used for healing wounds and for treating mental disorders and relieving tensions.

Stellaria alsine var. alsine (syn. Stellaria alsine var. undulata (Thunb.) Ohwi.) (Caryophyllaceae): Eng.- Chick Weed / Bog Stitchwort [W]. A perennial herb with slender stem growing on moist alpine and sub-alpine slopes. Leaves thickish, blue-green and used as vegetable. Flowers white. Fruit, oval, yellowish-brown, long capsule.

Stellaria aquatica (L.) Scop. (syn. Myosoton aquaticum (L.) Moench.; Cerastium aquaticum L.; Malachium aquaticum (L.) Fr. (Caryophyllaceae): Eng.- Water Chickweed; Ladakhi- Pushtingtse [W]. Perennial herb of damp shaded places. Leaves opposite, oval with wavy margins and glandular hairy and dark-green. Flowers white. Leaves powdered and mixed with oil are applied on the belly of expected mother to relieve muscles for an easy delivery(Fig. S 27).

Stellaria media (L.) Vill. (Caryophyllaceae): Eng.- Common Chick Weed; Kash.- Losdhi [W]. Annual weak, ascending and glabrous herb, inhabiting shaded and moist sub-alpine slopes. Leaves ovate, acuminate. Flowers white. Seed powder is given to children with milk to cure skin infections and allergy. Leaf paste applied to heal burns and frost bites.

Sternbergia colchiciflora Waldst & Kit. (Amaryllidaceae): Eng.- Yellow Autumn Crocus [W]. A bulbous herb inhabiting orchards,graveyards and karewa lands.Leaves dull green,ciliate. Flowers golden yellow, appearing in early spring. Also cultivated as ornamental.

Sternbergia vernalis (Mill.) Gorer & Harvey (syn. Sternbergia fischeriana (Herb.) Roem.) (Amaryllidaceae): Eng.- Winter Daffodil; Kash.- Taenke Batne [W]. A bulbous ornamental herb growing wild in sub-alpine forest slopes and grasslands, especially on moist and shady locations. Leaves dark-green and strap shaped. Flowers bright yellow and appear in early spring. Seen frequently planted on grave yards (Fig. S 28).

Stipa sibirica (L.) Lamk. (Poaceae): Kash.- Gumin Ghasse [W]. A perennial caespitose rhizomatous herb, very common in pasture areas. Flowers green. Leaf blades scaberulous, puberulous, hairy adaxially. Plants of fodder value.

Strobilanthes wallichii Nees (syn. Strobilanthes alata (Wall.) Nees): (Acanthaceae) [W]. A wild perennial herb upto 50 cm tall inhabiting wet areas in deciduous forests.Stem glabrous, erect arising from woody root-stock. Leaves slightly unequal, ovate, toothed. Flowers blue purple. Used in woodland gardens as an ornamental (Fig. S 29).

Styphnolobium japonicum (L.) Schott. (syn. Sophora japonica L. (Fabaceae): Eng.- Japanese Pagoda Tree; Kash.- Chathre Kul [NI]. A deciduous tree with spreading branches forming an umbrella type of round head. Leaves pinnate, leaflets 7-17, stalked, ovate-lanceolate, acute, dark-green and lustrous above, glaucous and pubescent beneath. Cream white pea like flowers are borne on pendent racemes. Var.

Pendula is the common type grown in the valley characterized by its pendulous branches and slow growth.

Swertia angustifolia Buch Ham. ex D Don. (Gentianeceae): Eng.- Felwort [W]. Annual herb of open alpine,sub-alpine slopes. Stem erect, sub-quadriangular. Leaves sessile, linear, elliptic-lanceolate. Roots yellow, fibrous. Flowers bluish white, to greenish with purple dots.Whole plant medicinal and used against fevers.

Swertia chirayita (Roxb.) Buch. ex Ham. ex C.B. Clarke (Gentianaceae); Eng.-Chiretta; Hindi – Kiryat/Charayata; Ladakhi-. Gyatik [W]. An erect herb distributed, in moist situations, across temperate Himalayas. Leaves sub-sessile lanceolate. Flowers greenish-yellow, tinged with purple. Capsule egg shaped, many sided, sharp pointed. Plant used in treatment of chronic fever and blood purifier.

Swertia petiolata D. Don (Gentianaceae): Ladakhi- Zantik. [W]. Tall perennial of moist places of alpine/ sub-alpine grasslands above 3000 m altitudes. The decoction of whole plant, in milk, is used against headache and body aches. Flowers bluish white, fruit a capsule (Fig. S 30).

Swertia thomsonii C. B. Clarke (Gentianaceae): Ladakhi- Chirayata [W]. Perennial, unbranched, erect herb growing in alpine pastures, near moist situations. Whole plant used for headache and fevers. Basal leaves ovate-spathulate, cauline leaves connate. Flowers greenish-yellow with bluish tinge. Fruit a capsule.

Syringa emodi Wall ex Royle (Oleaceae): Eng.- Himalayan Lilac; Kash.- Yausmin [W]. A large deciduous shrub upto 5 m tall inhabiting alpine slopes at 2100- 3600m.Leaves elliptic-oblong with entire margins. Flowers borne in dense branched clusters of white fragrant flowers at the end of branches (Fig. S 31).

Syringa persica L. (syn. Syringa laciniata (L.) Mill.) (Oleaceae): Eng.- Cutleaf Lilac [NI]. An ornamental shrub. Leaves on the new growth pinnately cleftted or 3-9 lobed, summer leaves simple. Flowers light purple, in loose panicles, fragrant.

Syringa persica L. (Oleaceae): Eng.- Persian Lilac [NI]. A large, glabrous ornamental shrub, with upright or arching branches. Leaves oblong or ovate-lanceolate, acute or acuminate, base cuneate, entire or 3-lobed. Fragrant lilac coloured flowers are borne in pyramidal panicles.

Syringa pubescens subsp. microphylla (Diels) M.C. Chang & X.L. Chen. (syn. Syringa microphylla Diels.) (Oleaceae): Eng.- Little Leaf Lilac; Kash.- Yausmin [W]. A small, bushy, broadly upright shrub, branches thin, soft, pubescent. Leaves rounded ovate to elliptic-ovate. Flowers in 4-7 cm long, finely pubescent panicles, corolla, lilac, very fragrant (Fig. S 32).

Syringa vulgaris *L.* **(Oleaceae): Eng.- Common Lilac. [NI].** A shrub or small tree; leaves ovate to broad ovate, acuminate, glabrous. Flowers are fragrant borne on long panicles. Three cultivated cultivars include *Clarkes Giant* (fls single, soft blue), *Lamaertine* (light lilac pink flowers) and *Vestale* (early flowering, fls pure white).

Fig. S 1. Sagittaria sagittifolia

Fig. S 2. Sagittaria sagittifolia (tubers)

Fig. S 3. Salix caprea

Fig. S 4. Salix fragilis

Fig. S 5. Salix viminalis

Fig. S 6. Salix wallichiana

Fig. S 7. Salvia moorcroftiana

Fig. S 8. Sambucus wightiana

Fig. S 9. Saussuria bracteata

Fig. S 10. Saussuria costus

Fig. S 11. Saussuria roylei

Fig. S 12. Saussuria simpsoniana

Fig. S 13. Saxifraga jacquemontiana

Fig. S 14. Scutellaria heydei

Fig. S 15. Senecio graciliflorus

Fig. S 16. Senecio vulgaris

Fig. S 17. Setaria italica

Fig. S 18. Silene latifolia

Fig. S 19. Sinopodophyllum hexandrum

Fig. S 20. Sisymbrium loeselii

Fig. S 21. Skimmia anquetilia

Fig. S 22. Solanum americanum

Fig. S 23. Solanum melongena var. Kashmiri long

Fig. S 24. Sorbaria tomentosa

Fig. S 25. Sorbus lanata

Fig. S 26. Stachys sericea

Fig. S 27. Stellaria aquatica

Fig. S 28. Sternbergia vernalis

Fig. S 29. Strobilanthes wallichii

Fig. S 30. Swertia petiolata

Fig. S 31. Syringa emodi

Fig. S 32. Syringa pubescens

T

Tagetes erecta L. (syn. Tagetes patula L.) (Asteraceae): Eng.- French Marigold; Hindi- Genda / Gule jafre; Kash.- Jaffre / Bata poshe [NI]. A herbaceous annual, naturalized in N W Himalayas and grown for border as well as bedding plant. Both double and single forms are cultivated. Double forms are golden yellow and carnation like.

Tagetes minuta L. (syn. Tagetes glandulifera Schrank.) (Asteraceae): Eng.- Southern Cone Marigold [NI]. An autumn flowering, strong smelling, grooved annual herb. Leaves odd pinnate, sharply serrate. Heads many, golden yellowish.

Tamarix chinensis Lour. (syn. Tamarix juniperina Bunge.) (Tamaricaceae): Eng.- Tamarisk [NI]. An ornamental shrub or small tree, with slender spreading branches. Leaves oblong lanceolate, acuminate; flowers pink, borne in racemes.

Tamarix indica Willd. (Tamaricaceae): Eng.- Indian Tamarisk; Ladakhi- Daktampat [W]. Tall hairless shrub, 2-6 m tall, with reddish-brown bark. Leaves tiny, sessile. Flowers light blue, in racemes. Ash of the burnt roots is applied externally to cure skin eruptions. In Ladakh its wood used as fuel.

Tamarix parviflora DC. (Tamaricaceae): Eng.- Athel Tamarisk / Small Flowered Tamarisk [W/UC]. A large spreading shrub or small tree with reddish brown arching branches. Leaves ovate, acuminate. Flowers pale pink, very short-pedicelled, in lateral racemes.

Tanacetum cinerariifolium (Trevir.) Sch. Bip. (syn. Chrysanthemum cinerariifolium (Trevir.) Vis. (Asteraceae): Eng.- Dalmatian Pyrethrum. [W]. A herb growing in grasslands of Kashmir. Leaves blue-green. Flowers white with yellow centres. Its extract is used as biopesticide with low mamalian toxicity. In Ladakh it is used in the preparation of a drink- *Chhangu*e.

Tanacetum coccineum (Willd.) Grierson (syn. Pyrethrum roseum Tzvelev.) (Asteraceae): Eng.- Pyrethum [NI]. A hardy herbaceous perennial producing daisy like flowers and popularly planted at sunny locations.

Foliage bright green. Variants with wide range of colour- white, salmon, crimson pink, crimson red, white and silver are grown in gardens.

Tanacetum gracile Hook. f. & Thomson (Asteraceae) [W]. An alpine desert aromatic herb of Ladakh Himalaya inhabiting stony slopes. Stem slender, branched. Leaves pinnately dissected, densely hairy.

Tanacetum griffithii (Clarke) Muradyan (syn. Chrysanthemum griffithii Clarke) (Asteraceae): Ladakhi- Chackerkeh [W]. Shruby, viscid herb, on dry alpine slopes and rocky crevices from 3000 – 4000 m. Plant with several, erect to ascending green arachnoid hairy branches arising from woody rootstock. Leaves hairy, bi- or tri-pinnatisect with small segments. Capitula solitary at apex of branches, whitish. Powdered plant used as poultice to relieve joint pain.

Tanacetum longifolium Woul ex DC (syn. Athanasia linifolia Burm. f.) (Asteraceae) [W]. A perennial hairy aromatic herb growing on alpine slopes at 3300- 5000 m altitudes. Heads yellow. Plant having anti-spasmodic, stimulant and carminative properties.

Tanacetum tenuifolium Jacquem ex DC. (Asteraceae): Ladakhi- Antang [W]. A perennial glabrous herb of high altitude areas. Flower heads yellow.Whole plant used for nasal congestion, renal colic and bodyache (Fig. T 1).

Taraxacum campylodes G E Haglund (syn. Taraxacum officinale (L.) Weber ex Wiggers) (Asteraceae): Eng.- Common Dandelion; Hindi- Dulal, Barau; Kash.- Mauedan Handh. [W]. A perennial low, scapigerous herb growing wild throughout Kashmir Himalayas in grasslands and meadowlands. Radical leaves form a basal rosette. Fruits called cypselae are straw coloured. Silky pappi form the parachutes. Foliage of the plant is used as vegetable in fresh from as well as after sundrying and especially given to ladies after child birth to supplement their nutritional deficiency (Fig. T 2).

Taxodium distichum (L.) Rich. (Cupressaceae): Eng.- Bald Cyperus [W]. A large, strikingly beautiful, columnar, conifers tree of Himalayas with shallowly fissured reddish-brown bark and strongly buttressed trunk. Leaves linear and flattened, grass-green, pectinate on short, deciduous shoots. Flowers unisexual; male catkins are in groups of 3-4 at the tip of branches; female flowers in small catkins at base of male.

Taxus baccata L. (Taxaceae): Eng.- Common Yew; Kash.- Posthal [W]. A medium sized evergreen tree (female clone) of erect habit, forming a dense, compact, broad column of closely-packed branches. As a young specimen it is narrowly columnar. Leaves black green, radially arranged. A very popular yew of alpine forests. Plant has been used in the preparation of beverages.

Taxus wallichiana Zucc. (syn. Taxus baccata L. subsp. wallichiana (Zucc.) Pilger.) (Taxaceae): Eng. Himalayan Yew; Hindi- Birmi; Kash.- Chikri [W]. An evergreen tree of Kashmir Himalaya, as an element of coniferous forests, at 2100- 2900 m. A broadly conical tree with spreading, horizontal branches, scaly, purple brown bark, and shoots that remain green for several years. Leaves linear oblong leathery and dark glossy green, above and yellow green beneath. Flowers dioecious, male flower in axillary oval yellow catkins, female flowers solitary and green. Bark and leaves anti cancerous. Extract from bark and leaves is used against breast and utrine cancers and also to relieve fevers and muscular pain. Bark of the plant was being used as tea by Kashmiris (Lawrence, 1895) (Fig. T 3).

Thalictrum cultratum Wall. (Ranunculaceae): Eng.- Knife like meadow Rue; Kash.- Chautra [W]. A herbacious plant of alpine Himalayas above 2500 m altitudes. Flowers small greenish-white in large much branched clusters. Stamens purple, protruding out. Leaves divided into oval leaflets. Root extract is used for urinary irritation.

Thalictrum foetidum L. (syn. Thalictrum vaginatum Royle.) (Ranunculaceae): Ladakhi- Chachoo [W]. A small perennial herb with radical and pinnately compound leaves. Stem branched, glandular-pubescent. Leaves ovate, sub-sessile, tri-pinnate. Extract of the roots is used to treat cojuctivitis.

Thalictrum minus L. (syn. Thalictrum minus var. majus (Crantz.) Hook.f. & Thomson) (Ranunculaceae): Eng.- Lesser Meadow Rue; Ladakhi- Chak-achoo/ Changu [W]. Tall rhizomatous perennial herb, on humus rich scruby slopes/forest margins, at 1800- 2800 m. Flowers yellowish white. An extract of the aerial parts of the plant is used as an eye sterilizer and to cure gout and rheumatism (Fig. T 4).

Thalictrum platycarpum Hook.f. & Thoms. (Ranunculaceae): Ladakhi- Maniroon [W]. A high altitude perennial, glabrous and erect herb. Leaves mostly cauline, petiolate, ternately compound. Inflorescence terminal panicle. Flowers whitish to pinkish. Achenes in a globose head. In Ladakh dried plants used as a supplement to fodder.

Thalictrum secundum Edgew. (Ranunculaceae): [W]. Tall, much branched,glabrous alpine, sub-alpine herb endemic to Kashmir Himalaya. Leaves sub-sessile, 2-ternate,to 2 to 3 pinnate. Flowers small, purplish. A medicinal herb.

Thelypteris confleuns (Thumb.) C.V Morton (syn. Dryopteris thelypteris (L.) A. Gray. (Thelypteridaceae): Eng.- Maiden Fern [W]. Perennial, long creeping fern, usually along drains and damp places at 1700-2500 m. Plant has long running rhizomes. Fronds deciduous, monomorphic, up to 75 cm long. Grown as an ornamental (Fig. T 5).

Thlaspi arvense L. (Brassicaceae): Eng.- Field Penny Cress; Ladakhi- Treka [W] A hardy annual, 30 -60 cm tall, branching occasionally. Usually inhabits alpine/ sub-alpine wetlands. Stems terminate in erect racemes of small white flowers. Leaves simple, oblong-ovate, shallowly lobed. Pods round, flat. Seeds used for rheumatism.

Thlaspi cornuticarpum Naqshi (Brassicaceae): Eng.- Bastard Cress [W]. Dainty hairless, rather fleshy annual of sub-alpine grasslands. Basal leaves in rosettes. Flowers white with purple veins in terminal clusters.

Themeda anathera (Nees ex Steud.) Hack (syn. Anthistiria anathera Nees. ex Steud.) (Poaceae): Kash.- Wula Ghaussa [W]. Tufted tall perennial, gregarious in habit. Occurs on slopes, on Karewa lands of Kashmir in abundance. Rhizomes elongated. Culms erect or geniculately ascending. Inflorescence a compound raceme. Flowers purple. Important fodder plant.

Thuja plicata Donn. ex D Don. (Cupressaceae): Eng.- Western Red Cedar; Kash.- Sarwa Kul [NI].A tall, columnar-conical ornamental tree, developing billowing lower branches with fissured, red-brown bark, horizontal or hanging sprays of foliage scale-like ovate, mid to dark glossy green leaves above, with whitish triangular spots beneath. Leaves have small dorsal glands, a pleasant fruity odour when crushed. Oblong-ellipsoid female cones. Cultivated in gardens and parks for hedges.

Thuja occidentalis L. (Cupresaceae): Eng.- Northern White Cedar; Kash.- Sarwa Kul [NI]. An evergreen ornamental tree planted for hedging in gardens. Branches are fan like with scaly leaves.Seed cone slender, yellow-green, ripening brown.

Thymus linearis Benth. (Lamiaceae): Eng.- Himalayan Thyme [W]. A spreading aromatic shrublet inhabiting rocky slopes of Himalaya. Leaves tiny, oblong; flowers pink-purple in dense whorls. Leaves were being used as substitute to tea.Used as antisceptic on skin, as carminative, antispasmodic and expectorant (Fig. T 6).

Thymus serphyllum L. (syn. Thymus serphyllum L. subsp. quinquecostatus (Clesk.) Kitamura (Lamiaceae): Eng.- Creeping thyme; Hindi- Ban ajwain; Kash.- Jungli Javend [W]. An aromatic spreading herb, growing in grasslands and on karewa slopes at 1700-3100 m. Flowers tiny, bell shaped, purplish. Seeds and flower heads used in preparation of many medicines. Plant used for coughs, catarrh and gastro-intestinal disorders (Fig. T 7).

Tilia platyphyllos Scop. (Malvaceae): Eng.- Broad leaved Lime [NI]. A large, vigorous ornamental tree of columnar habit, with downy shoots. Leaves orbicular-ovate dull green above, light green and lightly pubescent beneath, with axillary tufts of hairs turning yellow in autumn. Produces pendent cymes of 1-5 pale yellow flowers.

Toxicodendron acuminatum (DC.) C.Y Wu. & T.L. Ming (syn. Rhus acuminata DC (Anacardiaceae): Hindi -Tatri; Kash.-Arkhol/ Wutil [W]. A decidous tree of temperate forests. Tannins and extraction from bark is used as astringent and expectorant and for treating swellings and wounds. Exudation for the plant induces skin rashes and allergy.

Toxicodendron vernicifluum (Stokes) F.A. Barkley (syn. Rhus succedanea var. himalaica Hook. F. (Anacardiaceae): Eng.- Sumach/ Japanese Wax Tree; Hindi – Kakarasingi; Kash. – Arkhol [W]. A deciduous shrub or small tree growing in ravines at 1700- 2300 m. Leaves pinnate, leaflets 9-11, elliptic oblong to lanceolate. Flowers very small, yellowish, in axillary panicles. Fresh juice from twigs causes blisters (Fig. T 8).

Trachelospermum jasminoides (Lindl) Lamk. (syn. Trachelospermum divaricatum Thunb.) (Apocynaceae): Eng.- Star Jasmine [NI]. A woody, evergreen, twining climber. Leaves dark- green, elliptic or obovate, short-petioled, glabrous or pubescent beneath. Flowers very fragrant, white becoming creamy with age, in axillary and terminal cymes. Grown as an ornamental for its white fragrant flowers.

Trachycarpus fortunei (Hook) H. Wendl. (syn. Trachycarpus caespitosus Becc.) (Arecaceae): Eng.- Fan Palm [NI]. A remarkable unbranched, single-stemmed, evergreen palm, thickly clothed with the fibrous remains of the old leaf bases. Leaves large, fan shaped, 30-45 cm borne on long stout petioles in a cluster from the summit of the trunk. Small yellow flowers are borne in large, pendent panicles borne near leaf bases. Fruit 3-carpelled and deeply angled, bluish-black, each part about as large as pea. Grown as a speciment tree in gardens/ lawns.

Tradescantia fluminensis Vell. (Commelinaceae): Eng.- Wandering Trad [NI] A tender perennial flowering cum foliage plant grown as house plants. Leaves elliptic ovate, short stalked, pale-purple umbel side and with yellow-green variegations.

Tragopogon dubius Scop. (Asteraceae): Eng.- Yellow Goat's Beard; Ladakhi- Raiskun; Kash.- Budege [W]. Perennial herb, usually in dry locations – temperate sub-alpine woodlands, foothills. Achenes with long beak. Decoction of the whole plant is used by *Amchis* to cure peptic ulcer, dysentery and blood and liver diseases.

Tragopogon gracilis D. Don. (Asteraceae): Eng.- Slender Salsify; Ladakhi- Raiskun; Kash.- Dude Kund [W]. Perennial herb, usually in open dry locations, mountain slopes. Stems simple, slender erect, hairless. Flower heads borne singly on stalks, having few florets yellow with pink spots. Decoction of the plant used against abdominal problems and dysentery.

Tra gopogon kashmerianus G. Singh (Asteraceae): Eng.- Kashmir Salsify; Kash.- Dude Kund [W]. Biennial herb growing in temperate grasslands. Heads solitary on inflated peduncles. Achenes greyish brown. Decoction of the plant used for abdominal problems (Fig. T 9).

Tragopogon pratensis L. (Asteraceae): Kash.- Tharnoo [W]. A biennial herb of temperate meadows. Paste of the plant is used for rheumatism and gout.

Trapa natans L. (syn. Trapa natans var. natans; Trapa bicornis Osbeck.; Trapa quadrispinosa Roxb.) (Lythraceae): Eng.– Water Caltrop; Hindi- Singhada; Kash.- Gaure [W]. An aquatic herb occurring in most of the water bodies of Kashmir especially Walur Lake. Stem submerged. Leaves floating, triangular, leathery with sawtoothed margins. Flowers 4 petaled, white. Fruit a nut, with barbed spines. The fruit of the plant has been the source of excellent food from its kernels and a welcome fuel from its shells. In recent past the dried kernels used to be staple food of people living around Wular Lake. Local Hindus used to break their fast with porridge or bread made from its flour which still continues. Even today the edible kernels obtained after boiling or sweering of nuts are eaten and also sundried to be ground into flour. Of the chief varieties the best is called *Basmati* - having small nuts, small hornes, thin shells and high shelling percentage. Variety *Dogru* has larger fruit, thick shell, large hornes and with low shelling percentage (Figs. T 10, 11).

Tribulus terrestris L. (Zygophyllaceae): Eng.- Devils Weed /Devil's Thorn; Hindi- Khare-Khaske; Kash.- Meitcher Kaunde; Ladakhi- Zama [W]. Annual prostrate perennial herb, usually in orchards, wastelands and Himalayan dry lands. Fruit achenes with short bristles. Stem hairy. Leaves opposite, pinnately compound. Flowers with lemon-yellow petals. Extract of the fruit is given to treat kidney infections, and also applied externally to cure frost bites. In Ladakh its fruits are used for the preparation of a local drink - *Chhanuge* (Fig. T. 12).

Trigonella balansae Boiss. & Reut (syn. Trigonella corniculata (L.) L (Fabaceae): Eng.- Wild Fenugreek [W]. Branched glabrous perennial herb, on alpine/ sub-alpine slopes. Leaves compound with three leaflets. Flowers yellow. Fruit linear, acuminate, curved, glabrous with transverse anastomosing veins. Used as vegetable and as fodder. Seeds medicinal.

Trigonella foenum–graecum L. (Fabaceae): Eng. Fenugreek; Hindi-Methi; Kashmiri- *Meathe [UC].* An aromatic annual, 30-60 cm tall, cultivated throughout Kashmir as cold season crop for its foliage which is used for culinary and medicinal purposes. Leaves trifid, with oblong leaflets.

Trifolium alexandrinum L. (Fabaceae): Eng.– Egyptian Clover; Hindi-Berseem; Kashmiri - *Berseem* **[UC]**. Annual cultivated herb growing upto 60 cms height. Stems are hollow and succulent with dense rank growth from the lower axes. Leaves large, numerous, tender and trifoliate. Flowers heads yellowish white. Seeds egg shaped and greenish yellow in colour. Used mainly for green fodder, hay and pasture and for green manuring. Many varities are under cultivation yielding 60-90 q ha-1.

Trifolium pratense L. (Fabaceae): Eng.– Red Clover; Hindi – Tripatre; Kashmiri- Tripatre [W]. A perennial leguminous herb mostly growing wild in grasslands, and orchards and used as fodder in fresh and dried forms. Leaves long, leaflets three, and leaf blades marked with large spots. Flowers red-purple. Initially after having been introduced it was much grown for pasture, hay and green manuring**,** however it has now run wild. It is adapted to grazing better than other annual legumes (Fig. T 13).

Trifolium repens L. (Fabaceae): Eng.– Whilte Clover; Hindi– Shaftal/ Seburgi; Kashmiri- Batakh Nuer [W]. Low, creeping, glabrous perennial herb. Leaves long stalked dull green and trifoliate. Flowers white and fragrant. Three natural strains of white clover are found growing in Kashmir Himalayas namely, the small leaved *wild white*, the *medium leaved* or common type, the large leaved or *ladino type*. All the white types are growing wild in grass lands, orchards and pastures and used as fodder, as well as for green manuring/hay and silage (Fig. T 14).

Triglochin maritima L. (Juncaginaceae): Eng.- Seaside Arrow Grass; Ladakhi- Tzenaramba [W]. A perennial stoloniferous grass, growing in marshy lands and on river banks. Leaves fleshy. Flowers greenish, 3 petalled. In Ladakh its rhizomes are cooked; young aerial portions are used for pickles and soups. Underground parts are fed to cows to increase quantity of milk (Fig. T 15).

Triglochin palustris L. (Juncaginaceae): Eng.- Marsh Arrow Grass; Ladakhi-Narambo [W]. A slender perennial herb growing on damp grasslands of Ladakh. Emits an aromatic smell when bruised.Three petaled flowers are borne on long spikes. Fruit club shaped. Decoction of the whole plant used as an antiacid, antispasmodic and a light purgative.

Trillium govanianum Wall. ex D Don. (Melanthiaceae): Eng.- Birthroot / Wakerobin; Hindi- Nag Chhatre [W]. A perennial rhizomatous herb of alpine/sub-alpine shady slopes of Himalayas. Produce scapes, which are erect bearing three large bracts arranged in a whorl. Flowers solitary, trimerous with purple-brown petals. Fruits fleshy, red, berry like. Roots medicinal.

Triticum aestivum L. (Poaceae): Eng.- Wheat; Hindi- Gehoon; Kashmiri – Knakhe [UC]. An biennial grass, cultivated as a food cum forage crop. Wheat is one of the important grain cum forage crops in the valley but in the cold arid zone of Ladakh it the only important cereal crop cultivated both for grain as well as fodder purposes. Commonly cultivated traditonal and introduced and state released high yielding wheat varieties of Kashmir and Ladakh regions include:

- **Kashmiri Knakhe [NC]:** A traditional wheat variety of Kashmir Valley having been grown in villages of *Kandi* areas. Plants of short stature (70 to 105 cm) with amber redish grains, yielding about 2.0- 2.5 t/ha.
- **HS-240 [WC]:** Wheat variety developed by Indian Agricultural Research Institute (IARI), Shimla and released for North Hill Zone of the country during 1988. Plants with erect growth habit, erect flage leaf angle and semi erect angle of the ear. Ear density lax, tapering and awned. Grains amber coloured, hard, ovate and with medium crease.
- **HS-295 [WC]:**Wheat variety developed by IARI, Shimla and released during 1992 for North Hill Zone under rain fed conditions. Early growth habit is erect; flag leaf angle erect; angle of the ear semi- erect. Ears are white, tapering with white awns. Grains amber coloured, semi-hard, ovate and with medium crease.
- **VL-738 [WC]:** Wheat variety developed released by UPKAS, Almora during 1996 for cultivation in the wheat growing areas of North Hill Zone under timely sown, rain-fed as well as irrigated conditions. Early plant growth is semi-erect, flag leaf erect; ear angle erect. Ears tapering, white and awnless. Grains amber coloured, semi-hard, ovate, bold with narrow crease.
- **Mansarover (Sel-195) [RC]:** A popular wheat variety of Ladakh region cultivated both for grain as well as for fodder purposes. It was

rleased by SKUAST-K in 1999 specifically for cultivation in cold-arid wheat growing areas of Ladakh. It is semi-tall variety with small amber coloured grains. The variety matures within 115-125 days and is tolerant to biotic and abiotic stresses prevalent in cold climate of Ladakh. Grain yield ranges between 3.0-4.0 t/ ha and fodder yield ranges between 8.0-11 t / ha.

- **Singchen (Swl-8)** [RC]: Rleased by SKUAST-K in 1999 for wheat growing belts of Ladakh region. Plants tall with thick stem and amber coloured grains. Variety has medium maturity, shows resistance to lodging and yellow rust and ear cockle diseases. Grains are preferred to for chapatti making. Straw palatable to animals. Yield potential lies between 3.5 - 3.8 t / ha along with straw yield of 9.0 - 9.5 t / ha.
- **Kailash (Sel-194)** [RC]: A wheat variety developed and released by SKUAST-K in 2001 for cultivation in the cold-arid zone of Ladakh. Plants are medium tall with amber coloured, hard and bold grains; leaves are erect, waxy; ears compact and awned. Variety is tolerant to intermittent stress of high temperature, intense solar radiation and moisture stress. Matures within 120-130 days and yields 4.0 - 4.5 t / ha plus straw yield of 9.0 - 11.5 t / ha.
- **Shalimar Wheat-1 [WC]:** A high yielding wheat variety developed and released by SKUAST-K in 2004 for cultivation in Northern Hill Zone under rainfed/restricted irrigation, timely sown production conditions. Plants are erect, medium tall (76-115 cm) with tapering ears, bearing deep amber, semi-hard grains. The variety is resistant to stripe rust and leaf rust and has a grain yield potential of about 35 q ha-1. The variety has gained popularity in the valley as it matures early and fits well in the rice- wheat cropping sequence.
- **Shalimar Wheat** -2 [WC]: An early maturing genotype (earlier than Shalimar Wheat-1) with a maturity range of 220-225 days, thus fitting well in the wheat –rice cropping sequence under valley conditions. Plants medium tall, with larger leaf size, semi-erect flag leaf angle. Grains are deep amber coloured, hard, ovate with medium crease. Yield potential ranges from 30- 32 q /h.

Trollius acaulis Lindl. (Ranunculaceae): Eng.- Dwarf Globe Flower [W]. A perennial dwarf herb,5 -30 cm tall, with stout rootstock inhabiting alpine slopes at 3000- 4300 m. Leaves 5-7 lobed, incised/toothed. Flowers solitary, deep yellow; sepals 5-10, petals shorter than stamens. Centre of the flower has stamens and 12-16 oblong nectarines.

Tulipa clusiana DC (syn. Tulipa stellata Hook.; Tulipa clusiana var. stellata (Hook.) Regel.) (Liliaceae): Eng.- Himalayan Tulip [W]. A bulbous perennial herb, often growing on forest openings, dry karewa slopes. Leaves narrow and grey- green; flowers long, purplish outside, whitish within. Also blooms in April around saffron fields and on other Karewa slopes of Kashmir (Fig. T 16).

Tulipa fosteriana W. Irving (Liliaceae): Eng.- Emperor Tulip [NI]. A bulbous perennial herb grown for its large flowers. Flowers cup shaped with blunt pointed petals. Appears in various majestic colours and is cultivated widely in gardens (Fig. T 17).

Tulipa lehmanniana Merckl. (Liliaceae) [W]. A bulbous perennial herb of high altitude meadows of Ladakh. Flowers cup shaped and yellow. Grown as ornamental.

Tulipa lanata Regel. (Liliaceae): Eng. Bukhara Tulip; Kash.- Bandel Posh [W] A bulbous perennial herb. Leaves sessile, lanceolate, entire, often margins undulate, acute to sub-acuminate. Flowers bowl shaped, solitary terminal, scarlet red with black blotch at the base of tepals. Occurs in wild in saffron fields around Papmpore. It used to be grown on muddy roofs of houses and Shrines of Kashmir (Fig. T18).

Tulipa tubergeniana Hoog. (Liliaceae): Kash.- Bandal Posh [W]. A bulbous perennial herb. Leaves are lanceolate and grey green. Flowers, are long red with black blotches at the base and have wavy edged tepals.

Used to be grown on mud lined roofs and around sacred groves for beautification.

Tussilago farfara L. (Asteraceae): Eng.- Colt's Foot / Clay Weed; Kash.- Watpan [W] A perennial rhizomatous herb on sub-alpine slopes and stream banks. Flower heads yellow and appear before leaves. Leaves resemble colt's foot, have angular teeth on margins, and are produced after seed set. Aerial parts used for wound healing. Leaves are edible and used as herbal vegetable; are also used traditionally to treat cough, common cold, flue,,rheumatism and gout (Figs. T 19, 20).

Typha angustifolia L. (Typhaceae): Eng. –Lesser Reedmace / Lesser Bulrush; Hindi- Pater, Bora; Kash.- Zab [W]. A rhizomatous perennial growing in marshy lands and shallow water bodies throughout Kashmir. Rhizomes thick, stoloniferous. Leaves grass like, spongy, trigonous, deep green and smooth. Male spikes longer than female ones; anthers yellow with green tips. Mature leaves after harvesting and drying are traditionally used for preparations of soft mats, ropes for thatching houses, bundling of vegetables and for wrapping packets. Fresh leaves used as fodder for horses (Fig. T 21).

Typha domingensis Pers. (syn. Typha angustata Bory & Chaub.) (Typhaceae): Eng.- Southern Cattail; Hindi- Pater; Kash.- Petch./ Kala Roon [W]. A tall, marshy rhizomatous perennial growing in most of the marshy lands and shallow water bodies of Kashmir. Rhizomes creeping, branched; leaves semi cylindrical narrow above the sheath, with spongy stalks. Flowers borne on complex spikes on the same vertical stem. Seeds minute and attached to a thin hair, which helps in its dispersal. At maturity leaves are harvested, sundried and mostly plaited into mats (locally called *wagoo).* Silky flouts of spikes are used for stuffing and mud plastering. The Anchar Lagoon, to the north of Srinagar, used to be the great home of *Petch* leaves and all houseboats deweling on its waters were thatched with its mats. Its rhizomes have been used as famine food (Fig. T 22).

Fig. T 1. Tanacetum tenuifolium

Fig. T 2. Taraxacum campylodes

Fig. T 3. Taxus wallichiana

Fig. T 4. Thalictrum minus

Fig. T 5. Thelypteris confleuns

Fig. T 6. Thymus linearis

Fig. T 7. Thymus serphyllum

Fig. T 8. Toxicodendron vernicifluum

Fig. T 9. Tragopogon kashmerianus

Fig. T 10. Trapa natans

Fig. T 11. Trapa natans (dehulled kernels)

Fig. T 12. Tribulus terristris

Fig. T 13. Trifolium pratense

Fig. T 14. Trifolium repens

Fig. T 15. Triglochin maritima

Fig. T 16. Tulipa clusiana

Fig. T 17. Tulipa fosteriana

Fig. T 18. Tulipa lanata

Fig. T 19. Tussilago farfara (flowering heads)

Fig. T 20. Tussilago farfara (foliage)

Fig. T 21. Typha angustifolia

Fig. T 22. Typha domingensis

U

Ulmus parvifolia Jacq. (Ulmaceae): Eng. – Chinese Elm; Ladakhi Shampoo Tree / Yumbuk [W]. A small to medium deciduous tree growing upto 12 m tall, across the region as a temperate woody element. Leaves leathery, toothed and lustrous green. Trunk has a handsome, flaking bark of mottled greys with tans and reds. Also used as a landscape tree. In Ladakh the bark of the plant used as shampoo to promote hair growth.

Ulmus pumila L. (Ulmaceae): Eng.- Siberian Elm; Ladakhi- Nyum-buk [W/UC] A medium sized deciduous tree growing in Ladakh at 3500-4000m. Bark dark-grey. Branchlets greenish. Leaves alternate, small, elliptic, toothed. Flowers greenish, in compact drooping clusters. Sammara sub-orbicular, pale-buff with seed central. In Ladakh the bark of the tree is used as a scalp conditioner, andi-dandruff and to promote hair growth.

Ulmus villosa Brandis ex Gamble (Ulmaceae): Eng.- Cherrybark Elm; Hindi- Bhamri; Kashmiri- Bren [W/UC]. A large tree, bark rather smooth, dark grey-brown, young shoots yellowish red, finely rough, later smooth and brown. Branches pendulous. Leaves obovate or more oblong, upper surface more or less rough, lighter beneath, soft pubescent. Flowers 10-15 in dense, sessile clusters. A very attractive elm with delicate, partly pendulous young shoots grown as specimen tree mostly in sacred groves, graveyards and parks. Wood used for furniture and construction works and for making fruit boxes.

Ulmus wallichiana Planch. (Ulmaceae): Eng.- Kashmir Elm; Kashmiri- Brari [UC]. Deciduous tree, well distributed in broad leaved forests and moist ravines, at 1700 -2800 m. Bark greyish, longitudinally furrowed. Leaves elliptic, acuminate. Wood used for furniture works and in house building. Bark used in cutaneous diseases; also used for washing and dying clothes.

Urtica dioica L. (Urticaceae): Eng.- Stinging Nettles; Hindi- Bichu Booti; Kashmiri– Soi [W]. A robust perennial, rhizomatous herb growing throughout Himalayas in grass lands, waste places and road sides. Leaves soft, strongly serrated with cordate base and acuminate tip. Stem and leaves heavily covered with stinging hairs. Tender shoots are used as vegetable. Rootstock used as stomachic and diuretic (Fig. U 1).

Urtica hyperborea Jacq. ex Wedd. (Urticaceae): Eng.- Northern Nettle; Ladakhi –Zaachout [W]. Perennial, velvety and hairy sub-shrub, growing in alpine ranges of Kashmir and Ladakh above 3000m altitudes. Plant sparsely covered with stinging hairs. Leaves ovate- heart shaped Flowers greenish purple. Dried leaves used as vegetable and also fed to cattle, during winter, to give them warmth. Crushed leaves are also applied on boils to relieve pain and oedema (Fig. U 2).

Urtica parviflora Roxb. (Urticaceae): Eng.- Stinging Nettles; Hindi- Bichu booti; Kash.– Soyei [W]. A perennial herb, usually in forests and within tall herbaceous vegetation at 1700 - 2800 m altitudes. Plant covered heavily with stinging hairs. Roots used medicinally for fevers.

Utricularia minor L. (Lentibulariaceae): Eng.- Lesser Bladderwort; Ladakhi- Lingna [W]. A small carnivorous plant inhabiting shallow waters. Leaves compound,made of 2 or more discrete leaflets. Flowers bright yellow. In Ladakh the dried leaves are used in the preparation of a local intoxicating drinks called *Lingeatzish*. Roots and leaves are also used for fishing. Extract of the plant also used to treat rheumatism and arthritis.

Fig. U 1. Urtica dioica

Fig. U 2. Utrica hyperborea

V

Vaccaria hispanica (Mill.) Rauschert. (Syn. Saponaria vaccaria L.) (Caryophyllaceae): Eng.- Cow Soapwort/ Cow Basil [W]. Annual herb,as weed of orchards.Plant has blue-grey, waxy herbage and pale pink flowers. Used traditionally to relieve body swellings.

Valeriana hardwickii Wall. (Caprifoliaceae): Eng.- Indian Valerian; Hindi – Tagger [W]. Tall glabrous, rhizomatous herb inhabiting sub-alpine slopes and stream margins. Leaves pinnate; flowers pale –pink, borne in clusters at the end of branches. Ladakhis use it for scenting hair.

Valeriana jatamansi Jones (syn. Valeriana wallichii DC) (Caprifoliaceae): Eng.- Jatamansi Valerian; Hindi– Mushqi-bala./Sumble-jable [W]. A perennial herb with pubescent stem on shady alpine and sub-alpine slopes. Leaves toothed, heart shaped at base. Flowers snow-white or pink tinged borne on flat topped clusters. Dried rhizomes used with oil for perfuming hair. Also used for hysteria and nervous problems (Fig. V 1).

Valeriana pyrolifolia Decne. (Caprifoliaceae): Kash.- Sugandhi bala [W]. A perennial rhizomatous herb of forest areas. Stem solitary, sparsely pubescent. Flowers white, in compact corymbose heads. Leaves dark green pyrolifom. Root is used as herbal sedative.

Verbascum thapsus L. (Scrophulariaceae): Eng.- Common Mullein; Hindi- Ban Tambaku; Kash.- Wan tamokh /Buder taunt [W]. Erect woolly biennial herb of temperate Himalaya from 1800 - 3000 m altitudes on waste places, orchards, roadsides. Basal leaves clothed with a selender tomentum all over. Flowers yellow on long apical spike. Flowers, leaves and roots are used to treat asthama, coughs, pulmonary diseases and bleeding bowls. Seeds are narcotic. An essential oil from herb is used for frost bites, bruises and piles. Seeds are used as fish poison (Fig. V 2).

Verbena hortensis Vilm. (Verbenaceae): Eng.- Garden Vervain [NI]. A perennial border plant producing bushy plants. Leaves mid to dark green, ovate and serrated. Tight clusters of white, pink and red to blue and lilac flower heads appear from June to August. A number of varieties representing groups- *grandiflora* and *compacta* are cultivated in gardens and parks.

Verbena officinalis L. **(Verbenaceae): Eng.- Common Vervain / Holy Herb: Kash.- Hathe mool [W].** A perennial upright herb growing at 1700 – 2300 m amsl, usually in grasslands and in cultivated fields. Leaves lobed and toothed. Flowers June- October, pinkish. Used as febrifuge, astringent and anthelmintic.

Veronica anagallis-aquatica L. (Plantaginaceae): Eng.- Blue water Speedwell [W] A perennial herb of marshy lands. Flowers pale purple. Leaves edible. Entire plant is diuretic in action and its decoction promotes menstruation in suffering women.Roots and leaves are appetizers and induce urination.

Veronica arvensis L. (Plantaginaceae): Eng.- Common Speedwell; Kash.- Lichi Katche [W]. Annual herb, usually a weed of cultivation, also in wastelands. An important fodder plant. Leaves opposite, ovate, with coarse teeth. Flowers pale-blue. Fruits capsule, heart shaped.

Veronica beccabunga L. (Plantaginaceae): Eng.- Baccabunga / European Speedwell [W]. Perennial succulent herb, usually along water drains and borders of rice fields. Flowers bluish-pink. Leaves ovate, round tipped, fleshy and edible, used as vegetable. Fruit spherical capsule (Fig. V 3).

Veronica cashemerica Gandoger. (syn. Veronica himalayansis D Don. (Plantaginaceae): [W]. Low, pubescent herb, on alpine meadowlands. Leaves sessile, ovate-lanceolate, pubescent with toothed margins. Flowers mauve coloured. Leaves edible. Capsule compressed, ovoid, hairy. Leaves edible.

Veronica lanosa Royle ex Benth. (Plantaginaceae): [W]. A clump forming, erect stemed densely hairy herb of alpine pastures and rocky cliffs. Leaves narrowly elliptic, sessile, toothed. Flowers blue in terminal spike like racemes.

Veronica speciosa R. Cunn. ex A. Cunn. (syn. Hebe speciosa (R. Cunn. ex. A. Cunn) Andersen) (Plantaginaceae): Eng.- Newzyland Hebe [NI]. An evergreen, ornamental shrub with upright purple tinged shoots and broadly elliptic glossy, dark green leaves, purple beneath when young. Flowers dark-purple to blue-purple in axillary racemes.

Viburnum cotinifolium D. Don. (Adoxaceae): Eng.- Smoke Tree leaved Viburnum; Kash.- Kulmache [W]. A large deciduous shrub, common in forests and on scrubby slopes at 1700 - 2400 m altitudes. Twigs and inflorescence more or less stellately tomentose. Flowers April- May. Leaves ovate to orbicular-ovate, obtuse or shortly acuminate, greyish stellate tomentose beneath, slightly so above, short stalked. Flowers in terminal dense corymbose cymes; corolla white tinged- rose and sweet scented.

Viburnum grandiflorum Wall. ex DC (syn. Viburnum foetens Decne.) (Adoxaceae): Eng. –Guelder; Kash.- Kulmache [W]. A large deciduous woody shrub, occasionally a small tree in its habitat across sub-alpine forest slopes. Leaves elliptic-oblong, tapered to both ends, pubescent beneath. Flowers in dense, 5 - 7 cm cymes, corolla tubular, white turning rose- pink; fruits red to black, edible (Fig. V 4).

Viburnum opulus L. (syn. Viburnum opulus L. var. roseum.) (Adoxaceae): Eng.- Gueldar Rose / Highbush Craneberry [NI]. An upright, deciduous shrub, bark thin and light grey. Leaves rounded maple like 3-5 lobed, irregularly dentate, light green grey-green, and pubescent beneath. Flowers double snow bal, scented on globose cymes, greenish at first, fading to slight pink. Fruit orange-red berries, edible.

Vicia faba L. (syn. Faba vulgaris Moench.) (Fabaceae): Eng.– Broad Bean; Hindi - Bahla; Kash.- Baugla [RC]. A hardy erect simple stemmed annual, 60 - 90 cm tall, cultivated in the valley for its pods. Leaves pinnate, leaflets 1-3 pairs, elliptic to oblong, obtuse but apiculate; flowers in small racemes, white with large purplish blotches; pods variable in size and up to 10 seeded. Local as well introduced broad bean varieties are grown by vegetables growers for the production of pods while as in rural areas it is cultivate for seed only. In Ladakh it is grown as an inter crop with naked barley.

Vicia sativa L. (Fabaceae): Eng.- Common Vetch; Hindi– Ankra; Kash.- Sigri/ Gantae mougthe; Ladakhi- Bugsukhang [W]. A sprawling annual herb, climbing by tendrils, in cultivated fields and wastelands. Flowers purple. Leaves stipulate, alternate and compound ending in a branched tendril. Flowers pea like, axillary, light pink-purple. Seeds brown. Used as green manure and fodder. In Ladakh whole plant is used for treating urinary infections (Fig. V 5).

Vigna aconitifolia (Jacq). Marechal (syn. Phaseolus aconitifolius Jacq.; Phaseolus aconitifolius (Jacq.) Marechal (Fabaceae): Eng.- Moth Bean; Hindi- Moth bean; Kash.– Mougthe/ Gaebe Moughte [RC]. An annual herb; stems sub erect, slightly hairy, leaves trifoliate, leaflets deeply 3-5 lobed; flowers yellow, in long racemes, pods cylindrical 2.5 – 6.0 cm long, glabrous, 6-12 seeded; seeds small, yellow to brown or mottled black. It is cultivated both for food and forage in Kashmir. Seeds mostly, after boiling, fed to cattle during winters. The tender pods are sometimes used as vegetable.

Vigna mungo (L.) Hepper (syn. Phaseolus mungo L.; Phaseolus radiatus Roxb. non L.) (Fabaceae): Eng.- Black Gram; Hindi- Urd; Kash.- Mah [RC]. An erect, hairy plant, varying in height from 30-90 cm, cultivated as a pulse crop. Leaves trifoliate, leaflets entire ovate to rhombic- ovate in outline, acuminate 5-10 cm long; fls. small, yellow on short but later elongating peduncles; pods cylindrical erect or spreading, hairy. Seeds oblong with square ends, generally black with a white hilum protruding from the seed.

Vigna radiata (L.) R.Wilczek (syn. Phaseolus radiatus L.) (Fabaceae); Eng.-Green Gram; Hindi –Moong; Kash. – Mounge [RC]. An erect or sub erect, annual cultivated as a pulse cum fodder.Commonly cultivated varieties include:

- **Kashmiri Moonge:** Stems 4-120 high, leaves trifoliate, leaflets entire, ovate in outline; fls yellow or yellowish-green and crowded in clusters of 10-25, on long pedicles. Pods 5.5-10 cm long, cylindrical, with pubescence, seeds globular, smaller and mostly green in colour. Cultivated as a kharif crop as a pure or as intercrop.
- **Shalimar Moong-1:** High yielding variety released by SKUAST-K during 2004 for pulse growing areas of the state. The plant has determinate semi-erect growth habit, dark green foliage and uniform pod colour. The seed is round, bold, shining green with 9-10 seeds pod-1. The variety is medium in maturity (105-115 days), resistant to leaf spot, pod blight, immune to yellow mosaic virus (YMV) under field conditions. Protein content ranges from 23-24 per cent. Seed yield potential is 7.5 to 6.5 q ha -1.

Vigna umbellata (Thunb.) Ohwi and Ohashi (syn. Dolichos umbellatus Thunb.) (Fabaceae): Eng.- Rice Bean; Hindi- Sutri; Kash.- Roomah. [RC]. A slender, hairy twiner- 30-90 cm, long. Leaves trifoliate, leaflets entire or 3 lobbed, ovate to broad ovate, shortly acute. Flowers brightly yellow, borne on axillary racemes; pods slender pubescent, bluntly beaked, 6-9 cm long; seeds 6-8 cm long, oblong with rounded ends, brown or black/redish brown. Rice beans are grown as kharif crop both for food and fodder.

Vigna unguiculata (L.) Walp. (syn. Vigna sinensis (L.) Savi. ex Hassk.) (Fabaceae): Eng.- Cowpea: Hindi– Lobia; Kash.- Gabe Moth / Wari Moth [RC]. Annual twining herb. Leaves alternate, pinnately trifoliate. Pods cylindrical. Flowers whitish pink and showy. Cowpeas have been important grain as well as forage crop especially in Kashmir Himalaya. After harvest the plants are sundried and subsequently used as fodder especially for sheep and goat. Two types of cowpeas are under cultivation, one with pale yellowish and other with black seeds however the former has been the most popular. The seeds of the plant are also being fed to sheep after boiling especially during winter as a nutritious supplement. Two types of cowpeas are under cultivation.

- **Kashmiri Wari Moth:** A sub erect, trailing or climbing, bushy annual with glabrous stem. Leaves pinnate, leaflets 7.5 - 15 cm long, ovate rhombiodal or broad; flowers in racemes, white/ pale violet coloured with an yellowish pods variable in size, with 10-20 seeds; seeds varying in size, shape and colour. Marginally cultivated as a commercial crop.
- **Shalimar Cowpea-1:** A high yielding variety developed by SKUAST-K through selection. The plant type is characterized by indeterminate growth with climbing habit when grown as an associated crop with maize. The leaf is green with glabrous leaf blade. The pod bears 10-12 seeds on an average that are oval in shape and black in colour. The average seed yield potential is 8.5 q ha-1.

Vigna unguiculata subsp. sesquipedalis (L.) Verdc. (syn. Vigna sesquipedalis (L.) Fruwirth. (Fabaceae): Eng.- Yard Long Bean; Hindi - Chowli; Kash.- Roongi [RC]. An annual climber, with weak branches, pods pendent elongated 30 to 90 cm long curved fleshy, more or less inflated and flabby. Seeds elongated kidney shaped. Cultivated for its tender pods.

Vinca major L. (syn. Vinca major var. argento-variegata Loudon) (Apocynaceae): Eng.- Greater Periwinkle [NI]. A small evergreen ornamental shrub, with long trailing stems. Leaves broadly ovate, entire, ciliate or finely serrate on margins. Flowers purple- blue, tubular, solitary and axillary.

Viola biflora L. (Violaceae): Kash.- Bunafsha [W]. A low stemless alpine herb, with rosette leaves, on shady moist alpine slopes at 3000 - 3500 m altitudes. Leaves reniform, long petioled, crenate, stipules ovate-oblong, spur very short. Flowers straw-yellow. Leaf extract is used for high fever.

Viola betonicifolia Sm. (Violaceae): Eng.-Arrowleaf Violet; Kash.- Maidani Bunufsha [W]. A rhizomatous, perennial, glabrous herb found in grasslands and as forest under growth. Leaves arrow shaped. Flowers bright purple. Dried flowers are used as purgative amd mixed with tea during colds and coughs. Also grown as ornamental.

Viola kashmeriana W Becker (Violaceae): [W]. A low stemless perennial herb with rosette leaves, commonly growing on rock surfaces and solpes at 1700 - 2200 m. Flowers bluish. Leaves long petioled., cordate. Used in treating respiratory troubles.

Viola odorata L. (Violaceae): Eng. Sweet Violet; Hindi– Banafsha; Kash.- Bunafsha [W]. Perennial rhizomatous stemless, tufted herb, growing in Kashmir Himalaya, from 1500-2000 m amsl, mostly wild in alpine and sub-alpine grasslands, preferring moist and cool situations. Leaves are heart shaped. Flowers of purple or white are produced from Febuary to March. Plant is used as the strong febrifuge in acute and chronic fevers. Flowers are emullient and demulcent. Roots purgative, febrifuge, tonic, expectorant and removes inflammations. Also used for ornamental purposes for bedding out and for edging paths (Fig. V 6).

Viola patrinii Ging. (Violaceae): [W]. Low, stemless pernnial herb, growing under forest shade. Leaves oblong-ovate, long petioled, cordate, cuneate, serrate; spur cylindric, obtuse. Flowers lilac. Dried flowers used for colds.

Viola pilosa Blume (syn. Viola serpens Wall ex Ging.) (Violaceace): Eng.– Sweet Voilet; Kash.- Nuna posh / Bunafsha. [W]. Perennial herb growing under forest shade at 1800-2300 m. Flowers bluish or white and are used in perfumery. Medicinally used as demulcent, in biliousness and lung troubles (Fig. V 7).

Viola tricolor L. (syn. Viola tricolor var. hortensis DC.) (Violaceae): Eng.- Garden Pansy; Kash.- Panzy [NI]. Most popular garden plant grown for its showy spring attractive flowers. Number of single and hybrid variations in different colour and shades are grown in home gardens and parks. Spring flowering varieties are mostly preferred to and cultivated.

Viscum album L. (Loranthaceae): Eng.- Mistletoe; Hindi- Banda; Kash.- Ahul [W] A semi-parasitic shrub of temperate Himalaya, generally seen growing on walnut trees. Stem greenish yellow, branched. Leaves simple, obovate to oblong with wavy margins, opposite at nodes. Flowers like round balls. Berries used for epilepsy, in uterine haemorrhages and in liver ailments (Fig. V 8).

Vitis vinifera L. (Vitaceae): Eng. Wine Grape; Hindi- Angoore; Kash. – Dache [UC]. Deciduous woody climber. About six popular varieties of grapes are presently growing in Kashmir which are listed below:

- **Kishmish Dach:** Fruit bunches small, sweet and green coloured; disease resistant with uneven ripening pattern.
- **Kruhon Dach / Kawa Dach:** Bushy, compact and large seedless; berries dark-blue coloured; heavy bearing; grown in Dachripur and Laur areas of Ganderbal.
- **Hussaini Dach:** Fruit bunches compact and large with large berries; leaves penta-lobed; berries oblong, sweet and whitish green.
- **Inabi Shahi:** Fruits in small and loose bunches, with uniform ripening;

fruits oblong and large, whitish green; grown mostly around Kralbabh, Ganderbal.

- **Thomson Seedless:** Fruits medium oblong in large bunches with uniform ripening, assuming yellow green colour at maturity.
- **Jungli Duche:** A wild deciduous climber with long bifid tendrils. Leaves ovate dark-green, orbicular-cordate, irregularly toothed, glabrous above tomentose beneath, long stalked. Flowers small, in leaf-opposed panicled cymes, petals greenish white; berries variable in size, greenish or bluish black. Growing across roadsides and ravines.

The grape varieties of Ladakh include *Chhor Gon, Naq- Gon* and *Batyr Gon.*

Volvariella volvaceae (Bul) Singer (Pluteaceae): Eng.- Straw Mushroom; Kash.- Saeki-Bube [W] An edible straw mushroom growing in moist shady places and gregariously in clusters in gardens. Also cultivated on agricultural wastes. Caps egg shaped when young, expanding convex or conical, flesh white. Gills free from stem, white becoming pink

Fig. V 1. Valeriana jatamansi

Fig. V 2. Verbascum thapsus

Fig. V 3. Veronica beccabunga

Fig. V 4. Viburnum grandiflorum

Fig. V 5. Vicia sativa

Fig. V 6. Viola odorata

Fig. V 7. Viola pilosa

Fig. V 8. Viscum album

W

Waldhemia glabra (Decne.) Regel. (Asteraceae): Eng.- Smooth Ground Daisy [W] A tufted mat forming, aromatic high altitude plant of drier areas of Kashmir Himalaya at 4000-5000 m altitudes. Leaves wedge shaped and bluntly 3-5 lobed, occur in rosettes. Flower heads solitary on stems, pink to purple ray florets, disc florets yellow. Whole plant used for rheumatism.

Waldheimia nivea (Hoook. f. & Thomson ex Clarke.) Regel (Asteraceae): Ladakhi- Palla [W]. Loosely mat forming perennial growing in open stony habitats at 3900- 5400 m altitudes of Ladakh. Leaves cuneate with 3-5 lobes, densely white fetted. Flower heads sessile with dark-brown disc florets and white ray florets. Decoction of the whole plant is used for hyperacidity and as stomachic.

Waldheimia stoliczkai (Clark.) Ostenf. (Asteraceae): Ladakhi - Solomarpo [W]. A perennial loosely tufted glabrous perennial herb of alpine grasslands at 4000- 5000 m altitudes. Leaves bi-pinnate. Heads terminal with yellow disc florets and light pink ray florets.The decoction of shoots and leaves is used in headaches, fevers and bronchial troubles. In Ladakhi folk medicine the plant extract is claimed to act as a blood purifier (Fig. W 1).

Waldheimia tomentosa (Decne.) Regel.(Asteraceae): Ladakhi- Solokarpo [W]. A perennial herb of high altitude meadows of Ladakh. Whole plant used for rheumatism.

Weigela florida (Bunge) A. DC. (Caprifoliaceae): Eng.- Old Fashioned Weigela [NI] A deciduous ornamental shrub, with arching branches. Leaves light green short petioled to nearly sessile, oval-elliptic finely wrinkled, serrate, glabrous above, underside pubescent tomentose. Flowers usually 3-4, without a common stalk, axillary; calyx lobes half as the tube, corolla campanulate above, pink, interior lighter to white, exterior pubescent, limb lobes large and spreading.

Wisteria sinensis (Sims). Sweet (Fabaceae): Eng.- Chinese Wisteria / Blue Acacia [NI]. A deciduous high twining climbing shrub. Leaves pinnate, leaflets 7-13, ovate -oblong to more lanceolate. Flowers blue-violet, slightly fragrant, in dense 15-30 cm long racemes. Grown as a hedge plant.

Wulfeniopsis amherstiana (Benth.) D.Y. Hong (syn. Wulfenia amherstiana Benth.) (Plantaginaceae): Eng.- Himalayan Wulfenia [W]. A semi-glabrous perennial herb on alpine slopes above 2000 m. Leaves spathulate, coarsely crenate. Flowers blue- purple, tubular, drooping with 4 pointed petals, borne on one sided spikes. Whole plant used against high fevers (Fig. W 2).

Fig. W 1. Waldheimia stoliczkai

Fig. W 2. Wulfeniopsis amherstiana

X

Xanthium strumarium L. (Asteraceae): Eng.- Common Cocklebur; Kash.- Tcheere [W]. An annual herb usually inhabiting disturbed soils. Leaves simple, alternate. Flower heads are borne in leaf axils and have disc florets only. Fruits obovoid in a hardened involucres with hooked beaks and hooked bristles. Stem turn black whem it matures. An obnoxious weed whose clinging fruit stick to wool of all herbivores.

Xanthoxylum armatum DC (Rutaceae): Eng.- Indian Prickly Ash; Kash.- Timru [W]. A deciduous spiny shrub growing in alpine forest vegetation. Leaves trifoliate, with leaf-stalk winged. Leaflets sessile, elliptic to ovate, slightly toothed, sharp tipped. Flowers small, arising from leaf axils and are diocious. Leaves and flower buds used against fevers. Bark and fruits used as carminative and anthelmintic (Fig. X 1).

Fig. X 1. Xanthoxylum armatum

Y

Youngia tenuifolia (Willd.) Babc. & Stebb. (Asteraceae): Ladakhi– Neu [W]. A perennial herb of high altitude cold arid areas. Most leaves radical,rosulate, lobed, cauline sessile, entire. Flower heads terminal. Ligules orange yellow; pappus soft, white. Extract of the powdered tubers used to check bleeding of gums. Also taken for diarhhoea and dysentery.

Yucca aloifolia L (Asparagaceae): Eng.- Adam,s Needle / Dagger Plant [NI]. A slow-growing, rounded evergreen shrub or small tree with a simple or branched stem and densely arranged, linear to narrowly lance-shaped, toothed, dark green leaves. Bears stout, erect panicles of pendent, bell-shaped white purple-tinged flowers, held above the foliage. Introduced and grown in gardens.

Yucca gloriosa L. (Asparagaceae): Eng.- Spanish Dagger Plant [NI]. An erect ornamental shrub with a stout stem, bearing terminal tufts of narrowly lance shaped, stiffy pointed, arching and dark-green leaves. Produces pendent, bell shaped, sometimes purple-tinged, white flowers in upright panicles, up to 240 cm long.

Z

Zantedeschia aethiopica (L.) Spreng. (syn. Calla aethiopica L.) (Araceae): Eng.- Arum Lily; Hindi – Iram sousan [NI]. A rhizomatous perennial, suitably planted in moist and shaddy situations and under green house conditions. Leaves deep green, glossy arrow shaped. White spathes surround a yellow spadex, borne from May to June.

Zantedeschia rehmanii Engl. (Araceae): Eng.- Pink Arum [NI]. A deciduous evergreen perennial with lanceolate mid- green leaves, spotted with silvery white flecks. Spathes range from pale- pink to wine red.

Zea mays L. (Poaceae): Eng.- Maize, Corn; Hindi- Makka; Kash. – Makayi [UC] A tall annual grass grown as a food cum- fodder plant across Kashmir Himalaya. Presently the principal agricultural maize groups represented in the valley include large seeded Dent Corn (*Zea mays* subsp. *indentata* (Sturtev) Zhuk) and small seeded Flint Corn (*Zea mays* subsp. *saccharata* (Sturtev) L H Bailey). Both the types are single stalked and 2 to 3 m high, grown both for table purpose, flour and *sattu* preparation and also adapted for poultry feedings (Fig. Z 1). Various traditional cultivars/land races, exotic varieties and recently released high yielding varieties under cultivation in Kashmir region are:

A. Traditional Farmer's Varieties / Land races

- **Islambadi Makayei:** Plant medium tall (1.20 - 2.30 m), bearing 1 to 2 cobs. Yellow seeded, yielding 3.0 to 3.5 t /ha.
- **Qauzigund Makayei:** Grown in Kandi maize belt of Qazigund. Small seeded; orange type.
- **Pahalgam Makayei:** Grown on the hilly track of Pahalgam at altitudes of 2000 to 2500 m amsl. The variety is adapted to cold conditions, is early maturing (100 -105 days) and has semident gains. Yield potential is 2.0 to 2.5 t/ha. It is moderately susceptible to *Turcicum* leaf blight. Used for bread making.
- **Gurez Makayei:** Plant type is smaller with thin stem. It is the traditional land race of Gurez and Tilail areas and sown on the foot hills up to 3500 m amsl. Grains small, flint, predominantly white, although yellow lines

are also present; has cold tolerance, extra early maturity and moderate tolerance to *Turcicum* leaf blight.

- **Niver Makayei:** Grown in the hilly maize growing areas of the valley. It is characterized by tall stalks, early maturity, bold grains and cobs with higher number of kernel rows. The seeds are either of yellow or creamish colour and accordingly known as *Safed Niver* and *Zared Niver*.
- **Mishre Makayei:** It is the sweet corn or sugar corn type grown in some traditional maize belts. Grains of bright yellow to golden colour are highly shrunken due to soft endosperm. It is popular for the consumption as roasted or boiled. It is very early maturing and matures in 85-90 days.
- **Aru Makayei:** Grown in the higher valley belts mostly around Pahalgam upto 2800 m altitudes. Plants are medium tall, cob size is small, kernel is flint, available in different colours of orange, purple and yellow. It is early maturing and having very high tolerance to cold. Having initially evolved and cultivated by farmers of Aru, Pahalgam, hence the name.
- **Vyaloo Makayei:** Moderately flint and light yellow variety. It is generally single cobbed and tolerant to cold under altitude range of 2000 to 2500 m. Yield potential is about 2.0 t/ha. Matures within 100-105 days but is susceptible to *Turcicum* leaf blight. Having been adapted and mostly cultivated in the Vyaloo, Kokernag belt of South Kashmir, hence the name.
- **Poonch Makayei:** A local traditional fodder cum grain maize variety grown in the hilly areas of Poonch and adjacent areas. A tall variety with medium cob size and small dull whitish semi -dent grains.
- **Paigambri Makayei:** A maize variety grown in some hilly areas of Poonch and Doda. It has characteristic large cobs with bold grains of varied shades of dense white, and dark purple. The native *Gujjars* of the areas cultivate the variety and maintain its genetic identity with reverence as they believe that the cultivar has been gifted to them by Prophet himself, centuries before.

B. Released high yielding varieties

- **Hybrid Makka -123:** A Himalayan hybrid developed by State Dept. of Agriculture and released on All India basis, for northern hilly areas in 1964. The hybrid yields about 5.5 t /ha and is recommended up to an altitude of 2200m. It has orange yellow semi-dent grains and mature in about 135-145 days in the valley.
- **Composite-1:** Variety developed and released by State Dept. of Agriculture, during 1967 for cultivation in the temperate regions of the state above 1550 m amsl. Plants tall with orange-yellow flint grains. Matures within 125-130 days and has yield potential of 4.0 - 4.5 t/ha.

- **Composite-2:** Variety developed by State Dept. of Agriculture, in 1968 for cultivation in the temperate belts of the state. Grains bold, yellow and semi-dent type and ideal for *chapatti* and *satu* preparation. Matures within 130-140 days and has a yield potential of 4.0 to 5.5 t/ ha
- **Composite-3**:Variety developed for maize growing belts of the state by State Dept. of Agriculture, during 1970. Grains yellow and dent type. Matures with 130-135 days and yields about 4 - 5 t/ha. Recommended up to the elevations of 2000 m amsl.
- **Composite-4:** Developed for Himalayan zone as whole. The variety has bold, white dent grains and is suited for areas like Uri, Poonch, and Rajouri where there is a preference for white types.
- **Composite-5:** The variety developed by State Department of Agriculture during, 1974 for cultivation in intermediate zone of the state and in plains of Jammu (500 - 1500m altitudes). Its yield potential is 5.0 - 5.5 t/ha. Grains are orange, flint type and bold.
- **Composite-6:** A composite maize variety jointly developed and released by State Dept. of Agriculture, and SKUAST-K during 1991 for cultivation in the temperate and intermediate zones of the state at elevations of 1500 to 1800 m. Plants medium tall, vigorous with ear placement at reasonable height. Stilt roots strong. Ears medium thick with compact green husk cover. Tassel branched and semi compact. Silk predominantly light green. Grains bold, flint type, orange yellow. The variety is resistant to *Turcicum blight* under field conditions. Yields 4.5-5.0 t/ha under suitable management conditions. Matures within 155-160 days in temperate zone and 125 to 130 days in intermediate zone.
- **Composite-7:** A composite variety developed and released by State Dept. of Agriculture in the year 1984 for general cultivation in the hilly maize growing belts of the state. The plants are tall with lower ear placement, orange yellow semi-dent to dent grains, medium maturity and with a yield potential of 3.5 - 4.5 q/ha.
- **Composite-8:** Variety developed by the State Dept. of Agriculture and SKUAST-K and released in 1993 for cultivation in lower and irrigated foothill areas in the Kashmir Valley and higher reaches of Jammu region and mid elevations of Poonch, Rajouri and Udhampur districts of Jammu region. Plants medium tall, vigorous with air placement at reasonable height. Leaves dark green with narrow apex. Ears long and mediun thick with compact green husk cover. Tassel branched and semi compact. Silk predominantly light green. Grains bold, creamy white, semi-dent, conical with flat lateral surfaces. Matures within 150 to

155 days in Valley and higher altitudes of Jammu and 110-115 days in intermediate zone. Yields 4.0-5.0 t/ha (Fig. Z 2).

- **Composite-14:** Developed and released by State Dept. of Agriculture and SKUAST-K in the year 1996 for cultivation in high altitude areas of J&K State from 2000 to 3000 altitudes. Plants medium tall (1.5 to 2 m). Leaves narrow, green. Tassel lax and medium sized. Silk light yellow to purplish. Husk cover green, fully covered. Cobs medium sized with conical shape. Grains bold, orange to orange yellow, predominantly flint which makes them good for *chappati* preparations. Matures within 135-145 days in the valley, 100-110 days in intermediate zone and 95-100 days in plains. Average yield ranges between 4.5 to 5.0 t/ ha (Fig. Z 3).
- **Composite-15:** A composite variety developed by State Dept. of Agriculture and SKUAST-K in the year 1993 for cultivation in high hill maize growing regions of the state. Plants tall, vigorous with air placement at reasonable height. Ears medium thick with compact green husk cover. Silk predominantly light green. Grains bold, orange yellow. Yield potential is 4.5 - 5.0 t/ha. Variety has early maturity and matures within 155-160 days in temperate zone and 125-130 days in intermediate zone.
- **Composite Super-1:** Released during 1994 by State Dept of Agriculture, and SKUAST –K for general cultivation in Kashmir and hilly maize growing regions of Jammu. Plants tall, vigorous with air placement at reasonable height. Leaves deep green. Stilt roots strong which make the plant lodging resistant. Ears medium thick with compact green husk cover. Silk predominantly light green. Grains bold, orange yellow and semi –flint. Yield potential is 5.0 - 6.0 t/ha. Matures within 155-160 days in temperate zone, 125-130 days in intermediate zone and 110-115 days in plains.
- **Shalimar Kishan Ganga Maize Composite-1:** The fodder cum grain variety developed and released by SKUAST-K during 2004 for cultivation in the maize growing areas of Gurez and Machil. Recommended for Cold Hills of Kashmir (above 2500 m). Medium tall plants with strong internode anthocyanin pigmentation and resistant to lodging. Grains flint type, uniform light orange coloured with whitish yellow tip. Silk slightly purplish, cobs medium sized and conico-cylindrical. The variety is extra early and matures within 120-125 days. Tolerant to leaf blight, downy mildew and stem rot. Yields 4.5 to 5.0 t/ha (Fig. Z 4).
- **Shalimar Kishan Ganga Maize Composite-2:** An extra early maize variety released during 2004 by SKUAST-K specifically for cultivation in the maize growing belts of Gurez and Machil areas of Kashmir region.

Recommended for Cold Hills of Kashmir (above 2500 m altitudes). Medium tall plants with strong tassel and silk anthocyanin pigmentation. Resistant to cold, lodging and shattering. Leaf angle wide. Grains flint type, uniformly light orange coloured with yellow sides and white yellowish tip. Cobs medium sized and slightly conical. The variety is extra early and matures within 120-125 days. The variety is dual fodder cum grain type and has a yield potential of 4.0 to 4.5 t/ ha.

- **Shalimar Maize Composite-3:** The composite variety has been developed by SKUAST-K through varietal cross between Pahalgam Yellow, Local Lidroo, Pop 85, C-4 and C-15 and subsequent selections. It is a yellow grained maize variety with early maturity (135-145 days), higher grain protein and high yield, disease resistance and cold tolerance. It has an average grain yield potential of 4.5 t/ ha. Suitable for cultivation in high altitudes of Kashmir Division.
- **Shalimar Maize Composite-4:** Developed by SKUAST-K, from a varietal cross between the maize genotypes KH -5991, F-7012, SSFX-9199, and C-36 received from Directorate of Maize Research Hyderabad. The variety has good biomass, timely maturity and high yield. Grains flint type with orange –yellow colour and high protein content. It has an average yield potential of 6.0 t/ha and exhibits a yield superiority of 30% over C-6. Recommended for plains of Kashmir valley and mid altitudes of Jammu. Suitable for poultry industry because of small grain size.
- **Shalimar Maize Hybrid-1:** A commercial maize hybrid developed and released by SKUAST-K during 2010. Developed by crossing two inbred lines (CML 349, CML 354) received from CIMMYT Mexico and released for cultivation in high altitude areas of Kashmir. Variety has flint and white grains, early maturity (135 to 145 days), high yield potential of 6.5 t/ha with a yield superiority of 30% and 50% respectively for grain and fodder yield over standard check C-15 and C- 8.
- **Shalimar Maize Hybrid- 2:** Released in 2013 and KDM -347 X KDM -500 have been utilized for the development of the hybrid. KDM-500 has been derived from CM-128. Recommended for Mid altitudes of Jammu & Kashmir between 1650-1800m altitudes (irrigated/rainfed). The hybrid has a yield potential of 70.0 quintals/ha, it is yellow in colour, early in maturity.
- **Shalimar Maize Composite -5**: Released during 2013; developed from SO3HI HQ, A QPM maize population received from CIMMYT, Mexico. Recommended for High altitudes areas of Kashmir between

1800-2250m amsl. (rain fed). The variety has a yield potential of 50.0 quintals/ha, it is white in colour, early in maturity and is tall.

- **Shalimar Maize Composite-6:** Developed during 2013 from BH-2528, C-15 and KDM 914B materials. It has yield potential of 55-60 q/ha. It matures in 135-140 days in the Valley and is recommended for higher elevations up to 2200 meters. Its a yellowish orange bold grain composite.
- **Shalimar Maize Composite-7**: Developed from F-7012, C-6 and C-36 materials and released during 2013. Recommended for Mid altitudes of Jammu & Kashmir between 1650-1800 m altitudes (irrigated/rainfed). The composite has a yield potential of 65.0 quintals/ha, it is yellow in colour, early in maturity.
- **Shalimar Pop Corn-1:** Released at national level during 2016. HKI-PC-7, HKI-PC-8,KD-Local Pop-1, KD-Local Pop-3 were used for the development of the variety. Recommended for Zone-1, 2, 3 and 5 across India, it's a yellow flint maize with yield potential of 5 t/ha and a popping of 80%.It's early in maturity and disease resistant.
- **Shalimar Sweet Corn -1**: Released during 2016 at national level. WOSC, KDSC-3,Vega-2,Merit-4 materials have been used for the development of the variety. High yielding and extra early maturity sweet corn composite with distinct sweet corn traits, tender cobs and inbuilt tolerance to biotic and abiotic stresses prevalent in the mid altitude areas of Kashmir.
- **Shalimar QPM-1:** The variety was developed by crossing KDQPM-14 x KDQPM-13 and released during 2016. The single cross QPM hybrid is suitable for areas up to 1850 m in elevation in Kashmir Valley. It's an early maturing, orange yellow flint hybrid with a tryptophan content of 0.81 % of the total protein and a yield potential of 60 q/ha.It is resistant to TLB and common rust disease.

Zinnia angustifolia Kunth. (syn. Zinnia linearis Benth.) (Asteraceae): Eng.- Carpet Zinnia [NI]. A half hardy annual with upright stems and oblong light green leaves covered with bristly hairs. Bright orange single flowers are produced from July to August. A wide range of forms with varying colours are cultivated in gardens.

Zinnia elegans L. (Asteraceae): Eng.- Youth and Old age; Kash.- Zinnnia [NI] Most popular annual ornamental herb grown for its daizy like bright flower heads. Used for bedding out and for cutting. The light to mid green leaves are ovate and pointed. Course upright stems carry showy purple flowers heads from July to September. Several attractive varieties are cultivated with varying colour shades.

Ziziphus jujuba Mill (syn. Ziziphus mauritiana Lamk.; Ziziphus sativa Gaertn.; Ziziphus vulgaris Lam.) (Rhamnaceae): Eng.- Wild Jujubi /Indian Date; Hindi- Kandiari; Kash.- Brayei [W]. A deciduous, branched, medium sized tree/spiny deciduous bush found in wild state in Kashmir Himalaya on dry slopes and forest openings at 1700-3800 m. Leaves oblong-ovate to ovate lanceolate, crennate-serrate, glabrous on both surfaces. Flowers small with inconspicuous yellowish green petals, fascicled in the axils of leaves, collate; drupes globose, succulent, dark red or purplish black, wrinkled with hard 2-celled stone, edible. Leaves are laxative and used in scabies and throat troubles. Fruits are credited with emollient, pectoral and cooling properties and hence are used by local Hakims for treating various human ailments (Fig. Z 5).

Ziziphus oxyphylla Edgew. (syn. Ziziphus acuminata Royle.) (Rhamnaceae): Eng.- Jujube; Hindi- Kandyari; Kash.- Brayie [W]. An erect, glabrous, armed deciduous shrub with spreading, purple armed branches and germinate prickles. Leaves obliquely ovate, unequal-sided, acute, finely crenate. Flowers fascicled in the axils of the leaves, petals green, spreading;drupes edible. Grown as ornamental/avenue tree. Also stand planted around some sacred groves and Sufi shrines for shade and beautification (Figs. Z 6).

Fig. Z 1. Maize type cultivated in Kashmir

Fig. Z 2. var. Composite-8

Fig. Z 3. var. Composite-14

Fig. Z 4. var. Shalimar KG-1

Fig. Z 5. Zizyphus jujuba

Fig. Z 6. Zizyphus oxyphylla

Selected Bibliography

Bamzai, P. N. K. 1987. Socio –economic History of Kashmir (1846 – 1925). Metropolitan Book Co. (P) Ltd., Netaji Subash Marg, New Delhi – 11002, India: pp. 9 -32.

Bhat, G. M. and Navchoo, Irshad A. 1988. Ethnobotanical Exploration of Ladakh – J & K; J. Phytol. Res. 1 (1): 70 – 76.

Buth, G. M. and Navchoo, Irshad, A. 1988. Ethnobotany of Ladakh (India) – Plants used in health care; J. Ethnobiol. 8 (2): 185 – 194.

Dar, G. H., Bhagat, R. C and Khan, M. A. 2002. Biodiversity of Kashmir Himalaya. Valley Book House, University Road, Hazratbal, Srinagar, Kashmir, J&K,India- 19006.

Dhar, Uppendra and Kachroo, P. 1983. Alpine Flora of Kashmir Himalaya. Scientific Publishers, Jodhpur, India.

Kachroo. P. 1980. Ladakh – An integrated survey. Ladakh Cell, University of Kashmir, Srinagar – 19006, India.

Kak, Ab Majid. 1983. Ethnobotany of Kashmiris. Kahkashan XVI (1) 1981 -82: pp 1-15;Bansi Art Press, Srinagar, Kashmir.

Kaul, M. K. 1997. Medicinal Plants of Kashmir and Ladakh. Indus Publishing Company, FS – 5, Tagore Garden, New Delhi.

Lawrence, W. R. 1967. The Valley of Kashmir. Kesar Publishers, Residency Road Srinagar, Kashmir.

Navchoo, Irshad, A and Buth, G. M. 1989. Medicinal System of Ladakh, India; Jr Ethnopharmacology, 26 (1989): 137 – 146.

Navchoo, Irshad, A and Buth, G. M. 1995. Ethnobotany of Ladakh: Studies on the traditional medicinal plants and their uses. Oriental Science 1 (1995) ISSN 0971– 703 X.

Parray, G. A. 2001. Genetic divergence in local rice germplasm of Kashmir valley and stability of promising cultivars; Ph D Thesis, SKUAST of Kashmir,Srinagar, J&K.

Paul, T. M. 2001. Studies on the woody ornamentals for landscaping; Ph D Thesis, SKUAST of Kashmir, Srinagar, J&K.

Sharma, C. C. 1986. Kashmir Agriculture and Land Revenue System under the Sikh Rule (1819 – 1946). Rima Publishing House, ER – 10, Inder Puri, New Delhi; pp. 19 -31.

Singh Umrao, Wadhwani A.M. and Johri, B. M. 1990. Dictionary of Economic Plants in India. Publication and Information Division, ICAR, New Delhi - 12.

Singh Gurcharan and Kachroo, P. 1994. Forest Flora of Srinagar. Bishen Singh Mahendra Pal Singh, 23 – A, New Cannaught Place, Dehra Dun – 248001, India.

Wadoo, M. S. 2007. The Trees of our Heritage. Idris Publications, 121 – Firdous Colony, Bucha pora, Srinagar, Kashmir, J & K.

Wani, Shafiq Ahmad and Zeerak, N, A. 1999. Project Report of National Agri. Technology Project on 'Sustainable Management of Plant Biodiversity'. Division of Plant Breeding & Genetics, She-re Kashmir University of Agric. Sc. & Technology of Kashmir, Shalimar, Srinagar, Kashmir, J & K, India.

Zeerak, Nazir Ahmad and Wani, Shafiq Ahmad. 2012. Agri– Horticultural Biodiversity of Temperate and Cold – arid Regions (Indian sub-continent). New India Publication Agency, New Delhi – 110 088.

Wanchoo, P. N. 2000. Horticulture in Himalayas. Bishen Singh, Mahendra Pal Singh, 23 – A, New Cannaught Place, Dehradun, India: pp. 407 – 510.

Youngerband, F. 1970. Kashmir. Sagar Publications; Ved Mansion, 71, Janpath, New Delhi: pp. 194 – 221.

Index

I. English Names

C

D

E

F

G

H

I

J

K

L

M

N

O

P

T

U

V

W

Y

Z

II. Hindi, Kashmiri and Ladakhi Names

A

B

C

D

E

F

G

H

I

J

K

L

M

N

O

P

S

T

Z